T0177485

Oxford Series in Ecology and Evolution

The Comparative Method in Evolutionary Biology
Paul H. Harvey and Mark D. Pagel

The Cause of Molecular Evolution
John H. Gillespie

Dunnock Behaviour and Social Evolution
N. B. Davies

Natural Selection: Domains, Levels, and Challenges
George C. Williams

Behaviour and Social Evolution of Wasps: The Communal Aggregation Hypothesis
Yosiaki Itô

Life History Invariants: Some Explorations of Symmetry in Evolutionary Ecology
Eric L. Charnov

Quantitative Ecology and the Brown Trout
J. M. Elliott

Sexual Selection and the Barn Swallow
Anders Pape Møller

Ecology and Evolution in Anoxic Worlds
Tom Fenchel and Bland J. Finlay

Anolis Lizards of the Caribbean: Ecology, Evolution, and Plate Tectonics

Jonathan Roughgarden

From Individual Behaviour to Population Ecology
William J. Sutherland

Evolution of Social Insect Colonies: Sex Allocation and Kin Selection
Ross H. Crozier and Pekka Pamilo

Biological Invasions: Theory and Practice
Nanako Shigesada and Kohkichi Kawasaki

Cooperation Among Animals: An Evolutionary Perspective
Lee Alan Dugatkin

Natural Hybridization and Evolution
Michael L. Arnold

The Evolution of Sibling Rivalry
Douglas W. Mock and Geoffrey A. Parker

Asymmetry, Developmental Stability, and Evolution
Anders Pape Møller and John P. Swaddle

Metapopulation Ecology
Ilkka Hanski

Dynamic State Variable Models in Ecology: Methods and Applications
Colin W. Clark and Marc Mangel

The Origin, Expansion, and Demise of Plant Species
Donald A. Levin

The Spatial and Temporal Dynamics of Host–Parasitoid Interactions
Michael P. Hassell

The Ecology of Adaptive Radiation
Dolph Schluter

Parasites and the Behavior of Animals
Janice Moore

Competition Theory in Ecology

PETER A. ABRAMS

Department of Ecology and Evolutionary Biology, University of Toronto, Canada

OXFORD
UNIVERSITY PRESS

Great Clarendon Street, Oxford, OX2 6DP,
United Kingdom

Oxford University Press is a department of the University of Oxford.
It furthers the University's objective of excellence in research, scholarship,
and education by publishing worldwide. Oxford is a registered trade mark of
Oxford University Press in the UK and in certain other countries

© Peter A. Abrams 2022

The moral rights of the author have been asserted

Impression: 1

All rights reserved. No part of this publication may be reproduced, stored in
a retrieval system, or transmitted, in any form or by any means, without the
prior permission in writing of Oxford University Press, or as expressly permitted
by law, by licence or under terms agreed with the appropriate reprographics
rights organization. Enquiries concerning reproduction outside the scope of the
above should be sent to the Rights Department, Oxford University Press, at the
address above

You must not circulate this work in any other form
and you must impose this same condition on any acquirer

Published in the United States of America by Oxford University Press
198 Madison Avenue, New York, NY 10016, United States of America

British Library Cataloguing in Publication Data

Data available

Library of Congress Control Number: 2022936370

ISBN 978–0–19–289552–3 (hbk)
ISBN 978–0–19–289553–0 (pbk)

DOI: 10.1093/oso/9780192895523.001.0001

Printed and bound by
CPI Group (UK) Ltd, Croydon, CR0 4YY

Links to third party websites are provided by Oxford in good faith and
for information only. Oxford disclaims any responsibility for the materials
contained in any third party website referenced in this work.

Acknowledgements

This book arose from a planned review article that grew too long. Charley Krebs and Tom Schoener encouraged me to turn it into a book. It was an idea that I had resisted for quite some time, so the book would likely not have come into existence without their influence. Tom also used some preliminary chapter drafts in a graduate seminar course, which furnished some much-needed motivation for speeding up my writing schedule. Chris Klausmeier, Bob Holt, and Mark McPeek early on provided valuable feedback, and Michael Cortez sent comments on the longest chapter. I am grateful to Ian Sherman, Charlie Bath, and Giulia Lipparini at Oxford University Press for their input into this project. I am also very much indebted to my wife, Janet Pelley, for her forbearance as I spent an inordinate amount of time writing this work, and for providing feedback and copy editing on many of the chapters. The book was written during the first two years of the SARS-CoV2 epidemic, which limited access to libraries and people who could have made the writing process easier. Weekly discussions with Tom Reimchen, Don Kramer, and Larry Dill have helped keep me informed of recent developments in ecology and evolution during that period. The Natural Sciences and Engineering Research Council of Canada provided financial support through a Discovery Grant.

Contents

1

Introduction
Competition theory past and present

1.1 A short history

This book is a critical look at current competition theory, which constitutes a significant fraction of theoretical ecology. My general view of the purpose of ecological theory is similar to that expressed by Michel Loreau (Loreau, 2010, p. 268): 'It is my firm belief that ecological theory should be both a guide for basic research and a guide for action'. It is likely that very few professional ecologists would dispute Loreau's statement. Given the unprecedented rates of loss of species and natural biological communities, most ecologists hope that ecological theory will provide practical advice on altering the dynamics of natural populations undergoing undesired changes, such as declines towards extinction. Certainly, as Loreau states, it 'should' do so. Nevertheless, ecological theory in its present state has largely failed to make predictions or contribute to plans that could help to avert the collapse of ecological systems. This book will examine the reasons for these shortcomings by focusing on competition theory, a subset of ecological theory. Competition theory is central to most of the other branches of ecological theory and has a particularly long history in the field. Investigating its current state and why it has not progressed more rapidly may provide more general insights for ecological theory and for ecology as a whole.

Understanding ecology's current state requires some history. Ecology, the scientific study of the distribution and abundance of living organisms, developed out of natural history. Natural history had provided an accounting and description of the organisms in natural communities long before ecology arose as a scientific discipline. Ecology sought to explain the properties of those communities and predict their future. Accounting for or predicting either the population size or geographic range of almost any species on Earth represents a major challenge. It was clear to some biologists in the early twentieth century that the natural history knowledge of that period would not be sufficient to answer most of the major questions about abundance, even for a single species. Because distributions and abundances are constantly changing, those biologists recognized that the mathematics of dynamical systems was clearly important for understanding ecology. As the scope and magnitude of human impacts

Competition Theory in Ecology. Peter A. Abrams, Oxford University Press. © Peter A. Abrams (2022).
DOI: 10.1093/oso/9780192895523.003.0001

on natural systems have escalated, mathematical models have become increasingly important tools to provide insights and predictive power.

Ecological theory began to appear in the 1920s, some of it spurred by applied questions in fisheries. However, in spite of a few developments over the next three decades, largely in fisheries, pest control, and human demography, a mathematical framework for understanding ecological communities was still largely absent from ecological literature in the 1960s. At that point quantitative ecology started to receive renewed attention from biology departments in universities in North America. In part this was because of the perceived insufficiency of the theory that existed at the time; it certainly had a poor track record in applied fields like fisheries management. The 1960s were also a time of expansion in biology departments and increased interest in ecology among the general population. Now most major universities have a department devoted to ecology, evolution, and behaviour, and many have one or more faculty members engaged in developing ecological theory. The scientific literature on ecology and evolution has expanded enormously.

In spite of this expansion of the field as a whole, my impression is that theory in ecology has not lived up to its promise at the time I started graduate work half a century ago. It has definitely not provided the type of predictive and explanatory power that I and many others at that time were expecting to develop quickly. This book will examine both the history and the current state of theory on one of the major types of ecological interactions, competition. Competition theory provides a case study of some of the problems in developing theory in ecology more generally. It was the central interaction studied by many of the ecologists who were responsible for the revival of theoretical ecology in the 1960s and early 1970s. Focusing on competition theory can still be justified, because an adequate understanding of competition either requires or implies a good understanding of all other interactions.

Competition occurs within and between species. It therefore has an effect on the abundance of every organism. Competition between species sets limits on the number of species that can coexist. Competition between individuals within a species determines the maximum abundance of a species and how rapidly it is approached. Within-species competition is also the driver of natural selection and its resulting evolutionary change. Competition is at least part of the mechanism behind divergent selection that leads to speciation. Thus, competition is responsible for generating new species as well as for limiting the number of species that can be present locally. All of these considerations contributed to making competition a central focus of the mathematical approaches to ecological interactions that developed in the 1960s and 70s. Competition had been a major area of interest for decades before this, with major arguments in the 1940s and 1950s about how important it was in determining the abundance of species (see Hutchinson, 1978).

Most ecologists who have studied this interaction have come to the study of competition with some preconceptions. All humans have some personal experience with competition. It often begins with issues such as, 'Who gets to play with a favourite toy?', and continues throughout life, involving money and various other needed or desired 'items' that are in limited supply. Most of economics is about competition, and a Web of Knowledge search for articles about 'competition' from the past five years yields more publications from economics than from ecology. Thus, most students in an introductory ecology class have some preconceived notions about competition. In spite of (or perhaps because of) this, the meaning of competition in ecology has remained something of a grey area. To address this confusion, the next chapter in this book is devoted to documenting the history of definitions and arguing for one based solely on shared use of resources that influence population growth.

1.2 The need for resources in competition theory

One central thesis of this book is an idea that was more widely accepted four decades ago than it is today; i.e. that understanding competition requires an understanding of its underlying consumer–resource interactions. Resources are substances that are required for survival and/or reproduction. For example, in predators, resources always include their prey, but they may also include sources of water, shelter, or camouflage. This means that a proper ecological understanding of competition between predators requires, at a minimum, some study of prey population dynamics, the rates of consumption by predators, and the effects of that consumption on the predators' birth and death rates. The last of these requires some knowledge of the supply and workings of those other factors influencing the predator's birth and death rates. Competition between plant species involves exploitation of water, light, and mineral nutrients. However, all of these are tied to the space where they occur, and the nature of the consumer–resource interaction may differ greatly with the physical characteristics of the place where the water, light, and nutrients are found. Detritivores, parasites, pathogens, and other groups are also consumers, and they often require different types of models of competition. This is particularly true of the latter two groups, which inhabit their 'resource', and their survival or reproduction may be adversely affected by overly high consumption rates. Consumer–resource interactions are also an essential component of many mutualistic interactions between species. A more complete theory of competition, based on a better understanding of consumer–resource interactions should help to provide a foundation for predicting future changes in distribution and abundance of all of these groups. A theory that includes resources explicitly is also needed for understanding the evolution of ecologically important characteristics in any species.

Acknowledging the centrality of consumer–resource interactions in competition leads directly to the question of the adequacy of existing consumer–resource theory. Thus, another theme of this book will be that, for both competition theory and consumer–resource theory, there has been an over-emphasis on historical models and overly simplistic formulations for understanding the interactions. In competition theory this has meant a continuing reliance on the Lotka–Volterra model, which is discussed below. In consumer–resource theory, the legacy of simple models with a long history means, inter alia, a concentration on cases with a single resource, or just two resources in the minority of studies that consider more than one. For those branches of competition theory that explicitly include resources, the most common assumption is that there are just two consumers and two resources, and the resources are assumed not to interact with one another. Simple models also omit adaptive behaviour and phenotypic plasticity from the set of processes determining the shape of consumption rate functions. Existing consumer–resource and competition theories have additionally tended to ignore the impacts of the abundances of species occupying higher and lower trophic levels on a focal consumer–resource pair. Linear relationships between resource abundance and consumption rate are often assumed as the default, as are linear relationships between resource consumption and per individual birth and death rates.

In competition theory, there is still a large body of new work produced every year using the oldest and simplest possible model of competition in which each species has an instantaneous per capita growth rate that declines linearly and independently with the abundance of every one of its competitor species. The result is that many ecologists have expectations based on theoretical systems that are *likely* to have properties that are extremely uncommon, even within the vast domain of different natural communities.

1.3 The Lotka–Volterra model

Because this book will be mainly about theory, it is necessary to begin with a brief review of the foundational mathematical model of interspecific competition, even though that model lacks any representation of resources. This is the simple model independently developed by Alfred Lotka and Vito Volterra, in the mid-1920s (see Volterra, 1931). As late as the mid-1960s, this was still the only model of competition referred to in most of the literature (Hutchinson 1965). This model has played such a major role in past research I will begin by reviewing its basic form and properties for any readers who may not remember their undergraduate ecology course.

The Lotka–Volterra (LV) model notably lacks any explicit representation of resources. It assumes that the per individual rate of increase of each consumer (competitor) species declines linearly with increases in its own abundance and declines linearly (usually with a different slope) with increases in the abundances of the second consumer species. The effects of changes in abundance on per individual growth rates are immediate. The most common representation of the LV model of two-species competition describes the rates of change in their abundances (N_1, N_2) as follows:

$$\frac{dN_1}{dt} = r_1 N_1 \left(1 - \frac{N_1 + \alpha_{12} N_2}{K_1}\right)$$

$$\frac{dN_2}{dt} = r_2 N_2 \left(1 - \frac{N_2 + \alpha_{21} N_1}{K_2}\right)$$

(1.1a, b)

The parameters r_i and K_i are respectively the maximum per individual growth rate and the equilibrium population size ('carrying capacity') of species i when it is present alone. The effect of interspecific relative to intraspecific competition is measured by α_{ij}, which is the ratio of the effect of the abundance of species j on the per capita rate of increase of species i, relative to the effect of the abundance of species i on its own per capita growth rate. The competition coefficient α_{ij} increases with the similarity of the two species in their relative use of different resources; it also increases as the absolute ability of species j to harvest all resources increases relative to that of species i. This model can be expanded to encompass more species; in this case there is a summation over all other species j of $\alpha_{ij}N_j$ in the numerator of the fractional terms. Whether both species coexist can be determined by assessing whether the per capita growth rate of each species (the quantity in parentheses in eqs 1.1a, b) is positive when the focal species is at near-zero abundance and its competitor is at its carrying capacity. Thus, under the specific form of eqs (1.1), the maximum per capita growth rates, r_1 and r_2, have no impact on whether the two species will both persist. If the product $\alpha_{12}\alpha_{21}$ is greater than one, it is impossible for the species to coexist; i.e. no values of K_1 and K_2 allow each species to persist together indefinitely. It is possible that one species always excludes the other; this may occur for a small difference in the K values when the product of the competition coefficients is close to unity. When the product of the α's is small, a larger ratio of K's is required to bring about exclusion of the low-K species, but this is always a possibility.

One outcome that was not widely appreciated before the development of the LV model was that in some cases when interspecific competition is stronger than intraspecific ($\alpha_{12}\alpha_{21} > 1$), either species is capable of excluding the other, with the outcome depending on initial abundances. In the two-species LV model, it is possible to determine whether the species will coexist by examining the growth rate of each species when it is very rare and the other species is at its carrying capacity. If both of these 'invasion growth rates' are positive, the two species will coexist, and if not, coexistence is impossible. Both invasion growth rates are negative in the case of alternative outcomes. Alternative outcomes were observed in later laboratory experiments.

The sets of outcomes in the LV model are usually illustrated using 'isocline diagrams'. This involves setting the per capita growth rates equal to zero, solving each equation for one variable in terms of the other, and then plotting the resulting two lines on a graph whose axes are the two population densities (sizes). The equilibrium densities are given by where the two lines intersect. The equilibrium is stable if arrows originating at points near the equilibrium, and whose direction is given by the pair of growth rates at that point, are directed back towards the equilibrium. However, if the arrows indicate a cycle around the equilibrium (never true for eqs (1.1)), isocline analysis is inconclusive. Although 3-D diagrams can be drawn for 3-variable systems (e.g. McPeek 2019a), these are usually not sufficient for determining stability, and are certainly not the easiest way to do so.

Unfortunately, the conditions for coexistence in the two-species LV model do not apply universally to models that have explicit resource dynamics, or to models with temporal variability, or to a range of other models and the consumer–resource interactions that underlie them. There is no analogue to the two-species coexistence condition LV model with three or more competitors. In any LV model, each species' population growth rate responds immediately to a change in the other species, something that is usually not the case for real consumer–resource interactions. Even when the LV model is being used as a rough approximation, estimating the parameters of the model from first principles—particularly the strength of competition given by the α values—requires a consumer–resource model. These and many other limitations of models lacking explicit resource dynamics will be discussed in this book. However, it should be noted that Volterra did include resources in his conceptualization of competition, even though he did not use the term. This is evident in his idea that only a single species could persist indefinitely if there was only a single limiting factor, or, as it later came to be known, the 'competitive exclusion principle'.

The limitation to a single consumer on a single resource follows from the two-species competition model when the product of the competition coefficients is equal to or greater than one. If there is purely exploitative competition for a single resource by two consumers, the product of their competition coefficients is unity, and there will be either no equilibrium point with positive densities of both, or, for a unique ratio of carrying capacities, a line of neutrally stable points. Volterra pointed out that the species that persist at the lowest level of that factor would prevent the existence of (i.e., exclude) all the others. Hardin (1960) notes that some authors later attributed this idea to Gause (1936), who had carried out a laboratory experiment in which competitive exclusion occurred. Hardin also noted that a wide range of authors had proposed similar ideas before and after Volterra. Many more recent authors attribute the idea to Tilman (1982).

In any case, the issue of coexistence on a single resource has remained a topic of continued interest, largely because early authors did not devote much thought to what exactly constitutes a *single* resource. Haigh and Maynard Smith (1972) suggested that the same resource species (or substance) at different times could constitute different resources, as could different parts of an individual of the resource species. Stewart and Levin (1973) and Armstrong and McGehee (1976a, b) showed that the prohibition of coexistence on a single resource type did not apply to consumer species in seasonal environments or species undergoing sustained population cycles. There has now been a large body of work (summarized in Chesson 2020b) on the potential for environmental variation to allow coexistence on a single resource. However, this requires that the consumers differ in their temporal responses to changing resource densities, in which case resource items becoming available at different times could be divided into different classes of resources. It has repeatedly been shown that space and time are both involved in the separation of distinct resources. All of the work over the years that has actually demonstrated coexistence on what seemed initially to be a single resource depends on the competitors responding differently to resource items based on place or time (or simple failure of the author to consider all limiting factors). These issues will be discussed further in the chapters that follow.

In spite of its limitations, the LV model did reveal some properties of at least some two-competitor systems, which were not widely appreciated in the 1920s. In addition, Lotka and Volterra's works inspired some early experimental explorations of competition in the laboratory by Georgy Gause (1936), who studied competition between two protozoans, and by Thomas Park (1948), who did the same for a pair of flour beetle species. These were the first influential experimental approaches to understanding competition in biological communities. Each of these authors related their results to the LV model. However, they were mainly concerned with examining whether or not competitive exclusion occurred in pairs of species, and did not really provide a challenging test for the adequacy of the model.

While the LV model played a key role in the birth of competition theory, it was clearly too simple to understand the composition of natural communities, or predict changes in those communities. Its assumption of linear effects on abundance, its focus on only two competing species, and the absence of an explicit accounting of resources combine to greatly restrict the ability of the LV model to describe any natural system, or even the majority of laboratory systems. This became clear many years later, after Francisco Ayala (1969) published a study on competition between *Drosophila* species in the laboratory in the journal *Nature*. His assumption that the LV model applied exactly to his two-species competition experiments led him to conclude that his observation of coexistence invalidated the competitive exclusion principle. This misinterpretation was later corrected in Gilpin and Ayala (1973) and Ayala et al. (1973). It had been assumed that the *Drosophila* had only a single resource in the jars in which the competing larval flies were raised, but later work showed that the species differed in the use of drier and moister areas within the medium (Pomerantz et al. 1980).

Ecology is certainly not alone in attempting to understand very complex systems in which it would be unrealistic to develop a quantitative understanding of all of the dynamic components that could influence changes in variables of interest. The normal course of theoretical development of a scientific field would have seen an exploration of the consequences of the most basic elaboration of the model; in this case that elaboration is clearly to include the dynamics of the resources that are the subject of the competitive interaction. This second stage of development was initiated in the late 1960s and early 1970s, but it seems to have been largely abandoned before it was given a chance to restructure the field.

1.4 The first revival of competition theory

The theoretical approach to understanding competitive systems, initiated by Lotka and Volterra, was not advanced significantly in the biological literature until a young professor of biology at Princeton University, Robert MacArthur, revived interest in mathematical studies of competition in the 1960s and early 1970s. Richard Levins (1968) independently stressed the importance of resources for understanding competition, and, in joint work, these two authors tried to relate competition between species to their similarity in resource use. MacArthur's final (1972) book, *Geographical Ecology* contained a large section that was devoted to interspecific competition. Two years earlier he had published an article on this topic, which was the very first article in the first issue of a new journal, *Theoretical Population Biology*. Both works analysed the conditions allowing competing species to coexist. They proposed a general relationship between overlap in the two spectra of resource capture rates by two consumer species, and the resulting competition coefficient describing the relative strengths of inter- and intraspecific effects of abundance on per capita growth rate. The main innovation in these works was to examine the consistency of the LV model with more detailed mathematical models that contained descriptions of both resource and consumer dynamics. The per capita growth rates of all species were linear functions of the variables affecting growth. MacArthur's main conclusion was that these models were consistent with the LV model. Unfortunately, MacArthur had already published several influential works using the LV model, so his assessment of the ability of that model to reflect resource dynamics was somewhat biased. MacArthur's career was tragically cut short by cancer in 1972, but competition continued to be one of the main topics within theoretical ecology for the rest of that decade.

For some later authors, MacArthur's work seemed to justify continuing studies of the LV model, or closely related forms that lacked resources. For others, his work was an indication that relationships with resources were a key to understanding some competitive systems. This contradiction has persisted until the present time in part because research on competition fell out of favour for a decade or more at the end of the last century. In his 1997 book on competition, James Grover (p. 277) wrote, 'About fifteen years ago, the study of competition seemed moribund, if not dead'.

1.5 Recent competition theory

Fortunately, research on competition theory has revived, and has increased greatly over the past two decades. For example, Peter Chesson's (2000a) *Annual Reviews* article on competition has been cited at an accelerating rate over the years, and now has several thousand citations. Many highly cited review and synthesis papers dealing with competition have been published during the past three years alone (see Chapter 4). There have been significant advances in understanding spatial (Leibold and Chase 2018) and temporal (Chesson 2003, 2018, 2020b) aspects of competition.

In spite of this sizeable number of publications and citations, the theory seems to have lost direction. While a resource-based approach has continued to be used by some quantitative ecologists, particularly in systems involving planktonic algae (e.g., Huisman and Weissing 2001; Litchman and Klausmeier 2001), its use by ecologists as a whole has diminished (as noted by McPeek 2019a). Many recent papers on competition theory have concentrated on models that lack resources (Levine et al. 2017; Saavedra et al. 2017; Hart et al. 2018). Even when a resource-based approach has been used, the component functions have, in the vast majority of studies, been drawn from early literature, and have failed to reflect the range of possibilities that are likely to occur among consumer species exhibiting adaptive behaviour in a food web context. One consequence of this is that we still know very little about how somewhat more realistic models differ in their dynamic behaviour from the 'first generation' models. A further consequence of the concentration on historical functional forms is that little effort has been devoted to determining functional forms that actually exist in natural systems.

This historical inertia of the theoretical work means that there is still little ability to predict how measurable changes in the environment will alter population sizes in groups of competing species, and competition between species is still mostly absent from analyses of applied problems such as harvesting and conservation. Although interspecific competition in natural communities is unlikely to involve only isolated pairs of species, a large fraction of recent analyses are restricted to two-species competition, often with only two resources, and usually without any other foodweb context. In addition, a great deal of recent research seems to be focused on estimating descriptive statistics that do not contribute to making accurate predictions about population responses to environmental or evolutionary change. All of these generalized criticisms will be detailed in subsequent chapters.

Evidence for the uncertain standing of competition theory is not simply the opinion of a few specialists. Reiners et al. (2017) analysed a survey of over 1300 members of the Ecological Society of America, to determine which concepts practising ecologists found most useful. 'Lotka–Volterra Models' was the lowest ranked concept on their list, even though 'competition' was ranked in the top ten. In a separate analysis of the influence of a long list of ecological concepts using figures on citation frequency (McCallen et al. 2019), the top ranked concept was 'models'. It could be that these opinions reflect the viewpoint of this introductory chapter. However, the models employed in much of the recent literature share many of the same limitations as the LV model, and the LV itself continues to be the primary or exclusive model in many high-profile theoretical works. The problem seems to be twofold: (1) that most field workers only regard theory as useful if it is something that can be 'tested' in relatively short-duration experiments; and (2) the majority of theoreticians have not been interested in a systematic exploration of how the most likely modifications of the simplest models alter their dynamics and outcomes.

1.6 General themes of this book

I will argue that both the LV and many of the more recent models are indeed insufficient, in that they do not reflect many of the dynamic characteristics of large classes of slightly more realistic consumer–resource systems involving shared resources. The book will also discuss possible reasons why the range of mathematical models of competition in the literature continues to be so narrow. The literature on competition in recent years has become increasingly focused on the question of coexistence, but it has ignored development of a wider set of consumer–resource models that will be needed to predict not only competitive exclusion, but also how competitors that continue to exist will change in response to any environmental shifts that affects one or more of them. Models that are capable of addressing this second set of questions will also lead to superior prediction of the conditions that allow coexistence, and they are usually necessary to obtain meaningful predictions of, or explanations for, evolutionary changes in sets of competitors. Thus, this book is an attempt to promote the development of a much larger and more useful body of models and approaches to competition.

A current preoccupation seems to be determining to what extent coexistence is due to 'stabilizing' and 'equalizing' factors, neither of which is well defined. In considering the role of temporal variation in altering competition, the focus has been on cases when variation favours coexistence. Within that set of cases, the goal has been to classify the mechanism as being 'the storage effect' or 'relative nonlinearity', or some mixture of the two, following Chesson's (1994) categorization of coexistence mechanisms. While there is no doubt that useful insights have grown out of Chesson's approach, the goal of incorporating competition into applied disciplines like conservation or exploiting natural resources seems to have disappeared, and classification is of little use here. Ideally, we should have a body of theoretical work that explores how various aspects of consumer–resource interactions affect predictions of future population sizes and competition-related traits in species experiencing environmental change.

One of the problems contributing to the development of competition theory applies much more generally to theory in ecology. It is the idea that all theory needs to be tested by determining whether some prediction of the theory applies across a wide range of real (preferably field) systems. The problem is that the theory that exists is so simplified that, without additional work on slightly more complicated models, we will not know whether a particular prediction has any chance of being 'true' in some sense of the word. If a prediction from theory that includes some feature is not manifested by one or more natural systems, this does not mean that the new feature should be discarded from future models; it is more likely to signal that other missing features need to be added. In other words, more theory is needed to allow empirical scientists to have some idea of what should occur (or have occurred) in the system being studied. Coming up with intelligent hypotheses to test empirically requires a large theoretical infrastructure.

The enormous range of ecological communities also means that there will never be a model that applies universally. The logical way to proceed from the LV model would be to incorporate a range of real-world features that are missing, but are likely to be present in real systems. At some point, this should provide a robust framework for choosing a range of alternative explanations for some observed population dynamical result involving competitors. MacArthur's original consumer–resource system was an initial step in the direction of building towards such a body of theory. Modifying MacArthur's system to include nonlinear components and interactions between resources were initial steps towards expanding the model, begun in the 1970s. The full implications of these expansions have still not been worked out in detail, largely because most theoreticians ceased working on this approach. Even the simple result that efficient consumers can cause extinction of some resources is ignored by most competition theory.

Suppose a new invasive species begins to increase rapidly, and an ecologically similar resident species decreases at an even more rapid rate. Can we expect the same relative changes in abundance to continue in the future? The Lotka–Volterra model implies they would do so. Some of the more plausible consumer–resource models suggest that they should not, and that decreasing per individual effects on the resident are more likely as the invader population grows (Abrams et al. 2008b). A useful theory of competition should at least suggest what information is required to predict the most likely consequences for the native species as the invading one increases.

Ecological systems are both extremely complex and difficult to study without altering the system itself. To come up with a hypothesis about why one or more species is changing rapidly in abundance, one must have some idea of the important processes that could be operating. This must come from theory based on simpler systems. Models that incorporate the key processes of resource use are essential in generating basic hypotheses that might underlie the observed change.

Part of the problem with 'mechanistic models' that include more variables is that determining which one applies to a given system requires work that is not attractive to individual researchers. For example, what is the range of forms of relationships between food abundances and consumption rates (i.e., functional responses)? Which are most likely to apply to certain scenarios or species? A single study will only contribute a small amount towards answering these broader questions, but many such studies on a range of species would usually require much more time than that available during a graduate student career or during the duration of a single research grant. As a result the studies are not being done. When they are done, they most often involve a single type of resource and a single individual consumer. In part, this is due to the historical inertia from functional responses having been originally defined for single prey/food types. The limitations of functional response studies also reflect the historical idea that only prey abundance—not predator abundance—influenced the response. Neither of these assumptions is likely to apply to competitive interactions in natural communities.

There is still hope that these historical influences on both the study of consumer-resource interactions, and the science of competition more broadly, will diminish in the future. I am not arguing that ecologists should ignore past work; in fact, I believe the opposite is true. The problem has been that, when an area drops out of favour and then returns (as competition has done), there is a tendency to ignore much of that history, and concentrate on the earliest parts. The history of empirical and theoretical approaches to competition has been characterized by abrupt changes in the degree of general interest in the topic as well as the underlying assumptions made and questions asked. These seem to indicate that a change in focus is possible. Perhaps this gives hope for the future.

This book will return to these themes, and will end with a set of recommendations for producing a more useful competition theory in the future. Such a body of theory could provide the 'guide for action' called for by Michel Loreau in the quotation that introduced this chapter. This is likely to be my last publication in this area, to which I have devoted most of my career, and I wanted to provide a unified account of articles produced in many different decades and which appeared in many different journals. As a consequence, readers will note that rather large amounts of the text deal with 'old' work and also with my own work.

1.7 Necessary background and a look ahead

This book does not attempt to cover a large fraction of the mathematical models and methods that have been used to study competition. Rather it is an attempt to find ways to restructure the theoretical work being done on competition so that it is more likely to produce an empirically useful body of predictions and explanations. This involves mathematics, but all of the arguments are illustrated using the mathematically simple framework of ordinary differential equation models of homogeneous populations, and the examples all involve six or fewer species. (Models that include space are considered, but, in these cases, the dynamics within a patch are based on homogeneous populations that are well-mixed within the patch.)

While the book does not employ any advanced mathematics, some familiarity with differential equation models in ecology is assumed. Case (2000) or Otto and Day (2007) would provide more than enough background. The recent article by Grainger et al. (2022) describes most of the techniques used here. Some of the chapters focus more on history and philosophy than on specific models (Chapters 2, 3, 4, and 12). Others present fairly detailed analyses of simple models with resources, most of which were initially explored prior to this century (Chapters 5 and 6). Chapter 7 contains the main treatment of empirical work (and one model). Chapters 8 and 9 describe new analyses of models with seasonal variation in conditions, and they contain a large amount of numerical results. Chapters 9, 10, and 11 are broad reviews of subfields within competition theory, and Chapter 12 provides an overview.

2

Defining and describing competition

2.1 Introduction

One would have thought that the question, 'What is competition?' should have been satisfactorily answered long before the year 2000. After all, the term has been a major topic for ecological theory for nearly a century, since the work of Alfred Lotka and Vito Volterra, which was described in Chapter 1. Every modern ecology textbook gives competition extensive coverage. Current articles about competition seldom if ever define the term, which should indicate that the definition is almost universally agreed upon. In addition, one would think that research usually proceeds by agreeing on exactly what defines a process before developing theory to understand that process. Nonetheless, competition theory has been around almost as long as the interaction has been recognized, and it became one of the most popular topics for ecological research in the 1970s. Nevertheless, Paul Keddy felt compelled to open his book on competition (Keddy 2000, p. 4) with the statement, 'The right definition is like a sword that will clearly cleave nature into pieces that we can understand; the wrong definition is like a blunt instrument that only mashes the object of our inquiry into more confusion.'

A range of definitions were used before Keddy made this statement, and a similar range have continued to be used since. In fact, there was considerable disagreement on that definitional question among ecologists when I began my formal training in the field in the early 1970s. This surprised me, as a large number of ecologists at that time seemed convinced that competition was the most important interaction in natural communities. A variety of textbooks and articles from that era identified competition as one of the two basic types of interspecific interaction, the other one being predation. (A few added mutualism as a third category.) Predation was accepted to be an interaction in which individuals of one species consumed all or part of individuals of another species. There was an active debate about whether competition or predation had a greater effect on the abundances of species (e.g. Connell 1975). Resolving the debate was difficult, in part because the literature at that time contained several quite different definitions of competition. How could competition not have had a universally accepted definition in 1970? Even more surprising than the lack of a clear definition of competition at that time is the fact that the vast majority of ecologists have avoided an active debate on the definitional issue over the intervening half century. In spite of Keddy's admonition, there is no single widely accepted definition of

Competition Theory in Ecology. Peter A. Abrams, Oxford University Press. © Peter A. Abrams (2022).
DOI: 10.1093/oso/9780192895523.003.0002

competition. This chapter will document the definitions used by various authors over the years. It will also present reasons why the definitional issue needs to be settled, and will suggest a resolution.

The above paragraph has used the term 'competition' to refer to a process that involves two or more species; this is 'interspecific' competition. The same basic mechanism that results in interspecific competition also operates within species, resulting in 'intraspecific' competition. This within-species form is commonly referred to as 'density dependence' by ecologists. For evolutionary biologists, intraspecific competition is fundamental to understanding the outcome of natural selection on ecologically important traits. In addition to the definitional issue for interspecific competition, there is a strange lack of consistency in the theoretical and empirical work on the intra- and interspecific forms of competition. This inconsistency will be discussed in Chapter 5.

'Competition' is not the only problematic term in the classification of interspecific interactions that occur in ecological communities. Defining and classifying interactions of all sorts has been a long-standing problem for ecologists (Abrams 1987a). This chapter will therefore touch upon the more general problem of establishing a logical and self-consistent set of rules for categorizing interactions between species.

2.2 Historical definitions of competition

I will not attempt a complete historical account of the term 'competition' in the linked fields of ecology and evolution. For more details on the early history see Hutchinson (1965) or Kingsland (1995). The term was already in common usage when Lotka and Volterra independently introduced the first mathematical model of interspecific competition in the 1920s (presented in Chapter 1). I will begin this history with the early 1970s, corresponding to the beginning of my own formal study of ecology. At that time, most universities did not include departments devoted to ecology and evolution, and the vast majority of the current body of published work on competition had yet to appear.

The year before I began my graduate studies, a short book summarizing population biological theory for beginning students was published, 'A Primer of Population Biology', by two Harvard professors, E. O. Wilson and W. H. Bossert. Their definition of competition was, 'the active demand by two or more organisms for a common vital resource' (Wilson and Bossert 1971, p. 156). This may be a little too succinct, as there is quite a bit of leeway in what constitutes 'a common vital resource' and 'active demand'. Subsequent authors usually have referred to 'consumption' rather than 'active demand'. Nevertheless, this definition is close to the one that I will support here. Although Wilson and Bossert's definition of competition applies to both within- and between-species forms, these two types were discussed separately in their book, and within-species competition was, in most instances, referred to as 'density-dependence'. The mathematical models of both within- and between-species processes that they discussed failed to include any explicit representation of the abundance or dynamics of the resources. Their models expressed the population growth rate of the competing consumers solely as a function of their own abundances.

Shortly after Wilson and Bossert's book appeared, C. J. Krebs (1972) published the first comprehensive modern textbook covering the entire field of ecology. He endorsed L. C. Birch's earlier definition (Birch 1957), that 'Competition occurs when a number of organisms (of the same or different species) utilize common resources that are in short supply; or, if the resources are not in short supply, competition occurs when the organisms seeking that resource nevertheless harm one or other in the process'. This definition maintains the centrality of resources, but it introduces the possibility that there are additional mechanisms, beyond simple consumption, which may influence the effects of one resource consumer on another. Interestingly, Birch (1957) was motivated to publish the above definition because he felt multiple definitions of competition were being used, and that 'Some of these meanings are so ambiguous that the word has largely lost its usefulness as a scientific term' (Birch 1957, p. 5). Neither Birch nor Krebs has a detailed discussion of what constitutes a single resource. This is not really required for the definition, but it does have an important effect on the population-level consequences of resource consumption for the consumers and the resources. In addition, it is crucial for assessing whether species will be able to coexist, as will be discussed in later chapters.

The second part of Birch and Krebs' definition is the issue of 'harm . . . in the process'. This part of the definition is now usually described as interference competition. 'Harm' may sound like a well-defined term, but several different metrics could be used to distinguish between harmful effects. Does the harm decrease the immediate per individual population growth rate or the immediate resource intake rate, or can it increase one or both provided that it decreases the ultimate population size? As will be explained below, these different ways of quantifying 'harm' do not always result in the same classification of the interaction. Harm is usually considered to include interference with intake of resources, interference with reproduction, or the killing or injuring of other individuals.

Models of competition should reflect the definition of the term. Fifty years ago, most ecological researchers who thought about modelling competition would have employed the Lotka–Volterra (LV) model presented in Chapter 1. The defining feature of the interaction between two competitors in that model is that the immediate per capita growth rate of each species declines with an increase in the abundance of the other. By 1970, this had been the predominant model of competition for over four decades. It was the only model discussed by Wilson and Bossert (1971). Unfortunately, measuring the per capita growth rate at one point in time requires two population measurements that are close in time. This is difficult and entails large uncertainties. Thus, most investigators studied the long-term change in population size in one species following addition or removal of a second species. We will see later that both the linearity and the immediacy of the effects on population growth rate are problematic if one considers the dynamics of the resources that are the subject of the interaction. In any event, the LV model lacked resources, so it was at least somewhat inconsistent with definitions based on resource use.

As was noted in Chapter 1, Robert MacArthur was largely responsible for the major increase in the scientific study of competition that began in the late 1960s. In his final book in 1972 (p. 21), MacArthur proposed that, 'two species are competing if an increase in either one harms the other. Any machinery that can have that effect will be called competition.' The 'harm' was not defined, but it is clearly meant to apply at the level of the population rather than the individual. MacArthur's usage of the term strongly suggests that he was interpreting 'harm' as a decrease in the size of one species' population caused by an increase in that of another species. Notably, he mentions that the mutual negative effects of two species that came about by increasing the population of a shared predator would qualify as competition. This separates the definition from a requirement that it involve resources and puts the emphasis on the negativity of the effect on population size.

While on the subject of definitions, readers should be aware that 'predator' in this book will be used to describe any consumer of a living resource, with 'prey' being any living entity consumed by a predator. Thus, 'predator' should be understood to include herbivores, and 'prey' includes plants. This book will largely ignore parasites and diseases, which live within their host (resource), and therefore require a different modelling framework.

The status of the interaction via shared predators is still somewhat uncertain. MacArthur did not attempt to model this interaction directly or discuss its role in consumer–resource interactions. It was later named 'apparent competition' by Holt (1977) in an article that developed the first formal theory regarding the population-level effects of shared predation. Most recent articles seem to regard interactions via shared predators as a different category from 'normal' competition (see Holt and Bonsall 2017). Apparent competition will be treated in detail in Chapter 5.

Another definitional grey area in the 1970s was the situation in which individuals of each of two species kill and consume individuals of the other species, i.e. mutual predation. The problem with mutual predation is that it is possible for one species to benefit more from its consumption of its competitor than it loses from negative effects of that competitor. This could produce (+,–) effects of the two predators on each other's abundance. It is even possible that it could produce (+,+) effects on population size. Neither mutual predation nor apparent competition is included in most recent discussions of interspecific competition, although Kuang and Chesson (2008) have argued that apparent and resource competition are identical in many respects.

Despite his definition emphasizing population-level effects, MacArthur (1970, 1972) was the first biologist to explore how shared resource use by two consumers could lead to the model of competition suggested by Lotka and Volterra. He showed that if four conditions were met, this model could approximate the consumers' dynamics in a consumer–resource model. These conditions were: (1) all of the resources had logistic population growth; (2) none of the resources interacted directly with others; (3) resource dynamics were fast enough relative to the consumer that all resources quickly reached a quasi-equilibrium state following any change in consumer population size; and (4) the change in consumer population(s) did not cause extinction of any resources. However, MacArthur did not actually investigate when conditions (3) and (4) would be satisfied, and he did not even acknowledge condition (4). A 2-resource version of his model is explored in Chapter 3, to examine the consequences of these omissions. In any case, this quasi-consistency of the LV model with one very simple resource-based model led MacArthur to develop additional theory in which resources were not considered explicitly (May and MacArthur 1972).

The derivation in MacArthur (1972) may have contributed to the later popularity of definitions of competition based purely on negative effects on equilibrium population size. Negative effects are in fact the only possibility under the LV model of two competitors. MacArthur had previously published theory based on the LV model (MacArthur and Levins 1967), so his interpretation of the consumer–resource models may have been biased towards finding consistency between the two approaches. In any event, most subsequent authors adopted MacArthur's definition of competition in terms of effect on ultimate abundance rather than mechanism. Paired negative effects on abundance were assumed to be the definitive evidence for the occurrence of competition in subsequent experimental studies (Connell 1983; Schoener 1983) and were also thought to be a prerequisite for concluding that coupled evolutionary changes in two species were a consequence of competition (Schluter 2000).

MacArthur's focus on competition helped inspire many subsequent theoretical and empirical works. A large number of articles in the 1970s expanded our understanding of competitive interactions beyond the confines of the LV model. Laboratory systems involving *Drosophila* showed that competitive interactions were highly nonlinear (Gilpin and Ayala 1973; Ayala et al. 1973). In those systems, the effect of adding a fixed number of individuals of species 1 on the growth rate of 2 was relatively smaller when the initial abundance of species 1 was larger. This contrasts with the density-independent effect assumed by the LV model. A series of theoretical works by Schoener (1973, 1974c, 1976, 1978) explored the role of resource dynamics in producing different types of nonlinearity, as did Leon and Tumpson (1975) and Abrams (1975, 1977).

Stephen Levine (1976) called into question the assumption that mutually negative effects always occur as a consequence of the joint use of resources. He analysed simple mathematical models of two important cases in which shared use of resources could produce positive effects of a change in the number of individuals of one species on the abundance of another. The first case involved two consumers that shared two resources, but the resources had negative effects on each other's per capita growth rate. If consumer 1 had a strong negative effect on its preferred resource (resource 1), and that resource had a strong negative effect on consumer 2's preferred resource (resource 2), then the effect of increasing the abundance of consumer 1 would be an increase in consumer 2. In other words, they would not 'compete' according to a definition based on effects on long-term population size. Vandermeer (1980) provided a more detailed treatment of this case.

The second of Levine's examples used the LV model, but assumed three competing species. In this system, an increased abundance of consumer species 1 could have a positive effect on the long-term abundance of species 2. Positive effects occurred when greater numbers of species 1 caused a sufficiently large decrease in species 3, which was a stronger competitor of species 2. The indirect positive effect via competitor 3 could easily be larger than the direct negative effect on species 2, reversing the sign of the net effect of species 1 on species 2. Both of Levine's examples were common occurrences in the set of simple models he examined, and I am unaware of any subsequent work providing a good argument for why the mechanisms they represent should be rare in nature. However, there was no immediate response by the ecological community to resolve the problems that these cases represented for the definition of competition. Later work has revealed a large range of other mechanisms involving joint use of resources that have the possibility of $(+,-)$ or $(+,+)$ effects (e.g., Abrams and Cortez 2015a), as well as cases in which the interaction sign depended on the magnitude of the perturbation to the 'initiating' species.

Grime (1979) addressed the conflict between definitions based on mechanism vs those based on effect (directions of change in abundance), in a book focused on plant ecology. He suggested the use of shared resources as the basis for defining competition. The early 1980s saw other works that promoted a resource-based view of competition. Tilman's 1980 article and 1982 book were particularly influential, although he made a point of referring to 'resource' competition, implying the existence of other forms. A broader discussion of the proper definition and/or minimal model of competition was apparently cut short by a more prominent dispute, the debate over the 'importance' of competition for determining the composition of natural communities. Chapters for and against competition may be found in the book organized and edited by Strong et al. (1984). The contributing editors, all from Florida State University, were all on the 'anti-competition' side of the debate, but both sides were well represented in the individual chapters. The main issue of this controversy was whether geographic patterns in the co-occurrence of species could be interpreted as strong evidence for the operation of interspecific competition. A second issue was whether differences in the body sizes or feeding morphologies of coexisting species provided evidence that competition had caused these differences. The subsequent rapid decline in empirical research on competition (Grover 1997) suggested that the critiques of Strong and his associates might have discouraged studies of competition in the following years. Strong et al.'s book was based on a conference that took place in 1981; by the time the book finally appeared, two comprehensive reviews of field studies of competition (Schoener 1983; Connell 1983) had already appeared. Neither of these was based on the correlational evidence that was the subject of the controversy, and each of these reviews found that interspecific competition was observed in the vast majority of studies that were designed to assess whether it occurred. The evidence for competition in those studies was a significant decrease (increase) in population size (measured over some time frame) in response to an experimentally imposed increase (decrease) in the abundance of a co-occurring species on the same trophic level. This evidence of widespread competition was unfortunately not sufficient to prevent a loss of interest in studying the interaction.

The analyses in Connell and Schoener's review articles might seem to indicate a shift towards a definition of competition based on the effect on ultimate population size. However, Connell (1983) claimed to agree with Birch's (1957) definition of competition. Although this definition was resource-based, Birch seemed to assume, without explicitly requiring, mutually negative effects on population size. Schoener (1983) did not formally define competition, but did require that the species occupy the same trophic level and that at least one of a pair of competitors decreases in response to an increase in the other. This again seems to indicate a shift towards a double requirement of both mutually negative effects and a mechanism based on shared resource use. Schoener (1983) divided competition into six types; consumptive, preemptive, overgrowth, chemical, territorial, and encounter. Preemptive, overgrowth, and territorial interactions all involve space as a resource (or a proxy for the resource). Chemical and encounter both involve harm inflicted by individuals of one species on individuals of the other. Thus, this scheme is more of a subdivision of the two recognized categories of exploitative and interference competition, than a fundamentally new classification. Although it might seem that some interference effects are independent of resources, it does not make evolutionary sense for individuals of one consumer to kill, chase, or harm members of another species (which entails time, energy, and possibly risk of injury) unless there are increases in resource availability that subsequently result from these actions. Most of these scenarios argue that a proper definition of what constitutes a single resource needs to take location or time into consideration. It is possible that, in some cases of interference, the 'harm' is a side effect of some other activity, and has no real cost to the perpetrator. The latter possibility no doubt occurs in some cases (e.g., elephants stepping on insects), but is likely to be rare in systems with mutually negative effects.

While the definitions of interspecific competition assumed in Connell's and Schoener's 1983 reviews are not purely based on co-utilization of resources, it appears that almost all (if not all) of the studies they review that showed mutually negative effects would have also qualified as 'competition' based on such a requirement. However, in both of these reviews, the authors identified small numbers of cases that did not exhibit such effects. Were these few outcomes a consequence of Levine's or some other mechanism, such as limitation of abundance by separate predators? So far as I know, this was never investigated. These two reviews also do not shed any light on whether cases that lacked significant negative effects on population density also lacked effects on resource abundances.

Proceeding to the 1990s, the range of views on the definition of competition was still quite wide. Nicholas Gotelli (1991) published a general introduction to ecological theory, aimed at advanced undergraduates and beginning graduate students. His definition (Gotelli 1991, p. 112) was based purely upon mutually negative effects: 'Competitive interactions are those in which two species negatively influence each other's population growth rates and depress each other's population sizes'. James P. Grover, in his 1997 book 'Resource Competition', states on page 1 that 'Competition refers to mutually negative interactions among two or more individuals or populations'. If it is between populations, Grover specifies that 'population growth rate' is the quantity that is negatively affected. This is clearly an effect-based definition using population density as the perturbation. Nevertheless, his book is restricted to what he considers to be a subset of competitive interactions in which mutual use of limiting resources is the mechanism producing the mutually negative effects. Gurney and Nisbet's (1998) theory text lacks an explicit definition of competition, but only represents the interaction using consumer–resource models. The examples they consider did not exhibit positive effects. T. J. Case's (2000) theoretical ecology textbook stated that competition occurs when 'individuals of one species suffer a reduction in growth rate from a second species due to their shared used of limiting resources . . . or [due to] interference'. Case did not specify the time frame over which this reduction in population growth rate is to be measured. Keddy (2000, p. 5) defined competition as 'the negative effects that one organism has upon another by consuming, or controlling access to, a resource that is limited in availability'. This combination of effect sign and mechanism again neglects the time frame over which the negative effect is to be measured, as well as not specifying the range of properties that could be used to define the negative effect.

Textbooks from the current millennium have mostly adopted a definition that includes both a requirement that resources be involved and a requirement that mutually negative effects occur, although the operational definition of a 'negative effect' is usually not provided. For example, the 5th edition of C.J. Krebs' textbook (2001, p. 179) defines competition as a situation in which 'two species use the same limited resource or seek that resource, to the detriment of both', a definition that requires both a resource-based mechanism and a mutually detrimental outcome. Another widely used textbook from the same period, Ricklefs and Miller (2000, p. 384, Table 20-1) suggests that interaction types are classified by the 'effect of' one species on another. Competition is a case when 'interactions between individuals . . . have negative consequences for all species involved'. However, the chapter on competition theory in this book defines competition (p. 403) as 'use or defence of a resource by one individual that reduces the availability of that resource to other individuals'. Thus, both types of definition are supported, but the mechanistic one seems to ignore the Levine (1976) scenario with its positive net effects, and is generally vague about the time frame of measurement and about situations with more than one resource. Turchin's (2003) monograph on population dynamics separates indirect and direct interactions, and therefore equates (−,−) interactions to interference competition. This does make his classification of interactions more logically consistent than that in most treatments. However, because interference is almost always present as a result of exploitative interactions with shared resources, neither can be fully understood without considering the resource-mediated pathways.

Textbook definitions have not been converging towards a consensus during the past decade. The most theoretically oriented general textbook available currently (Pásztor et al. 2016, p. 23) defines competition as 'individuals sharing one or more regulating factors and thus having a negative effect on each other'. The current best-selling (as of 2019 on Amazon.com) ecology text is Bowman et al. (2017). Their definition (p. 318) is, 'a nontrophic interaction between individuals of two or more species in which all species are negatively affected by their shared use of a resource that limits their ability to grow, reproduce, or survive'. This again combines mechanism and effect, while adding the somewhat unclear requirement of being

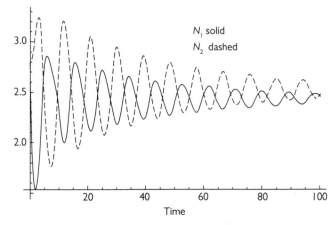

Fig. 2.1 Dynamics of two consumers in a MacArthur system (eqs 3.4), modified by having competition between the two resources. The competitors have opposite relative specializations on the resources, but are otherwise equivalent and the simulation is initiated at their equilibrium densities. The start of the simulation corresponds to an imposition of a 50% increase in the death rate of consumer species 1, from 1.0 to 1.5. The initial consumer abundances are both 2.784. The system will eventually come to an equilibrium in which N_1 has been reduced from 2.784 to 2.435, while N_2 has been reduced from 2.784 to 2.524. The basic parameter values are: $b_{ij} = 0.125$ for all i, j; $d_i = 1$; $c_{11} = c_{22} = 0.4$; $c_{12} = c_{21} = 0.1$; $r_i = 2$; $K_i = 0.02$; $\alpha_{12} = \alpha_{21} = 0.9$ for both resources.

non-trophic, and apparently implying a single resource. (The latter requirement was clearly not meant, given the subsequent examples provided.) The only theory referred to by Bowman et al. (2017) is the LV model, which lacks resources. A somewhat more advanced recent textbook on community ecology (Mittelbach and McGill 2019) cites the works by Grover (1997) and Case (2000), discussed in the previous paragraph, for possible definitions.

In summary, most relatively recent definitions of competition mention resources and they usually require negative effects, although the nature and timing of these negative effects is often not specified. Timing is important because there is often an oscillatory change in the sign of the change in both population growth rate and population size of a second consumer species following the addition or subtraction of individuals of the first consumer (in a two-consumer-species system). For example, Figure 2.1 shows the changes in abundance over time following an instantaneous 50% increase in the death rate of individuals of consumer 1 at time 0. The system represented here is the system described by Levine (1976). This is based on the MacArthur model (eqs (3.4) in the next chapter) modified by having the resources themselves exhibit LV competition in the absence of the consumer. The initial dynamics seem to exhibit competition by many definitions—one species goes up when the other goes down. However, the end result is that harming consumer 1 reduces the abundance of consumer 2, reflecting a positive interspecific effect of 1 on 2.

The continued and widespread use of competition models that lack resources certainly detracts from research into the role of resource dynamics in shaping competition. The most-cited article on competition between species published during the past five years (Levine et al. 2017) mainly considers models that lack resources. (Note that this is not the Levine of the 1976 work referred to above.) Other highly cited works published in recent years have also concentrated on models of competition that lack resources (Saavedra et al. 2017; Mayfield and Stouffer 2017; Hart et al. 2018). One might be able to justify this in cases in which resource dynamics were very fast relative to consumer dynamics, and those resource dynamics were used to derive the form of the consumer-only model. Yet in most cases, this has not been done. O'Dwyer (2018) is a recent discussion of some of the limitations of representing within-and between-species competition for resources based on models without explicit resource dynamics. Lack of resources is particularly problematic when population sizes vary continually.

2.3 What *should* the definition be?

This historical range of definitions illustrates three unsettled questions. The first question is whether the definition should be based on mechanism or effect, or perhaps a combination of both. The second is how narrowly or broadly to delineate either mechanism or effect. The third question is what the operational definition of the interaction should be; what perturbation of species A should be imposed and what response of species B should be measured to determine how one affects the other? Even if the effect on population size is not used to define competition, it is still obviously important for understanding the interaction and its consequences for other species in the community. The definitions of Grover (1997) and Case (2000) use an increase in population size as the perturbation and the per capita growth rate as the response, but this is not what is used to classify most indirect effects (Yodzis 1988; Abrams et al. 1996; see below). Two possible ways to resolve the first two questions would be: (1) to make the definition of 'competition' consistent with the definition of other interactions; or (2) to choose the alternative that is most consistent with developing a quantitative description (i.e., a mathematical model) of the process that can describe population dynamics. Option (1) is not very helpful, as there has been no consistency in how interspecific interactions have been defined, a problem that I wrote about many years ago (Abrams 1987a), but one that has yet to be addressed by the ecological community as a whole. The two other major categories of interspecific interaction recognized in most current textbooks are predation and mutualism (e.g. Mittelbach and McGill 2019). Predation (in the broad sense; this includes herbivory and parasitism) is defined by a mechanism (consuming living organisms), while mutualism is defined based on outcome (mutually positive effects on population growth and/or abundance). Thus, appealing to the manner in which other large classes of named interactions are defined does not help us decide on the definition of competition. On the other hand, criterion (2) for choosing a definition of competition clearly favours mechanism because mathematical models are not based on outcome; in fact, they are generally developed in order to reveal what the outcome should be in cases where it is not obvious.

A definition based on mutually negative effects of increased population sizes does not correspond well to the situations that are usually considered to be competition. Work on indirect effects within food webs has revealed that many different indirect pathways of effects between species can produce mutually negative effects on abundance (Yodzis 1988; Schoener 1989, 1993; Abrams et al. 1996). This includes 'apparent competition', but also includes indirect effects involving more than one intermediate species in the chain, most of which are never considered to be competition. Schoener (1993) suggested the possibility of labelling these cases as 'food web' competition, but that terminology does not seem to have been widely adopted. The theoretical findings of Levine (1976), and similar ones by Vandermeer (1980), had long ago shown that shared use of resources in limited supply was consistent with mutually positive effects of increased abundance of one consumer on the abundance of another. More recent work (including Matsuda et al. 1993, 1994; Abrams et al. 2003; Abrams 2003; Abrams and Nakajima 2007, Abrams and Cortez 2015a) shows that $(+,-)$ effects between consumers of a shared set of resources are also common in models of such systems, as are effects that change sign depending on the magnitude of the perturbation, and on the initial population sizes of both species. Figure 3.2 in the next chapter provides an illustration of these phenomena. The fact that initial abundances and perturbation magnitude can both alter the sign structure of interspecific effects between consumer species argues strongly against a definition based on mutually negative effects.

One could argue that the definition of competition is relatively unimportant. It certainly does not alter the nature of reality. On the other hand, an inadequate definition could narrow the range of systems studied. While ecological systems are infinitely varied, the number of scientists engaged in studying interspecific interactions is many orders of magnitude smaller. The interactions that are chosen for study tend to be those that fit within the framework of the defined interactions and effects. For instance, the first empirical system characterized by at least one positive interaction between species that share competing resources was studied shortly after the theoretical studies by Levine and by Vandermeer were published. This was a case in which ants and rodents competed for seeds, but the large-seeded plants, which were mainly eaten by rodents, outcompeted the smaller-seeded plants, used mainly by ants (Davidson et al. 1984). The net result was a positive effect of greater abundance of rodents on the abundance of ants, despite their joint use of intermediate seed sizes. However, the possibility of positive effects via shared resources seems to have received little empirical attention since then. During the intervening years, there has been no general theoretical explanation suggesting such positive interactions should be rare. Given this state of affairs, the most likely explanation is that ecologists are either ignoring the whole question or choosing not to investigate cases where positive effects between different consumers that share resources are deemed a likely possibility. Alternatively, ecologists may be failing to publish the 'negative result' of an increase or stasis in the population of one species following an increase in the population of another species that shares resources with the first. Without knowing the effects of species on resources, such results could be attributed to changes in uncontrolled or unmeasured environmental conditions during the course of the experiment. A final possibility is that systems having many competing resources seldom exhibit positive effects, but theoreticians have not systematically explored this possibility. A broad definition of competition based on shared resource use seems most likely to avoid the narrowed scope that has characterized the past several decades of studies on species that share resources.

Because of the arguments reviewed above, the definition that will be used here is: *Competition between two species occurs when a population (species) shares one or more depletable resources with the other population (species).* Resources are substances that contribute to population growth and whose availability to other consumers can be reduced as a consequence of being used (occupied or consumed). Substances that are too abundant to be altered by consumption are not included in those involved in competition. (Oxygen in most aboveground terrestrial environments is a substance that contributes to population growth but is typically too abundant for the consumption to affect its abundance.) Competition need not produce mutually negative effects, no matter what definition of 'effect' is used.

2.4 Implications of the definition

A few of the advantages of going back to such a strictly resource-based definition are:

1. It makes competition theory more consistent with general theory on indirect effects. The dynamics of the intervening entities (including biological species and resources) are important parts of the theory on indirect effects generally (Yodzis 1988, 1989); they have always been included in models and descriptions of the other interactions involving a single intermediate population/entity (e.g., apparent competition and top-down or bottom-up interactions in three-species food chains).

2. It provides a basis for describing and measuring the full range of factors that influence the effect that one consumer may have on another. These include changes in any of a consumer's traits that affect its interaction with its various resources, changes in the interaction between the resources, and changes in the abundance of one or both consumers.

3. It establishes a more productive approach to understanding the evolution of the traits that define the competition within and between species, as many of these traits affect both relative and absolute uptake rates of different resources.

Predicting the effects of environmental and evolutionary changes on species' abundances represents two of the main goals of ecology. In addition, the science of ecology should be useful in explaining some of the properties of existing ecological communities and how they and their component species have changed over time. All these goals require a quantitative description of the interactions between dynamic components of the system. If we accept that competition always involves effects on resources, then it seems that a minimal model would need to contain: (1) some description of resource dynamics in the context of the community in which they occur; (2) descriptions of rates of consumption of the various resources by the consumers; and (3) the effects of those consumption rates on the demographic rates (birth, death, growth) of the consumers. These are the components of a minimal consumer–resource model with a single consumer species as well (Case 2000; Murdoch et al. 2003; Otto and Day 2007).

Models of interspecific competition from the period 1973–1980 that included resources often had results that differed from those of the most comparable models that lacked resources (e.g., Schoener 1973, 1974c, 1976, 1978; Abrams 1975, 1977, 1980a). Calls for representing competition using resource-based models continued in the 1980s (e.g., Abrams 1983a, b; Schoener 1982, 1986; Tilman 1982, 1987), and have been repeated more recently (McPeek 2019a, among others). Even if a model is developed by assuming that resources are at quasi-equilibrium with respect to current consumer densities, deciding on the form of that model requires knowledge of the dynamics of the resources. Once a resource-based model is constructed it makes little sense to replace it with an approximation to it that leaves out resources (Abrams 1975). Letten and Stouffer (2019) is a recent analysis that nevertheless does so. They show that consumer-only models with fitted nonlinear terms can better approximate resource-based models than can the LV model. However, except as part of a critique of the LV model, it is unclear why one would need or want to replace a model with explicit resources by an approximation that lacked them. Chapter 6 will show that simple models with explicit resources often predict that the interspecific effect of one consumer species on another is positive rather than negative, at least for some ranges of population sizes. Chapters 8 and 9 demonstrate that explicit representation of resource dynamics is particularly important for understanding competition in variable environments.

At this point, a brief digression regarding the three component functions of consumer–resource interactions identified above is called for. It turns out that our current understanding of each of these is at best inadequate. This theme is developed in more detail in Chapter 3. First consider resource dynamics. The subsequent chapters suggest that we lack a good understanding of the most likely shapes of population growth functions when the resources are themselves alive ('biotic') and are therefore resource-consuming entities. For biotic resources, the same three features identified above must be known, which includes the dynamics of the resource's own resources, as well as its functional and numerical responses. If the resource is not self-reproducing ('abiotic'), the dynamics depend on input and output processes. Each one of these may be affected in many ways by the amount of resource present

(Abrams 1988a). However, virtually the only model of abiotic growth considered in the literature is one in which resources enter a system at a rate that is totally independent of the amount present, and resources leave the system at a rate directly proportional to the amount present. This is an accurate description for a laboratory apparatus that is widely used in biology, known as a chemostat, which involves constant input of resources into the system with a constant outflow of medium including resources (and often consumers as well). The chemostat was designed to deal with the problem of waste accumulation in small closed systems in the laboratory. Studies comparing chemostat resource dynamics to abiotic resources in natural systems are notably lacking. Abiotic resources include living organisms that have special properties and do not immediately reproduce themselves. Predators that can only capture limited classes of individuals from their prey (such as injured, newly born, or senescent individuals) can often be described as using abiotic resources. While on the subject of consumers and resources, note that predator–prey will be used synonymously with consumer–biotic resource; in other words, a herbivore–plant system is a predator–prey system. However, consumer–resource also includes abiotic (non-reproducing) resources.

The numerical response has been the subject of surprisingly little research, which has led to the assumption that it is linear in the vast majority of models. The functional response was made famous by C. S. Holling, in the 1950s, and has been an object of study for many years. However, that body of knowledge has not been very useful in studies of competition. Some of the reasons for this are given in Box 2.1 below. More details are provided in Chapter 3.

Box 2.1 Functional responses and their role in competition

One of the key elements of any consumer–resource system involving foods or nutrients as resources is the set of 'functional responses' of the consumers to the resources. A functional response is the relationship between the abundance of the resource and the amount 'consumed'. That relationship may depend on the abundances of other resources and on the abundances of a variety of other interacting species. Functional responses have been a key element of consumer–resource systems from the time that C. S. Holling (1959, 1965) proposed a set of four (later often reduced to three) common shapes for systems having a single type of resource. These relationships implicitly assumed no effect of abundance of other resources, predators, or any species other than the one resource in question, a problem discussed in Chapter 3. MacArthur (1970, 1972) assumed linear (Holling type I) functional responses for the consumers in his exploration of the connections between one particular consumer–resource model and the LV model. Type I responses imply no upper limit on the intake rate of the resource, a biological impossibility that is not present in Holling's types II, III, or the seldom-mentioned type IV. Oaten and Murdoch (1975) extended the type II response to systems with multiple resources, but did not discuss this in connection with competition. A type II response to resource 1 in a system with two resources (R_1, R_2) is represented by $C_1 R_1/(1 + C_1 h_1 R_1 + C_2 h_2 R_2)$. Here C_i is a per capita attack rate while searching and h_i is the handling time for resource i. Rosenzweig and MacArthur (1963) demonstrated that type II responses (but not type I) could generate sustained population cycles in

Box 2.1 *Continued*

1-predator-1-prey systems. Later, Armstrong and McGehee (1976a, 1980) showed that a type II response in one consumer species could allow coexistence of two consumer species on a single self-reproducing ('biotic') resource. Abrams (1980a) investigated systems with two consumers and two resources, and showed that similar type II consumer functional responses in both species made the strength of competition very sensitive to the absolute and relative abundances of the two consumer species; i.e., competition was a highly nonlinear function of abundances. Abrams' (1980a) analysis was confined to systems that did not cycle, but the dependence of measures of competition on the parameter values of the model is also nearly universal in cycling systems, as well as being significantly different from the corresponding measures in analogous systems having type I responses (Abrams et al. 2003; Abrams 2004b). Subsequent empirical work on competition has largely ignored functional responses, except for some studies of phytoplankton competing for abiotic resources; these have largely assumed the simplest form of the Holling type II response (see Abrams 1990c for other forms). The form of resource dynamics could be equally important in determining the strength of competition and the linearity of competitive effects (Abrams 1980b). Functional responses are the central topic of the following chapter, where more details are given. However, the simple early results reviewed in this box illustrate why considering these responses, and the resource dynamics they produce, are essential for understanding competition. This argues strongly for including resources in the definition of competition.

Ecologists should not be (and generally are not) only interested in the change in population size in one competitor that is caused by adding or subtracting some particular number of individuals of a single competing species (or adding or subtracting them at some specific rate). They should have a goal of being able to predict the effects of any feasible magnitude of change in the abundance of any competing species, as well as the effects of any environmental change that alters the dynamics of at least one of a set of competitors or their resources in some way. For example, what if the consumption rates of all resources are reduced for one consumer species, but unaffected in a second consumer? Will the first species decrease and the second increase? Resource-based models imply that this is often not the case (Abrams 2002, 2003, 2004a).

An understanding of interactions with resources is also needed to determine the relative fitness of individuals with different traits; this, in turn, is required to understand the evolutionary trajectory of that species and many of those that interact with it. By leaving resources out of a model, it becomes difficult to do any of this, and the lack of resources has already been a source of misunderstanding. This is true for intra- as well as interspecific competition. Leaving out resources in models of single-species growth has led to the idea that adaptive evolution will not decrease population size. This traditional view was buttressed by early models of r- and K-selection based evolution of the parameters of logistic growth, the traditional model of intraspecific competition. However, this evolutionary result was shown to be inconsistent with consumer–resource models by Matessi and Gato (1984). Evolution of the rate of resource consumption reduces population size in a large range of systems with self-reproducing resources. Nevertheless, the popularity of the logistic model of intraspecific competition (density dependence) has led to the persistence of the incorrect view that adaptation always increases population size, and maladaptation decreases it (see Abrams 2019). Similarly, the persistent belief that evolution favours divergence in the relative use of different resources when two competing species come into sympatry is inconsistent with many resource-based models (Abrams 1986a, 1987a, b; Vasseur and Fox 2011). More details on evolutionary responses to competition are provided in Chapter 11.

Even if we put aside the issue of resource dynamics, the definition of competition based on mutually negative changes in population size is inconsistent with the outcomes of many models having three or more competitors (even for models as simple as the Lotka–Volterra model). Most ecologists in 1970 would have admitted that competition between species was not confined to species pairs. Yet the bulk of theory on competition is still based on the two-species case. Levine's (1976) demonstration that an increase in one competitor's population could increase that of a second competitor by causing large changes in the abundance of a third competitor should have changed this narrow focus. This was followed by a more detailed analysis of sets of three-or-more species interacting according to the LV model (Gilpin and Case 1976; Lawlor 1979), which again demonstrated that net positive effects were possible in many cases. These results showed that mutually negative changes in population size cannot be the defining feature of competition, unless one places restrictions on the range of admissible systems. Allowing only two competitors would reduce, but not eliminate this problem, as shown by Levine's (1976) example with competing resources. In addition, there is the well-known example of rock–scissors–paper types of interaction in 3-species competitive systems (e.g., May and Leonard 1975; Laird and Schamp 2006; Soliveres et al. 2015). These involve intransitive competition in which, for example, species 1 would exclude species 2 in a strictly two-species system; 2 would exclude 3, and 3 would exclude 1. Long-term coexistence of all three can occur in many such systems. Such systems also are characterized by the fact that mortality applied to one competitor can increase its abundance (Cortez and Abrams 2016), a result that affects some sign-based definitions of interspecific interactions. A variety of views have been expressed on whether such intransitive competitive systems are common. Soliveres et al. (2015) make the case that they occur frequently. Even if this assessment is not correct, there is no doubt that intransitive competition does occur, and interactions in such systems cannot be described as mutually negative effects on abundance.

The above discussion does not imply that models lacking explicit representation of resource dynamics are always inadequate for answering all questions regarding the population-level effects of competition. Competition for space is one example, since the resource can reappear immediately when an occupying individual dies. However, even here, death may not immediately produce usable space; it may be necessary for most of the remains of the dead individual to decay or be otherwise removed before the space is again suitable for occupancy. Similarly, in a more-or-less closed aquatic system with competition for nutrients, the nutrients contained in dead or consumed individuals may be rapidly recycled, potentially allowing the approximation that the amount of resources in the system is functionally determined by the current number of consumers (e.g. Miller and Klausmeier 2017; Kremer and Klausmeier 2017). However, most recycling takes time, so the dynamics of space regeneration may be important. In any event, having a definition of competition based on shared use of resources makes it far more likely that such assumptions are assessed before making predictions using a simpler model. And the proper form of the consumer-only approximation should be determined using a reasonable model of the dynamics of the resource.

2.5 Competition within the framework of food webs

In the 1970s, competition and predation were regarded as the primary interactions between species. This classification of interactions was illustrated by several articles in the 1975 book edited by Cody and Diamond titled 'Ecology and Evolution of Communities'. This book contained chapters by many of the most prominent ecologists of the time. Several of those chapters discussed the relative importance of competition and predation for the dynamics of communities. Notably, Connell (1975), who had become well known for his field experiments on two competing intertidal barnacle species, argued that predation had a greater effect on community composition than did competition. Of course, exploitative competition for living resources is a consequence of the predation of the consumers on those living resources, so one of the necessary consequences of predation in any community whose predator species have overlapping sets of prey is to produce interspecific competition. A specialist predator experiences intraspecific competition as a direct result of its predator–prey interaction. These cases show that a comparison of the 'relative importance' of predation and competition is fundamentally meaningless. However, it may be useful to compare the magnitudes of indirect effects transmitted between two species on a given trophic level via higher vs lower trophic levels.

The most common classification of interactions in textbooks from the 1980s onward is that based on the three possible pairs of '+' and '−' signs: (+,+) is mutualism; (−,−) is competition; and (+,−) is predation. Mutualism was often ignored in the 1970s but is now included in all textbooks. The two signs are usually used to denote the effect of an increase in abundance of species 1 on the abundance or per capita growth rate of species 2. However, this classification suffers from ambiguity because the time interval is not specified, and both time interval and the magnitude of the initial change in abundance can alter the sign of the effect on the receiver. This is true of predation as well as competition. In addition, the change in short-term per capita growth rate may have a different sign than the change in equilibrium or long-term average abundance. In a system that does not come to a stable equilibrium, the sign characterizing the change in the equilibrium may differ from the sign that describes the change in the mean abundance. Adding more predator individuals does initially decrease prey abundance, barring the possibility of increased maladaptive fighting among predators. However, it is quite possible that, by the time the system reaches its new equilibrium, increased productivity of the prey's food more than offsets the predator's effect, and the prey actually increases (Abrams 1992b; Peacor 2002).

While it is possible to argue for a sign-based classification of interactions, if this is adopted, the nomenclature needs to be changed. Predation is usually defined as a mechanism of interaction, while mutualism is usually defined as a paired set of positive effects, each of which may be produced by many mechanisms. The inconsistency of using mechanism and effects for different interactions was discussed in Abrams (1987a). Another problem with the classification framework based on a fixed pair of signs is that it favours thinking about interactions purely in terms of species pairs. The interaction between almost any randomly selected pair of species in a natural community is likely to depend on the abundances of many additional species, and the sign of one or both effects may be changed by these additional species. Thus far, I have largely ignored the impacts of trophic levels higher than that of the competing consumers, but it has long been recognized that the presence of a specialist, food-limited predator on each of the consumers can eliminate the effect of competition between them on each other's equilibrium abundances. In spite of eliminating the impact of a neutral parameter change in one consumer on the population size of the other, the specialist predators in this case do not eliminate the evolutionary effects of competition between the consumers (Abrams and Cortez 2015b; see Chapter 11). Trophic level(s) lower than the resources (if they exist) are also likely to affect the interaction between consumers. This was implicit in Levine's (1976) model, although he represented the competition between resources using direct effects.

An even wider variety of more complicated food web connections between resources is possible in systems with two or more trophic levels below the prey. In predator–herbivore systems with potentially competing predators, the plants can have effects other than simply defining the food available to the herbivores. For example, plant species can alter the structure of the habitat in a manner that makes one or more herbivore species more vulnerable to their predators (for example, making the predators more difficult to detect; e.g., Holt and Barfield 2013; Pearse et al. 2020). This can alter the competition between different herbivore species. Effects of other trophic levels can arise when there are species at higher or lower trophic levels, whose abundance changes the behaviour of one or both species of a focal consumer–resource pair in ways that alter their interaction (Abrams 1984b, 1995; Preisser et al. 2005; Bolnick and Preisser 2005). Theory suggests that such behavioural effects can be transmitted through multiple trophic levels (Abrams 1992a). Examples of adaptive behavioural effects will be provided in the following chapter. In general, these effects broaden the range of potential outcomes for the interaction of any particular pair of species.

The currently proposed definition of competition based only on shared use of limiting resources means that a pair of species can be competitors while having mutually positive effects on each other's abundance (measured using neutral parameter perturbations of small magnitude). Matsuda et al. (1993) discussed a model of two predators sharing a single prey, in which the prey species exhibits adaptive defence with the property that defending against one predator species makes the prey more vulnerable to the other predator. This always entails short-term positive effects of increased abundance of one predator on the per capita growth rate of the second. However, the long-term effect on the abundance of the second predator may be negative or positive, depending on the magnitude of the change in prey population and the effectiveness and exclusivity of the two defences (Matsuda et al. 1993). Even with mutually positive effects on ultimate population sizes, the definition of competition proposed above would include this as competition because it involves a shared resource. The fact that the outcome of 'mutualisms' may depend on other species or environmental conditions has long been acknowledged in the literature dealing with that interaction (Bronstein 1994). However, this conditionality is seldom mentioned in discussions of competition.

The issue of sign-based vs mechanism-based classifications of interactions brings up the more general issue of how to quantify the 'effect of one species on another'. This seemingly innocuous phrase actually has many potential interpretations, as should be clear from the above. What is changed in the effector species and what is measured in the affected species? Over what time frame is that quantity measured? Bender et al. (1984) raised these and other ambiguities in measuring interspecific effects, and they were explored by Yodzis (1988, 1989). This topic will be examined further in the following chapter. That chapter will also provide a more detailed introduction to the nature of consumer–resource interactions.

3

Measuring and describing competition

A consumer–resource framework

3.1 The measurement of competition (and other interactions)

The previous chapter on defining competition leads to questions about how it should be described and measured. A definition based on resources means that there is a large body of theory about consumer–resource interactions more generally that can be applied to competition. Murdoch et al. (2003, p. 1) open their book on consumer–resource interactions, with the statement: 'The consumer–resource interaction is arguably the fundamental unit of ecological communities. Virtually every species is part of a consumer–resource interaction as a consumer of living resources, as a resource for another species, or as both . . . If we are to understand population regulation and its various manifestations, we therefore need to focus on consumer–resource interactions.' Intraspecific, as well as interspecific, competition depends on consumer–resource interactions. All of predation theory is based on a subset of consumer–resource theory as well. Host–disease interactions are a separate category of consumer–resource interactions. This chapter argues that measurement of almost all ecological interactions requires a good description of the consumer–resource components of the overall interaction. It also suggests that current consumer–resource theory needs to be expanded to take great account of the role of behaviour and other adaptive processes in interspecific interactions. This need has been stressed many times in the past (Abrams 1982, 1984b, 1995; Brown et al. 1999; Bolker et al. 2003).

Measuring the effect of one consumer (competitor) on another requires an appropriate experiment, whether it is carried out using the real system or a mathematical model. The 'experiment' using a model may be a simulation or an analytical calculation. Measuring the 'effect' of one species on a second requires a description of the perturbation to the first species and the temporal responses of both species following that perturbation. Bender et al. (1984) established the division of perturbations into 'pulse' and 'press' types. A pulse is a one-time and effectively instantaneous increase or decrease in the population size of the initiator species. This change will often die out over time; in a foodweb context, the perturbed population will often oscillate up and down around the original equilibrium during this process. However, knowing

Competition Theory in Ecology. Peter A. Abrams, Oxford University Press. © Peter A. Abrams (2022).
DOI: 10.1093/oso/9780192895523.003.0003

the short-term effects of a perturbed initiator population on other 'receiver' populations is useful because they can often identify quantities that need to be considered in constructing a dynamic model of the interaction.

The long-term effect on the abundance of a second species must be established using what Bender et al. (1984) called a 'press' perturbation. This involves a sustained change in some property of the initiator species. For example, instead of the one-time introduction of individuals in a pulse perturbation, the input of individuals is sustained over time in the corresponding press perturbation. The parameter that is changed in such a press experiment is ideally one that causes a unidirectional change in the per capita population growth rate of the initiator species, but that does not directly affect any other species in the system (Yodzis 1988, 1989; Schoener 1989, 1993; Abrams et al. 1996). Because of their lack of an immediate effect on the per capita growth rate of any other species, such parameters will be referred to as 'neutral' parameters here.

In a competitive interaction, the resources used by consumer species 1 will usually change following any alteration in the input of individuals of consumer species 1; these changes would eventually alter the abundances of other consumers that share at least some of the same resources. However, there might be fluctuations in all of the abundances as they approach the new equilibrium (dynamic attractor), as illustrated in Figure 2.1 in the previous chapter. In the case of a very small perturbation, it is unlikely that the change would produce sustained cycles. In most cases, the system would either be stable before and after the change of input, or be unstable in both cases. In the latter case, effects of the first consumer would have to be measured based on long-term average abundances after the system approaches its new limiting dynamics. It is possible for the initial change to be large enough to change the stability or to shift the system to a new attractor, assuming such an attractor exists.

If the perturbation is small enough, different neutral parameters will produce similar relative effects on the sizes of the two populations involved. However, this is not true of larger perturbations, or of non-neutral parameters. The neutral perturbation used to define interspecific effects in most previous theory is usually either the immigration or per capita mortality rate of the initiating species (Yodzis 1989). Both immigration and per capita mortality are quantities that can usually be manipulated in laboratory experimental systems. The same is true for at least some field scenarios as well. However, larger magnitude perturbations usually produce different effects per unit of parameter change than do small perturbations. In real experiments, very small perturbations do not produce large enough effects on populations to distinguish them from measurement error. For the two traditional neutral perturbations, greater immigration is usually a stabilizing factor, while lower mortality is often destabilizing. This may need to be taken into account in experimental measurements. If the system is nonlinear, the cycle amplitude often has a large impact on mean population size and is sensitive to neutral parameters. In this case, neutral perturbation size can have a particularly large impact on the measured effect.

In most food web models the majority of parameters are non-neutral; they directly affect the dynamics of two or more species. This is true for a change in any consumer's consumption rate of one or more resources. Such a parameter in one consumer will produce immediate effects on both the consumers and the affected resource(s), and will usually alter the magnitude of any interspecific effect because of this. The fact that two species have their growth rates affected simultaneously makes such non-neutral parameters inappropriate for assigning a sign to the effect of one species on another. This, in turn, means that a greater variety of effect signs are associated with non-neutral parameter change. For example increased consumption of a resource used by one consumer species may actually increase the equilibrium population sizes of both the resource and a competing consumer species, even in simple all-linear models (Abrams 2002, 2003). The direction of the response of another species after perturbing any parameter in the focal species may depend on the magnitude of the change. Thus, a complete picture of the interaction should consider the full range of potential magnitudes and parameter identities.

Nevertheless, if one wants to characterize an interaction by a single figure, it is reasonable to use a neutral parameter and consider a small magnitude change. A small magnitude parameter change is particularly useful from a theoretical perspective because analytical methods of determining the effect are possible when the equilibrium is stable (Yodzis 1988, 1996). Unfortunately, the standard approach in experimentation is to use a large magnitude change in order to obtain a statistically significant measurement of the response. This requirement has limited the application of the theoretical results. If the per capita dynamics are close enough to linear and the perturbation is not too large, a linear extrapolation of the 'small perturbation' outcome may provide a reasonable approximation to the actual change in abundance of the competitor. Another important consideration is that alternative equilibria or attractors exist in many models and real systems (e.g. Scheffer 2009); this was documented long ago for the LV model in systems having more than two competitors (Gilpin and Case 1976). The impacts of a parameter change on population sizes can be very different, depending on which equilibrium/attractor the system occupies. We still don't have any good estimate of how often alternative attractors arise in resource-based multi-species competition models, or in any type of natural communities.

In some cases, behavioural changes in resource species may produce a large part of the effects of a neutral parameter change in one consumer species on other consumers. Models and a large number of experiments have explored the possibility of such behaviourally transmitted effects in food chains (Abrams 1984b, 1995; Matsuda et al. 1993, 1994, 1996; Werner and Peacor 2003). Depending on the spatial locations of different consumer species and/or their foraging behaviours, it is possible that defending against one consumer entails greater risk from the others. Alternatively, defences may be generally applicable to both (or all) consumers, in which case the behaviour mobilized against one predator will reduce the capture rates of other predators. Both behavioural pathways imply rapid effects of altered predator abundance on the per capita capture rates. Changes in population growth rate due to such behavioural shifts are often referred to as trait-mediated effects, and are important, but largely ignored, components of consumer functional responses. They occur in plants as well as animals. For example, Ohgushi (2005) reviewed the effects of trait-mediated changes in plant defences against herbivores and finds that these usually cause positive rather than negative indirect effects between their consumers. This occurs because defences tend to be consumer-specific, so the initial effect of greater defence against consumer A is usually an increased intake rate by consumer B. This would be classified as a positive effect on the immediate per capita growth rate of species B. However, such a change need not mean an increase in the ultimate population of consumer B relative to its pre-manipulation state.

Models have shown that defensive behavioural/phenotypic changes in the resource species may enhance or reverse the effect that would have occurred with fixed behaviours (Matsuda et al. 1993, 1994, 1996; Sommers and Chesson 2019). Even in cases with large trait-mediated effects, there is likely to be some population-level component, so press perturbations to mortality or immigration can provide a measure of inter-consumer effects. However, behavioural change usually increases the non-linearity of interactions (Abrams 1984b, 1992c, 1995), so the ability to predict the effects of large-magnitude neutral perturbations is even more limited in these systems. The effects of both consumer and resource behaviours on the functional forms of consumer–resource models are discussed in more detail later in this chapter.

3.2 Methods of measuring and describing competition

The general principles outlined in the previous section will be translated into explicit formulas and illustrated for some simple models in this section. I begin by giving formulas for very small press perturbations in a neutral parameter. The formulas assume that species j is the initiator and they use the per capita mortality rate of consumer species j as the neutral parameter. Expression (3.1) below is the negative of the effect of the mortality of species j on the equilibrium population of species i. Measure (3.2) standardizes this relative to the change produced on species j itself. The competitive effect is then the change in recipient species i's equilibrium population $\hat{N}_i$ with a small increase in d_j divided by the corresponding change in the population size of the initiator species j. The two measures are:

$$-\partial \hat{N}_i / \partial d_j, \tag{3.1}$$

and

$$\frac{\partial \hat{N}_i / \partial d_j}{\partial \hat{N}_j / \partial d_j}. \tag{3.2}$$

These will be referred to as measures 1 and 2 below. The negative sign in expression (3.1) is present because increased mortality of species j has a negative effect on its own population growth, and interaction signs are usually defined by the impact of positive effects on (or addition of) the initiator. Both measures above represent the response of population size of species i to a very small press perturbation (Bender et al. 1984) in the per capita mortality of species j. The standardization involved in expression (3.2) makes it equivalent to the competition coefficient in the LV model given in Chapter 1. If increased immigration (rather than mortality) is used as the perturbation, the sign of expression (3.1) is reversed, as this represents a positive perturbation to species j. Using immigration rather than mortality does not change the sign of expression (3.2). For the moment, I will assume that there is at least one stable equilibrium point and that the population densities in the formulas are the densities at such an equilibrium point. The same procedure can be applied to long-term mean densities in systems with sustained fluctuations (see below), although there is generally no closed-form solution giving the temporal course of population size. As a result, the analogues of expressions (3.1) and (3.2) based on mean densities will generally need to be evaluated numerically, using a small but finite change in mortality (or immigration). The responses of population sizes to larger magnitude changes in d_j can be used to quantify interspecific effects of such larger perturbations; these measures will not, in general, be identical to the values given by the above formulas.

For characterizing the strength of interspecific competition, the second measure above seems preferable to the first in at least two respects. Firstly, as noted above, it is independent of the neutral parameter used, so it produces the same measure if immigration (or another neutral consumer parameter), rather than per capita mortality were used. That is not true of measure (3.1). Secondly, measure (3.2) is consistent with the common method of measuring competition in experimental studies (i.e., measuring the change in a recipient population for a given change in the initiator's population). In a two-competitor system, measure 2 corresponds to the competition coefficient α_{ij} in the traditional parameterization of the Lotka–Volterra model in which intraspecific effects of a consumer on its own per capita growth rate have been scaled to one (as in Chapter 1). Thus, measure 2 is a ratio of inter- to intraspecific effects.

On the other hand, because it is a ratio, measure 2 does not always identify cases where the change in the receiver species' population (measure 1) is large or small in magnitude. Measure 2 can be very large in some cases because the effect of a parameter of species j on its own abundance is very small. This makes the denominator of measure 2 small, which can make the ratio very large. Measure 2 produces a positive sign for the interaction in those cases in which both species increase in response to greater mortality of the initiator. These are usually cases when species j is characterized by a generalized 'hydra effect' (Abrams and Matsuda 2005; Cortez and Abrams 2016; Abrams 2019); i.e., the equilibrium population of the initiator increases in response to a neutral parameter change that lowers its own per capita growth rate (or the opposite: decreases in density in response to a parameter change that increases per capita growth rate). Such outcomes are not uncommon in situations with asymmetric competition, even in MacArthur's consumer–resource model (Abrams and Cortez 2015a). Abrams (2009a) and Cortez and Abrams (2016) provide a larger number of other theoretical examples. This possibility was not considered in Schoener's earlier (1993) review of indirect effects, where he stated (p. 366) that, 'In the most precise existing theoretical formulation, an effect is defined as a change in equilibrium population size caused by a change in the *input* or *abundance* of another species.' The italics have been added because of the problematic fact that increased input may cause decreased equilibrium abundance of the perturbed species. As noted above, hydra effects (where increased input decreases abundance) occur in a significant fraction of simple models of small groups of interacting species, and such effects can be large in magnitude (Abrams 2002, 2009a; Abrams and Matsuda 2005; Abrams and Cortez 2015a, Cortez and Abrams 2016). In systems with shared resources, hydra effects will often produce positive effects according to measure 2. Very small effects of the parameter on the abundance of the perturbed species can imply that measure 2 is large in magnitude even when measure 1 is relatively small.

Neither expression (3.1) nor (3.2), which are both defined for a particular equilibrium point, provides a full description of an interaction because both change depending on the initial mortality rates. In theoretical studies, this is not a problem, because the values can be determined for any set of mortality rates. Unfortunately, the dependence on the mortality rate makes it difficult to impossible to characterize the 'strength' of competition by a single number. As I argue below (see also Abrams 2001c), a full quantitative description of the consumer–resource dynamics is essential for understanding the relative change in abundances produced by any given perturbation to one consumer. The dependence of sign and magnitude of relative effects on both initial population sizes and the size of the perturbation makes single measures of competition insufficient for understanding the dynamics of most competitive systems. However, these are not problems with the Lotka–Volterra model, where competitive effects are simply assumed to be independent of the abundances of the competitors, and independent of any neutral parameter.

Even in a two-species system, the outcome of a press perturbation to the per capita mortality rate is usually significantly altered if there are sustained non-equilibrium dynamics in the system. In this case, expressions (3.1) and (3.2) should be reformulated as effects on long-term mean densities, i.e., equilibrium population sizes should be replaced by long-term averages of population size. Effects on long-term mean densities may differ in sign from those defined by the change in the equilibrium point in systems having endogenously driven cycles due to a saturating (type II) functional response in one or both consumers (Armstrong and McGehee 1980; see Box 2.1). This is also the case when sustained environmental fluctuations drive population cycles (e.g., Abrams 2004b). These issues have largely been ignored in the competition literature based on LV equations. This limitation is in part because of the focus on the 2-species LV model, in which cycles cannot occur without environmental forcing. However, even when the LV model is modified by seasonal variation in a parameter that affects a density-independent component of per capita growth rate (i.e., a neutral parameter), the linear dependence of per capita growth rate on population sizes ensures that the mean population size is equal to the equilibrium in an otherwise equivalent model that lacks environmental variation. This is not the case in models with nonlinear per capita growth rates or in cases when the environmental forcing affects a nonlinear term in the expression for per capita growth rate (see Chapter 8).

Changes in non-neutral parameters are likely to be involved in most responses of competitive communities due to climate change (Gilbert et al. 2014; Amarasekare 2015), as well as many of the changes produced by the addition of higher-level predators (Abrams 1995; Suraci et al. 2019). The presence of adaptive phenotypic plasticity (including behaviour) in one or more species means that changes in neutral and non-neutral parameters are often coupled (Abrams 2019). This implies a wider range of potential signs of interactions when changes in non-neutral parameters occur either as a direct response to the environment or as an indirect response to changes in neutral parameters or resource abundances. The absence of resources from many formal definitions of competition (and from the LV equations) has delayed the development of theory for understanding these non-neutral parameters. Thus, it is important that the impact of non-neutral parameters be more fully incorporated into theory of interactions transmitted by resources, i.e., that expressions (3.1) or (3.2) should only be two of many measures of potential interactions transmitted or influenced by shared resources.

Having reviewed a number of complications involved with measures (3.1) and (3.2) in consumer–resource models, it is useful to return to the Lotka–Volterra model, both to illustrate the use of these expressions, and to show their exceptionally simple properties in the case of this particular model. The normal LV model parameterization (see Chapter 1) is changed so that each species has a single neutral parameter. The inter- and intraspecific competition terms are represented by α_{ij} and α_{ii}, and dynamics are described by

$$\frac{dN_1}{dt} = N_1 \left(r_1 - \alpha_{11}N_1 - \alpha_{12}N_2 \right)$$
$$\frac{dN_2}{dt} = N_2 \left(r_2 - \alpha_{21}N_1 - \alpha_{22}N_2 \right)$$

$$(3.3a, b)$$

Here r_i (which is still the maximum per capita growth rate) can be understood as the maximum per capita birth minus death rate of species i, and it is a neutral parameter. The α_{ii} terms are strengths of intraspecific competition, which were given by $1/K_i$ in the traditional parameterization of Chapter 1. The α_{ij} terms are strengths of interspecific competition, which were given by α_{ij}/K_i in Chapter 1. The general formulas for effects of greater per capita mortality given by measures 1 and 2 above can be represented by examining the effect of a reduced r_i; alternatively, one can simply examine a positive change in the r_i and change the sign of expression (3.1). The interspecific effect of a small increase in r_2 on the equilibrium N_1 is given by $\alpha_{12}/(\alpha_{12}\alpha_{21} - \alpha_{11}\alpha_{22})$; this is always negative, as the denominator must be negative for the equilibrium to be stable. The two species will not coexist if this condition on the denominator is not met.

Measure 2 above is given here by the ratio of the effect of r_2 on the equilibrium N_1 relative to its effect on the equilibrium N_2; this quantity is $-\alpha_{12}/\alpha_{11}$. This is the same ratio of effects on the two population sizes if N_2 were maintained at a slightly higher value by some external interventions (and so was removed as a dynamic variable). In this case, eq. (3.3a) alone determines the abundances. The α_{ij} terms do not change with the addition or loss of another competing species in the LV model, although the formulas for the equilibrium abundances do change. In the two-competitor case, the increase in r_2 required to bring about extinction of competitor 1 can be linearly extrapolated from the effect of a very small change in r_2. There are no abrupt jumps in equilibrium population size with any small perturbation. Thus, the standard theoretical approach for investigating indirect interactions—a press change in a neutral parameter—yields the same measures of interaction strength and conditions for coexistence as would be obtained by manipulating population size directly and holding it at the new density. In addition, the magnitude of the perturbation does not affect the relative population changes of the two species. These claims are true provided that the LV model accurately describes the system. As shown below, this simplifying feature does not characterize most consumer–resource models.

3.3 Arguments against resource-based definitions and models

The idea that resource dynamics should be part of most models of the dynamics of competing species seemed to be more widely accepted 40 years ago than it is today. At that time I thought that the Lotka–Volterra model was likely to be largely abandoned in the future, as its form had been shown to be a very special case that was inconsistent with almost all consumer–resource models. This abandonment never happened; Chapter 4 examines a number of influential recent theoretical articles that either use the LV as their primary model or fail to include resource dynamics. It is useful to ask why methods of defining and measuring interactions have not been reformed.

There are two main arguments against requiring resource dynamics. The first is that inclusion of resources is just one of many ways one could make the LV model more realistic. Representing resources is no more important than representing age or size structure or intraspecific variability in resource use or evolutionary change, or a number of other factors left out of most models in community ecology. The second argument asserts that some forms of competition do not include resources, so the necessary inclusion of resources would rule them out.

Bolker's (2008) book is often cited in support of the first argument above. He pointed out that there was no clear division between 'mechanistic' and 'phenomenological' models; there was no obvious point at which you should begin or end with the addition of real-world features that were not represented in simpler models. The difference between resources and the other omitted details, however, is the near-universal presence of resources, even under the negative effects definition, and the completely universal presence of resources in the historical definition of competition used here. There is no other class of indirect interactions in which the intervening species/entity is left out of the definition or left out of models. This is largely because the dynamics of intervening species/entities plays a major role in determining the nature of the interaction. The only difference between competition and other indirect interactions involving 3 dynamic entities is that the intermediate entity need not be a biological species under competition; it can be detritus or a nutrient or sunlight or space (or one of many other entities that are not self-reproducing). Although some of these intermediate entities can reach equilibrium with respect to the consumer more rapidly than most biological species, they are equally likely to have dynamics that are slow enough that they cannot be ignored in systems where the system does not come to rest at an equilibrium (i.e., has the significant fluctuations in abundance that characterize most natural populations (Pimm 1991)). This result will be justified in Chapters 8 and 9 on competition in variable environments. This is not to say that age or size structure, inter-individual variation in other traits, time lags between resource intake and reproduction, or other features lacking in the LV model are not in need of exploration. However, they are important in all aspects of ecology, and are not part of the defining properties of the competitive process.

The second argument against universal inclusion of resource dynamics is the idea that competition often does not involve resources. For example, the textbook, *Community Ecology*, by G. G. Mittelbach and B. J. McGill (2nd ed., 2019) argues (p. 146) that 'consumer–resource models poorly represent some common types of competition (e.g. interference competition and apparent competition . . .). Even for exploitation competition, consumer–resource models depend on the notion of distinct resource types.' The first sentence in this quotation is countered by points made above; 'interference' usually produces effects on resources, and, in most, if not all, cases, it would not exist if interference had no effect on subsequent resource abundance. The second point about exploitation is certainly true, and neither I nor any other ecologist that I know of has suggested a restriction of the definition of competition to systems with only a single resource. It is true that interference can come in many forms, at least some of which need to be distinguished to be properly represented. Impacts on mortality and those on feeding rates are different forms of interference and have different consequences, both in reality and in consumer–resource models. However, there is no particular difficulty in representing both in a consumer–resource model; feeding interference makes the resource capture rate a decreasing function of the competitor's abundance, rather than having competitor abundance increase the mortality rate. In contrast, models that lack resources do not have a way of representing this difference. If the competitor causes mortality directly it can in theory be modelled equally well under frameworks with and without resources. However, in such a case, the consumer species that can be killed by a competing consumer is likely to exhibit behavioural avoidance of that second consumer, and this is likely to alter its resource consumption behaviour as well. Such behavioural effects cannot be modelled without explicit consideration of the resource. Moreover, direct mortality or other 'interference' between consumers that does not entail consumption (i.e. that is not 'mutual predation') is very unlikely to occur unless it increases subsequent resource availability. This requires some spatial localization of resource items, and it therefore means that an understanding of the competitive interaction requires models that explicitly represent the spatially distinct populations of both consumers and resources.

The conclusion from this section is that understanding the dynamics of the resources that underlie competitive interactions is usually necessary to describe the interaction of two or more consumer species. The next section examines MacArthur's 'all-linear' consumer–resource model, which has been used as the main logical justification for using and/or believing results from the LV model.

3.4 MacArthur's connection of LV to consumer–resource models

MacArthur's (1970, 1972) consumer–resource system, described very briefly without equations in Section 2.2, features logistic resource growth and linear consumer functional responses. MacArthur used the model to justify going back to the Lotka–Volterra model in some of his final work on the number of species that could coexist using a common set of logistically growing resources. It has continued to be used in theoretical analyses (e.g., Chesson 1990; Kuang and Chesson 2008). MacArthur implicitly assumed that resources never went extinct. However, a sufficient press perturbation to one consumer will often cause extinction or re-emergence of one or more resources. This is particularly likely when resources compete with each other (Abrams and Nakajima 2007; Abrams and Cortez 2015a). Hsu and Hubbell (1979) first identified the possibility of resource extinction, it was noted independently in Abrams (1980b), and its consequences for the relationship between the similarity in resource use and interspecific effects on abundance were explored in Abrams (1998, 2001a). However, the possibility of resource extinction has largely been ignored in the subsequent competition literature, even though it is a simple case of exclusion via apparent competition. This will be discussed again in Chapters 4, 5, and 6. Here I will just focus on how it changes the predictions of the LV model.

A 2-resource version of MacArthur's (1972) consumer–resource model is given by:

$$
\begin{aligned}
\frac{dR_1}{dt} &= r_1 R_1 \left(1 - \left(\frac{R_1}{K_1}\right)\right) - c_{11} R_1 N_1 - c_{21} R_1 N_2 \\
\frac{dR_2}{dt} &= r_2 R_2 \left(1 - \left(\frac{R_2}{K_2}\right)\right) - c_{12} R_2 N_1 - c_{22} R_2 N_2 \\
\frac{dN_1}{dt} &= N_1 \left(b_{11} c_{11} R_1 + b_{12} c_{12} R_2 - d_1\right) \\
\frac{dN_2}{dt} &= N_2 \left(b_{21} c_{21} R_1 + b_{22} c_{22} R_2 - d_2\right)
\end{aligned}
\qquad (3.4\text{a, b, c, d})
$$

The only new parameter introduced here is the conversion efficiency of captured resource type j into numbers or biomass of consumer i, given by b_{ij}. An LV model (using the traditional form from Chapter 1) can be derived under MacArthur's assumption of resources always at their quasi-equilibrium with respect to current consumer abundances. It has the following parameters, where the prime is used to distinguish them from the resource r and K. For consumer species i:

$$r_i' = b_{i1} c_{i1} K_1 + b_{i2} c_{i2} K_2 - d_i$$
$$K_i' = r_1 r_2 \left(b_{i1} c_{i1} K_1 + b_{i2} c_{i2} K_2 - d_i \right) / \left(b_{i1} c_{i1}^2 K_1 r_2 + b_{i2} c_{i2}^2 K_2 r_1 \right)$$

The two competition coefficients are given by:

$$\alpha_{12} = \left(b_{12} c_{12} c_{22} K_2 r_1 + b_{11} c_{11} c_{12} K_1 r_2 \right) / \left(b_{12} c_{12}^2 K_2 r_1 + b_{11} c_{11}^2 K_1 r_2 \right)$$
$$\alpha_{21} = \left(b_{22} c_{12} c_{22} K_2 r_1 + b_{21} c_{11} c_{21} K_1 r_2 \right) / \left(b_{22} c_{22}^2 K_2 r_1 + b_{21} c_{21}^2 K_1 r_2 \right)$$

The above formulas are only valid provided that both resources have positive abundances at the equilibrium point. They simplify considerably if the resources are all characterized by equal values of the pairs of subscripted parameters K, r, and b. Under the equal parameter assumption, the formula for the competition coefficient can be generalized to the case of an arbitrary number of resources, giving the following formula for the effect of consumer j on consumer i relative to i on itself:

$$\alpha_{ij} = \frac{\sum_{k=1}^{\omega} c_{ik} c_{jk}}{\sum_{k=1}^{\omega} c_{ik}^2}. \tag{3.5}$$

The sums here are over all ω resources in the system. In the case of two resources, if a consumer with significantly different consumption rate constants (c) on the two resources also has a sufficiently low mortality, one resource will go extinct when that consumer alone is present. Resource extinction can also happen in a 2-consumer–2-resource system, but this will imply extinction of one of the two consumers as well, given the above model. If both resources persist at an equilibrium point, the two consumers must also have different relative consumption rates of at least one of the two resources to coexist for any range of mortality rates. In a multi-resource system, a sufficiently low mortality of either consumer may produce extinctions of one or more resources.

MacArthur also carried out some analyses that assumed an infinite number of resources that could be ordered using a single continuous variable. The attack rate of each consumer depended on that variable (here denoted x), which was usually exemplified by resource size. He showed that if each species i had a Gaussian $c_i(x)$ curve having a different mean value, the competition coefficient was given by the analogue of eq. (3.5) with sums replaced by integrals over the variable x. Unfortunately, the weighting of different resources by b, K, and r was usually ignored in later work, despite an early article by Schoener (1974a), which stressed the large effects that these terms could have.

Under the two-species LV model, the per capita effect of one species on the growth of another is independent of population sizes, and there is a linear reduction in each species' equilibrium abundance with increases in its own per capita death rate or decreases in the other consumer's per capita death rate. A sufficient change in mortality will result in exclusion of one species, but this happens continuously;

the equilibrium abundance of the disfavoured consumer species decreases linearly and continuously with increases in its mortality until abundance hits zero. Under MacArthur's consumer–resource model, this remains true if all resources are present, but sufficiently large or small death rates may cause extinction of one or more of the resources. Because of the equal numbers of consumer species and resource types in the 2-consumer–2-resource system, extinction of one resource always results in extinction of one consumer (i.e., the one with a greater requirement for the single remaining resource).

Figure 3.1 illustrates a case in which either consumer (in the 2-consumer–2-resource system), if present alone, would cause extinction of one resource. Figure 3.1 assumes that the two otherwise similar consumers have opposite preferred resources. In the case illustrated, $c_{12} = c_{21} = 1/4$, and $c_{11} = c_{22} = 3/4$; each species consumes its 'preferred' resource at three times the rate of the non-preferred resource. I will later consider other parameter sets, but all are characterized by $c_{11} = c_{22}$, and by $c_{ii} + c_{ij} = 1$. In the figure, the two consumers have equal abundances when consumer 1 has a death rate equal to that of consumer 2 ($d = 0.1$). This basic equality combined with different resource utilization rates means that the two species are able to coexist for a range of values for consumer 1's death rate. However, imposing a press perturbation

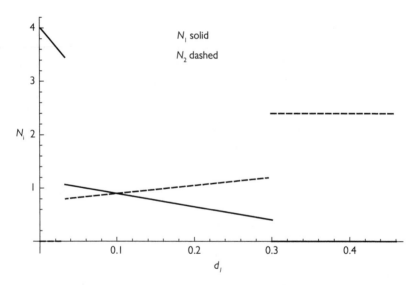

Fig. 3.1 Consumer populations as a function of the death rate of consumer 1 (d_1) in a symmetrical 2-consumer–2-resource MacArthur system. The lines give the equilibrium abundances of consumer 2 (dashed) and consumer 1 (solid) as a function of the mortality of consumer species 1, d_1. The other parameters are: $b_{ij} = 1$ for all i, j; $d_2 = 1/10$; $r = K = 1$; $c_{11} = c_{22} = 3/4$. The extinction of consumer 1 at $d_1 = 3/10$ corresponds to extinction of resource 2, while extinction of consumer 2 at $d_1 = 1/30$ corresponds to extinction of resource 1.

to the per capita death rate of consumer 1 has the effects shown in the figure. Lowering consumer 1's death rate causes an increase in its own density and a reduction in the abundance of consumer 2. However, with any continued directional change in species 1's death rate, there is an abrupt loss of this equilibrium, at which point the system shifts to one having a single consumer and a single resource. If d_1 is decreased to the point where consumer 2 and resource 1 go extinct, the result is an abrupt jump in the abundance of consumer 1. Note that this disappearance in Figure 3.1 occurs when the population of N_2 is only approximately 11% lower than its equilibrium when both species have identical death rates and abundances. This extinction corresponds to a more than threefold jump in the abundance of consumer species 1.

The condition for such a discontinuous change in equilibrium abundance of a consumer species as the mortality rate of its competitor is reduced, is $d < 2c_{ii} - 1$ in a symmetrical 2-consumer–2 resource model with the other parameter values used in Figure 3.1. For example, if $c_{ii} = 3/4$, as in the figure, a discontinuous change in the equilibrium of both consumers at a sufficiently low d_1 occurs when the initial equal mortalities of both species are less than 1/2. Considering that a mortality rate of 1 implies extinction of either species, even in the absence of competition (given the other parameters assumed in the figure), species having mortalities much greater than 1/2 should be relatively unlikely to persist in a variable environment. This suggests that, if the model were a good description of the system dynamics, discontinuous responses to an altered mortality in one species should be quite common. There is a second discontinuity in Figure 3.1 if d_1 is increased to a value of 3/10. This is characterized by the loss of both resource 2 and consumer 1, and by a discontinuous jump in the equilibrium density of consumer 2, which more than doubles its population size. Note that much of the literature in fisheries management is based on single-species logistic growth models. It suggests that reduction to half of the pre-fishery equilibrium abundance is a safe level of exploitation and is one that often produces close to the maximum sustainable yield. In this 2-consumer–2-resource system, the abrupt extinction of species 1 occurs when the added mortality to that species reduces it to slightly less than half of its original equilibrium.

The model behaviour described above and illustrated in Figure 3.1 is unusual primarily in that the change in the mortality rate d of either species of consumer has a linear effect on equilibrium consumer abundances over some range of mortalities. In essentially all models other than MacArthur's, the magnitudes of the population changes produced by a small press perturbation are quite sensitive to the initial equilibrium abundances of both consumers and resources (Abrams 1980b, 1983b). As a consequence, measures of the per capita effects of larger perturbations will depend on the initial densities and the magnitude of the imposed parameter change. So far as I know, the phenomenon of discontinuous change illustrated in Figure 3.1 has not been demonstrated for the MacArthur model in any previously published work, although the possibility of resource extinction was discussed in Abrams (1998, 2001a). Given the long history of subsequent studies using or referring to MacArthur's work, this figure demonstrates a lack of appreciation for the possibilities inherent in even this model that are inconsistent with the LV model.

Adding one or two features that differ from MacArthur's model gives rise to a range of additional phenomena. Figure 3.2 looks at the same type of mortality perturbation as in Figure 3.1. However, it plots one of the competition coefficients, α_{21}, as a function of d_1, rather than the abundances of the two species. It assumes a system that is similar to that explored in Figure 3.1 except that it has three resources, and the consumers have Holling type II functional responses, which saturate at high prey abundance (see Box 2.1). Each consumer utilizes one exclusive and one shared resource. The measure of competition given by expression (3.2) is calculated based

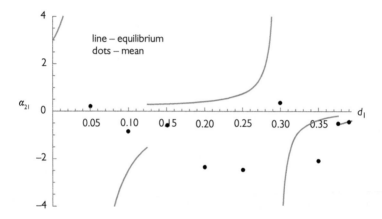

Fig. 3.2 The lines show the competition coefficient (expression (3.2)) based on equilibrium densities (line segments) and mean densities (dots) for a system similar to eqs (3.4). This system differs in form from that in Figure 3.1; it has three resources rather than two; consumer 1 uses resources 1 and 2, while consumer 2 uses resources 2 and 3. It also differs in that the two consumer species have Holling type II functional responses with a common handling time for all consumer–resource pairs. See Chapter 6 for a more complete description of the model. The line gives the values of measure (3.2) calculated at the equilibrium point; both positive and negative effects occur. Most of the equilibria are unstable, and the dots give the analogue of measure (3.2) calculated based on change in long-term mean abundance following a 1% increase in the mortality rate from the value given on the x-axis. Positive values mean that the two consumer species change in the same direction in response to increased death rate of species 1; they change in opposite directions for negative values. The parameter values are: $r_1 = r_2 = r_3 = 2$; $K_1 = K_3 = 1$; $K_2 = 2$; $c_{11} = 2/3$; $c_{12} = 1/3$; $c_{22} = 1/3$; $c_{23} = 2/3$; $b_{ij} = 1$; $d_2 = 0.25$. All handling times are $h_{ij} = 2$. This graph has four discontinuities corresponding to where a resource becomes extinct ($d_1 = 0.125$ and 0.3928), and where a mortality of consumer 1 does not alter its density, at $d_1 = 0.05$ and $d_1 = 0.2954$. The system exhibits cycles at all of the mortality values when mean density effects were calculated, except for the two highest values.

on the *equilibrium* densities (solid line). For a selection of different initial mortality rates, the effect on *mean* densities due to a small finite change in mortality is given by the solid dots. Most parameters result in sustained population cycles and as a result the two measures (dots and line) differ significantly. There are pronounced changes in both measures as the mortality of one consumer is increased or decreased. Positive effects and very large magnitude effects are both possible. The figure illustrates the more general result that calculations based on an equilibrium point can be quite misleading as an indicator of the effects on average densities. As Chapter 6 will show, large-magnitude changes in measures of competition as neutral parameters are changed are common, even in systems that have relatively few differences from MacArthur's model.

The remainder of this chapter will examine the structure and properties of consumer–resource models in greater detail, and will highlight how little we know about them, even in the case of a single consumer species. This implies that there exists a large unexplored frontier of different but biologically plausible models of competition involving two or more consumer species.

3.5 What do coexistence and exclusion mean?

The definition of coexistence may seem obvious. For example, in his book on resource competition, Grover (1997, p. 11) simply states that 'two or more populations may persist, in which case we speak of coexistence'. One problem is that nothing on Earth persists indefinitely; the vast majority of species that have ever existed on this planet are now extinct. Chesson (2020b) is one of the few accounts of competition theory that acknowledges this fact. Thus, coexistence needs to be defined as long-term persistence by all the species concerned, given the existence of some regime of environmental conditions (which may include constantly present variation). The other problem is that continued persistence of one species in a spatially distinct community may be terminated by large environmental perturbations such as exceptional weather, diseases, evolution of other species, or changes in the reproduction or input of some subset of resources. All these types of perturbation are possible in almost all naturally occurring sets of competitors. Is it necessary that each species is able to grow back from extreme rarity, and what constitutes that 'extreme rarity'? 'Exclusion' of a species has analogous definitional issues; is it a total inability to achieve a continued positive abundance, or an inability to increase when rare enough when the remaining species are at equilibrium?

A large body of theory now defines coexistence as the ability of each species to increase from near-zero abundance when the rest of the consumer species have reached their limiting dynamics (usually, but not always assumed to be a stable equilibrium point); e.g., Grainger et al. (2019) and Chesson (2020b). This is termed 'invasion analysis'. Invasion, however, need not imply continued coexistence (Abrams and Shen 1989; Mylius and Diekmann 2001; see examples in Chapters 8 and 9). And long-term persistence is possible even when neither species can increase when very rare (Gilpin and Case 1976; Barabás et al. 2018). These cases either require a large number of individuals in the initial propagule or some large perturbation away from the original equilibrium of the original community at the time of invasion. Pimm's (1991) review had already shown that most communities experience frequent large perturbations. Given the temporary nature of all biological communities, it seems more reasonable to adopt the less restrictive definition of coexistence as the possibility of continued persistence without directional change over many generations. For the typical differential equation model of competing species, this implies the existence of an attractor with all species present and bounded away from zero. The attractor should have a large enough basin of attraction that multi-generational persistence of all species is likely over such a time frame. Meszéna et al. (2006) give a similar definition of coexistence as the existence of '. . .a fixed point of the community dynamics with no population having zero population size'.

The definition of coexistence becomes more complicated in the context of a metacommunity, consisting of two or more spatially distinct 'patches' connected by migration of some or all of the species. As early as 1974, Simon Levin had pointed out that, for the LV model with contingent outcomes (species 1 excludes species 2, or 2 excludes 1, depending on initial abundances), a two-patch environment has a stable equilibrium with both species persisting. This is the equilibrium with only species 1 in one patch and only species 2 in the other (actually there are two such outcomes, the dominant species having reversed locations). There are also two additional potential outcomes with species 1 in both patches or species 2 in both patches. To apply the 'invasion criterion' to this case, one has to decide whether the resident species is present in one or both patches; only systems with an empty patch can be invaded. If both are initially empty and small groups of colonist individuals arrive randomly with long inter-arrival intervals and with an equal probability of an invasion by either species, then the end result is likely to be coexistence (always with a different species in each patch) about half the time. In any case, exclusion and coexistence are both possible in the system as a whole. Given a large number of identical patches, with some movement between most of them, and occasional extinction events, it is likely that the two consumers in this scenario will persist for a long time when the single patch model says that they cannot.

This case of alternative exclusion outcomes in a spatial context also presents a challenge to the belief that stronger competition (as measured by the product of the two competition coefficients at the equilibrium point in a spatially homogeneous system without temporal variation) makes coexistence less likely to occur. Competition coefficients whose product is greater than one usually involve some form of interference competition. For initially identical exploitative competitors with competition coefficients both equal to one, coexistence would not occur given some change in system-wide conditions that favoured one species or the other. Such a change, slightly favouring one species, would not cause system-wide exclusion in the multi-patch, alternative-exclusion case.

Defining coexistence also requires a definition of exclusion. This is usually taken to mean that the excluded species is completely absent. However, most communities come into existence by receiving immigrants from other communities, and it is illogical to think that this immigration ends at a point where that species becomes very rare. The same is true of biotic resources. In a more realistic metacommunity context, exclusion is more appropriately considered to include situations in which a species has a very low abundance that is maintained only by immigration. Even if the topic of interest is the interaction between consumers in one patch, it is important to include a low level of resource immigration. This will only affect the interaction between two consumers in cases where some subset of the potential consumer species results in near-zero abundance of one or more resources. With resource immigration these resources may be able to return from near-zero abundance following introduction of one or more new consumer species (see Chapter 5 on apparent competition). In the absence of resource immigration, those resources would be permanently lost. The question of how to deal with immigration in real-world competitive communities has been avoided by the concentration on laboratory systems, in which immigration can often be prevented.

3.6 What distinguishes a single resource from others?

Haigh and Maynard Smith (1972) was the first attempt to take a rigorous approach to answering this question. They noted that separation in time or space, and different parts of a single biological entity could constitute different resources. The necessary condition for being distinct was that abundance of the resource at one point in space or time did not functionally determine the abundance at another point. That is certainly true of the spatial system discussed in the previous section. Abrams (1988b) later added a condition that the consumers must be capable of distinguishing the different categories of resource for them to qualify as different resources for the purposes of coexistence. Resources that differ only in spatial location or the time at which they are present are generally connected to each other. Time has a directional property that is not universally present with space (although it may be in the case of flowing water systems). In any event, the interaction between resource items that only differ in when or where they are present is quite different from the interaction of resources that have distinct properties when present at the same time and place.

Even when resources are physically distinct, different resources interact in a number of ways. Haigh and Maynard Smith (1972) pointed out that different life stages or different parts of a particular biological resource constituted potentially distinct resources for consumers. Distinct resource species may compete for their own resources, but may also interact positively. The absence of resources in the LV model not only ignores these possibilities, but also suppresses thought about their potential role in interactions between consumers. That is probably why MacArthur never considered this possibility in his consumer–resource models.

3.7 Functional forms for the model components

Here I will use the term 'consumer–resource model' to denote mathematical models that contain descriptions of the dynamics of the relevant consumer species and all of the resources they consume. However, the model may contain a larger number of dynamic entities. There may be species or entities that are not consumers and are not consumed themselves, but still have negative effects on one or more resources.

This would, for example, include inedible plants that negatively affect edible ones in a scenario involving competition between herbivores. Whenever the resources are biological, it might be necessary to include their own resources, particularly when they share one or more of those resources. Similarly, it may be necessary to include predators of the consumers, or still higher trophic levels, if these affect resource exploitation by the focal consumers (e.g., behavioural effects on consumer foraging, as in Abrams 1984b, 1992c, 1995). Consumer–resource interactions are central to all competitive interactions. However, the mathematical forms used in most previous consumer–resource models of competition have come from a very narrow part of the range of potential forms. These models have largely employed a standard (small) set of simple relationships for their three basic components.

After distinguishing the resource populations based on individual properties, spatial location, and/or temporal availability, the basic elements of each consumer–resource interaction are the expressions for the resources' population growth rates, and the consumers' functional and numerical responses (Murdoch et al. 2003). These three components are the minimal set of functions required for any consumer–resource model. Ideally, the set of resources in a model should include all species/types that are consumed plus those that are not consumed, but still affect the dynamics of one or more resources that are consumed. The latter have usually been omitted, in part because most models of competition have simply ignored any inter-action between resources. However, this set of relationships is a minimum because competitors in natural communities are embedded in food webs. The question of how many other species must be considered to understand the interaction between com-petitors has yet to be resolved. Yodzis (1988) and Schoener (1989, 1993) discussed this problem of how many additional species had to be considered to properly assess the effects of two particular species on each other. This topic does not seem to have received much attention from empirical ecologists in the subsequent three decades. Many of the models considered in the rest of this book will unfortunately also leave out much of the range of potential indirect effects impacting consumers and their resources, as they have yet to be explored. However, if one is studying competition for biotic resources, it seems inconsistent to ignore the possibility that the resources interact with each other, and that one of the most likely interactions is competition.

When the resources are themselves living entities, an accurate description of their interaction likely requires consideration of their interactions with their own resources. Such multi-trophic level systems will again be largely ignored in this sec-tion, as there are so many possibilities, even without them. However, some indication of the potential for the resource species' own interactions to affect the interaction between their consumers was provided by early articles using linear per capita growth rate functions (Levine 1976; Vandermeer 1980). Both of these articles emphasized the potential for mutually positive effects of consumers on each other's equilibrium population size. This multi-trophic level approach has subsequently been extended to systems with nonlinear per capita growth rate functions (Abrams and Nakajima 2007) or highly asymmetric resource consumption abilities (Abrams and Cortez 2015a). These two works have revealed a range of between-consumer interaction signs (including (+,–) effects) and magnitudes compared to similar systems with symmetrical consumer abilities and non-interacting resources. Models in which these lower level resources are modelled explicitly with nonlinear resource growth functions must exhibit an even wider range of possible between-consumer interactions. In any case, sufficiently large negative interactions between resources are likely to imply one or more positive interactions between consumers, and these deserve to be part of competition theory.

One of the additional foodweb components that may often need to be included in models of competition is the set of the consumer species' predator(s). Although the between-consumer interaction via changes in predator abundance is considered to be a different interaction (apparent competition (Holt 1977); see Chapter 5), predators have regularly been shown to have effects on their prey's resource consumption behaviour (Lima and Dill 1990), and this needs to be included in models of systems in which the consumer species are subject to predation.

It is now recognized that a significant fraction of terrestrial plants have multiple pathways of effects between species, many of which are positive. Positive effect pathways include such processes as providing shelter for predators of herbivores and physical protection from adverse environmental conditions (Callaway et al. 2002; Brooker et al. 2008). Beneficial effects also include sharing of resources by fungal connections (Pither et al. 2018, Birch et al. 2020). These positive effects of consumers on each other are likely to alter conditions for coexistence; Gross (2008) argued that they always promote coexistence, but this was due to his unjustified assumption that there were interspecific but no *intra*specific positive effects. Nevertheless, it is true that positive effects within and between species that also share resources do occur in a significant number of cases; when they do, they should be included in dynamic models of between-consumer interactions (i.e. competition). Such positive effects and the potential effects of trophic levels higher than the consumer are not considered in detail in this chapter, mainly to avoid excessive complexity.

Even if we ignore some of the complications mentioned in the preceding paragraphs, there still exists a wide range of functional forms for model components. The following subsections describe the variety of functional forms that may characterize the three basic components of consumer–resource models, and provides some history on when they were first discussed in the literature. This shows that existing competition theory incorporating resource dynamics has thus far only considered a small fraction of possible interaction mechanisms arising from consumer–resource models that were introduced long ago. Thus, there has long been a need to expand our understanding of consumer–resource-based models of competition.

3.7.1 Resource growth

The resource growth process is a partial exception to the previous generalization that linearity is assumed in most consumer–resource models. In cases having an externally produced ('abiotic') resource, its per capita rate of increase is always nonlinear, and MacArthur (1972) acknowledged this. The most common abiotic resource model is that of chemostat dynamics, which has a constant input rate of resource and an exit rate that is linearly proportional to resource abundance; this gives a hyperbolic density dependence in the per capita growth rate of a consumer (Abrams 2009b, c). The simplest case of chemostat dynamics (with a zero resource exit rate) was used in Volterra's (1931) seminal work on competition, and was explored for some two- and three-resource systems in Schoener (1974c, 1976) and Abrams (1975). But resources cannot exist indefinitely; thus, non-zero exit rates were examined in similar models of competition by Abrams (1977), Schoener (1978), and Tilman (1980, 1982). The first two authors stressed the nonlinearity of the resulting competitive relationship between two consumers. Unfortunately, Tilman (1982) erroneously suggested that this model was consistent with LV dynamics (Abrams 1987c). The chemostat model for resource dynamics has now been used in many models of competition (e.g., Grover 1997; Huisman and Weissing 1999, 2001; Klausmeier and Litchman 2012). However, many subsequent studies failed to stress its inconsistency with LV dynamics of consumers.

The relative lack of research on the shape of density dependence no doubt has helped the assumption of logistic resource growth to persist in competition models that include explicit biotic resources. Nevertheless, both the theory (Abrams 2009a, b, c, d) and the empirical evidence we have (Sibly et al. 2005), suggests that near-linear density dependence is relatively rare. Abrams et al. (2008a) extend results in Abrams (1980b) to show that the curvature of resource density dependence determines whether heavily exploited resources are weighted more—or less—heavily in determining competitive effects. Lesser weights for a given resource mean that, even if both species use that resource at high rates, their competition coefficients need not be large. In a simple generalization of the logistic growth model, the density-dependent reduction in resource (R) per capita growth rate (R/K), is raised to the power θ. Concave density dependence ($\theta < 1$) implies lower weighting of heavily exploited types, while convex density dependence ($\theta > 1$) implies higher weighting of these types. Abiotic resource growth is likely to be characterized by lower weighting of heavily exploited types, and that is always true for the chemostat model of dynamics (see Abrams et al. 2008a).

Unfortunately, the detailed dynamics of single-species growth has not been a popular topic of research among empiricists, whether those species are plants, predators, or prey. Sibly et al.'s (2005) review of the shape of density dependence relies on time series analysis of species, most of which were not studied specifically to determine the shape of density dependence. Data from fisheries yields and population size estimates have been used to estimate the shape of density dependence in fish (Hilborn and Stokes 2010). This approach is again less than ideal for determining the functional nature of feedbacks via resource depletion. The underlying observations are not designed for determining the form of density dependence; the resources involved are often unknown or unstudied (and usually both); and the potential effects of competitors and/or higher-level predators are routinely ignored. The various factors that influence the nature of single-species growth will be examined in Chapter 5.

The effect of resource growth functions on the nature of competition was largely ignored in part because MacArthur's (1970, 1972) work on his consumer–resource model did not consider alternative (or nonlinear) functions for biotic growth, and only had a brief treatment of chemostat growth for the special case of a zero exit/loss rate of unconsumed resources. Although MacArthur did not acknowledge this possibility, efficient consumers that share a number of biotic resources can cause some of those resources to go extinct (Hsu and Hubbell 1979); this reflects apparent competition between resources. The consequence of such extinctions is usually a sudden change in the strength of competition between consumers (Abrams 1998, 2001a; Abrams et al. 2008a). If the number of resources is initially equal to the number of consumers, a resource extinction event will necessarily be followed by at least one consumer extinction in the standard scenario of a spatially homogeneous system without temporal environmental fluctuations. An additional problem with MacArthur's (1970, 1972) assumption of equivalence of LV models to consumer–resource models with independent logistic resources is that the equivalence only applies to systems in which the resources have fast enough dynamics to approximate a quasi-equilibrium with respect to consumer density. The same is true of derivations of logistic growth from a consumer–resource model with a single 'linear' consumer of a logistic resource (O'Dwyer 2018). Figures 2.1 and 3.2 suggest that this requirement for fast dynamics is often not satisfied. If there are large population fluctuations driven by environmental stochasticity or internally driven cycles, it is generally impossible to approximate the consumer's dynamics or determine its mean population size without a full model that includes resources. This conclusion applies to intraspecific as well as interspecific competition (Abrams et al. 2008b; Reynolds and Brassil 2013; O'Dwyer 2018). Chapters 8 and 9, dealing with competition in seasonal environments, provide many examples of qualitative differences between such temporally variable models and the constant-environment LV model.

3.7.2 Consumer functional responses

MacArthur's (1970, 1972) assumption of linear functional responses was examined not long after his seminal works first appeared. Nonlinear (type II) functional responses were shown to result in competition coefficients that change greatly in magnitude with changes in the abundances of each of two competitors by Abrams (1980a), who only examined stable 2-consumer–2-resource systems with no possibility of resource extinction. Armstrong and McGehee (1976a, b, 1980) explored unstable coexistence equilibria with a single biotic resource and a type II functional response on the part of at least one of two consumer species. This allowed coexistence in cycling systems due to the species' different ranking of per capita growth rates at high and at low resource abundance. The consumer having the more strongly saturating functional response also had a lower resource requirement for zero growth rate, but its population cycles caused the mean resource abundance to be much higher than its equilibrium when that consumer was the only consumer present.

Other than some further work on the logistic resource-type II functional system with one or two resources, most articles adopting type II functional responses in studying competition have used models that also assumed chemostat resource growth, which does not allow cycles in the 2-consumer–2-resource case with substitutable resources. The relative lack of research on competition models with nonlinear functional responses is unfortunate because both logic and available functional response measurements (Jeschke et al. 2002, 2004) suggest that a saturating response (the type II or some modification of it) must be a better model of natural systems than a strict linear response.

Even with traditional type II responses, the appropriate multi-species form is unclear when prey species are characterized by different values of B/h (energy gain per unit handling time). Type II responses were the basis of much of optimal foraging theory (Stephens and Krebs 1986). The Holling disc equation implies that the consumer's fitness decreases with its consumption of a food with a sufficiently low B/h relative to that of other food items. The resource(s) characterized by a smaller value of B/h in a 2-resource system should be dropped from the diet when the higher B/h resource is sufficiently abundant. In this case, the disc equation formula must be changed. Even with equal B/h for all resources, functional responses are altered when there is a trade-off between increasing attack rates on different resources. This applies to cases where resources are located in different places. Various ways of incorporating food choice into a dynamic model have been developed (Fryxell and Lundberg 1998; Abrams and Matsuda 2004; Abrams et al. 2007), but the effects of these dynamics of choice on competitive interactions between consumers has yet to be explored. Different consumers will usually differ in their values of the ratio B/h for different resources. For each consumer, it is obvious that increased abundance of the higher B/h resource can at least temporarily decrease predation on the lower B/h resource, and this may result in a very large effect of the better on the poorer resource(s) for various types of perturbation. Exploring the impacts of diet choice for competing consumers is still largely a task for the future. If the values of B and h are both roughly proportional to the caloric content of the food item, adaptive diet choice may have little impact on functional response forms, as food values will be equivalent, and diet choice will not be exhibited. However, this proportionality condition was not satisfied in most of the early behavioural ecology studies of diet choice (Pyke et al. 1977).

Another likely feature of most functional responses is consumer dependence (Abrams 1984b, 1992c; Abrams and Ginzburg 2000; empirical examples in Skalski and Gilliam 2001), but this has received almost no attention in the literature on competition. McPeek (2019a) is a recent exception, although he did not include consumer dependence in any of the models he analysed numerically. Consumer dependence is commonly expected to arise due to adaptive prey defence, but such defence is unlikely to depend purely on the abundance of a single predator species in a system with two or more competing predators. In such a case the existence and nature of interspecific interference or facilitation depends on whether the same type of behaviour reduces the risk from both, or whether distinct defences are required (see Matsuda et al. 1993, 1994, 1996; Sommers and Chesson 2019). Interference can also arise by predator individuals attempting to interfere with the foraging of other individuals. This range of possibilities seems to have inhibited use of interference terms rather than producing a variety of alternative models. The most used model is that of DeAngelis et al. (1975), which is based on a single predator type. In this model a term proportional to the density of the higher-level predator is added to the denominator of a type II functional response of each consumer.

Nonlinear functional responses may also arise when consumer behaviour can increase the consumption of one resource (or group of resources) at the expense of decreased consumption of another. This was first studied empirically by Murdoch (1969) and was termed 'switching'. The possibility that such choices influence functional response shape was incorporated indirectly in a paper on food webs by McCann et al. (1998), and more directly by Matsuda et al. (1993, 1994, 1996), Fryxell and Lundberg (1998), Abrams and Matsuda (2004), and Abrams et al. (2007). Nevertheless, consumer choice is still ignored in most of the literature on functional responses. The most comprehensive review of empirical work on functional responses (Jeschke et al. 2002, 2004) had few examples with more than a single prey type, so the nature of switching or other forms of adaptive behavioural trade-offs based on diet choice have received almost no empirical attention. Abrams (1987c, 1989, 1990a, b) explored some of the effects of nutritional interactions between resources, as well as toxins and gut capacity constraints, for both functional responses and competition. Both functional responses and consumer–resource dynamics were greatly altered by these factors.

Researchers pointed out long ago that adaptive behaviour should create functional responses of consumer species that are often influenced by the abundances of species at trophic levels higher than the consumer and by species (or resources) at trophic levels lower than those of the consumed resources (Abrams 1984b, 1992c, 1995; Bolker et al. 2003). While many of these behavioural influences have been confirmed to exist and be large in many food chains/webs (Werner and Peacor 2003; Bolnick and Preisser 2005; Preisser et al. 2005), they have seldom been incorporated into models of competition. Their effect on functional responses has also received little attention. Thus, while these rapidly acting short-term indirect effects may significantly alter competitive interactions, it is unclear if or when they will be included in empirically based models of competition. However, it does seem likely that short-term interactions between two competitors are often affected by species that have no direct consumptive relationship with either of the competitors.

In many species of prey, only individuals in certain habitats or from a limited size range are susceptible to their predator. If functional responses are expressed in terms of total population size, this usually produces a consumer-dependent form for those responses (Abrams and Walters 1996). However, the exact processes involved in the flows between more- and less-available prey categories are usually not studied, again leaving considerable uncertainty about the appropriate form of the functional response if it is expressed in terms of total prey density.

All the factors influencing functional responses reviewed above will turn any otherwise linear functional response into a nonlinear one. They also alter the forms of already nonlinear functional responses. In addition, these factors should often change the per capita growth rate functions of consumers and resources, either by making them nonlinear, or by altering their already nonlinear form. The wide range of topics covered in this section suggests that we have much left to learn about the effects of adaptive behaviour on competitive interactions.

3.7.3 Consumer numerical responses

The numerical response of consumers to resources has received much less attention than have resource growth functions and consumer functional responses (Abrams 1997). Even within the MacArthur (1970) framework, which assumes a birth rate that increases linearly with the intake of nutritionally substitutable resources, the conversion efficiencies of consumed resources into additional biomass are usually assumed to be identical across resources. This assumption is used in many of the models considered here. However, most capture and processing costs are known to be different for different resource species (Stephens and Krebs 1986). Adaptive evolution will often lead consumer species to be better at capturing those resource types that they can more efficiently convert into new biomass (Abrams 1986a). This should result in a positive correlation between conversion efficiencies and capture rates, and this correlation reduces interspecific competition relative to intraspecific. However, such a correlation need not hold. Chase and Leibold (2003) reviewed some of the effects of different resource valuation by different consumers in determining the outcome of competition. If such differences exist, they should of course be included in any applied consumer–resource model. However, the argument for the necessity of explicit representation of resources is only strengthened by this possibility.

Numerical responses in community models are usually linear functions of the weighted total resource intake rate (i.e., the sum of the functional responses weighted by the value of the resource), or are assumed to be directly proportional to the intake rate of the single limiting resource. In some models, particularly with plant nutrients, the numerical response has the same Michaelis–Menten form as the type II functional response. However, this does not account for the possibility of adaptive change in the relative intake rates of different resources (Abrams 1987b, c).

Even if resources do not differ nutritionally, the numerical response is not likely to be linear as a function of total intake. This is true whether one or several resource intake rates affect the response. The numerical response functions used in single-resource models include relatively few nonlinear forms, although Abrams (1995, 1997) and Getz (1993) have provided arguments for nonlinear forms with accelerating per capita death rates when resource intake becomes very low. There has been very little empirical study of numerical response forms for consumer species, as noted by Abrams (1997). The consumers' numerical response shapes need not alter the equilibrium resource densities, and this has probably contributed to lack of attention to the shape of such responses. However, numerical response shape does affect the functional response when there is adaptive adjustment of costly foraging behaviour (Abrams 1992a, c). In addition, nonlinear numerical responses affect mean resource densities in systems with fluctuating population sizes (Abrams 1997).

The most common reason for a nonlinear numerical response in published competition models is that there are multiple resources with nutritional interactions between the different resources that determine consumer birth and/or death rates. A significant amount of literature deals with the case of two nutritionally essential resources (Chase and Leibold 2003). 'Nutritionally essential' resources almost always imply that the consumer's numerical response is a nonlinear function of the different intake rates. The usual assumption is limitation solely by the resource that is present in the smallest amount relative to a fixed requirement (i.e., Leibig's Law of the Minimum). The result is a numerical response that abruptly levels off as the intake rate of a single resource increases past the point where it limits growth; this response shape was used in simple 2-resource models by Leon and Tumpson (1975) and Tilman (1982). Subsequent work on essential resources has also concentrated on per capita population growth rates that depend solely on the intake rate of the resource that is smallest relative to its requirement based on an ideal ratio of intake rates. Adaptive balancing of intake rates of different resources was shown to lead to very different forms of both functional and numerical responses (Abrams 1987c), which were shown to dramatically affect competitive dynamics in some simple models (Abrams 1987b; Abrams and Shen 1989). However, this possibility has been largely unexplored since then. Harpole et al. (2011) suggested that co-limitation by two resources is common in plants, something that is hard to account for unless there is some adaptive adjustment of relative uptake rates of different resources. Competitive systems having three or more essential resources were considered by Huisman and Weissing (1999, 2001), but have not received much further attention. Huisman and Weissing also assumed Leibig's Law. More complicated interactions between resources, based on resource toxins or consumer gut-capacity constraints also affect the forms of adaptive functional and numerical responses in two-or-more resource systems (Abrams 1989, Abrams and Schmitz 1999). As a result, these interactions with resources affect the competitive interactions between different consumer species (Abrams 1990a, b).

Direct effects of consumer density on the numerical response were shown to be able to produce coexistence of two or more consumers on a single resource by Levin (1970). Such effects have been included in consumer–resource models sporadically since then (e.g., Schoener 1976, Abrams 1986a, McPeek 2019a). In most cases, the analysis has concentrated on intraspecific effects, which have almost always been linear. However, as in the case of functional responses (Abrams and Ginzburg 2000, Abrams 2010a), several different mechanisms can lead to both inter- and intraspecific effects of consumer densities on their numerical responses, and these effects are likely to differ depending on the mechanism involved.

3.8 Analysis of models of competition

If one accepts the need for some representation of resource dynamics, then, for coexistence at a stable equilibrium, the minimum number of dynamic equations in a 2-competitor model is four; i.e. there must be at least two resources. Without simplifying assumptions, having four variables means that a much greater range of non-equilibrium dynamics and alternative outcomes are theoretically possible than is true of ordinary differential equation models having two dynamic variables. Determining the long-term dynamics of such models can be carried out by standard techniques of analysing dynamic systems (e.g. Strogatz 1994; Hirsch et al. 2004; Ellner and Guckenheimer 2006). In the majority of cases, a full analysis of non-equilibrium dynamics requires numerical methods, and large systems with unstable dynamics are difficult to analyse completely. Even the case of the 3-species Lotka–Volterra competition model provides a good illustration of this; see Gyllenberg et al. (2006). Thus, the generality of some results may be difficult to establish. However, even some numerical exploration of dynamics over a range of plausible parameter values provides a better approach than any that are based on an unlikely assumption of the attainment of unique stable equilibria. Even in the case of 'MacArthur systems' with linear resource density dependence and linear consumer functional and numerical responses, asymmetries in utilization curves and mean uptake rates have been largely ignored (Abrams and Cortez 2015a). Exhaustive parameter space exploration in more complicated systems may not be feasible, and establishing the biological likelihood of different sections of parameter space is even more problematic (see Abrams et al. 2003). While some potential outcomes may be missed in models requiring numerical analysis, concentrating exclusively on unrealistically simple systems and approximate methods is certain to provide a misleading picture of potential outcomes.

Analysis using a graphical method based on isoclines (lines where consumer growth rate is zero in a space defined by resource densities) represents an example of an overly simplified method that is inapplicable to many-species systems. The general isocline approach had been used in a seminal study of predator–prey relationships (Rosenzweig and MacArthur 1963), and it allows some insights into the dynamics of systems with two variables. Leon and Tumpson (1975), and later Tilman (1982), promoted isocline analysis for analysing interspecific competition for two resources. However, the methods are not easily extended to systems with three or more dynamic variables, where visualization of intersections is unclear, and simple rules-of-thumb for stability do not apply. Visualization is impossible with four or more variables. And nonlinear models often require numerical methods to plot isoclines, even in two dimensions. Because of these limitations isocline analysis has not been a major tool for understanding predator–prey relationships in recent decades. If the ultimate behaviour of the model is some form of sustained fluctuations, the graphical analysis of isoclines does not offer insight into dynamics and effects on average abundances. Thus, it is surprising that this method has been repeatedly resurrected (e.g., Chase and Leibold 2003), and recently reaffirmed by Letten et al. (2017) and McPeek (2019a). Given the range of software currently available for numerically analysing differential equations, there does not seem to be any argument for relying on isocline approaches, even when the number of variables is small enough to apply it.

McPeek (2019a) recommends a consumer–resource approach and models with type II functional responses, which frequently produce cycles. However, he uses isocline analysis without acknowledging its problematic features for the analysis of competition in such models (his figures 4–8). He ignores all cases in which there are cycles in the absence of a second consumer. His one example of limit cycles in a 2-consumer system does not acknowledge the difference between mean and equilibrium abundances in determining competitive effects. Competitors may affect each others' mean densities by altering cycle amplitude, so the shift in the equilibrium point with a parameter change usually differs significantly from the shift in mean (or maximum or minimum) density (see Armstrong and McGehee 1980; Abrams et al. 2003; Abrams 2004b, Abrams 2006a, b). Finally, McPeek's (2019a) isocline analysis of consumer coexistence in a 2-consumer–2-resource system requires that both resources are able to persist in an isolated system having either consumer species alone. This was not required in the Abrams (1998) article that is referenced by McPeek (2019a). The outcome in this case depends on assumptions about immigration of resources and invasion timing of consumers, as explained in detail in Chapter 6.

Invasion analysis is an increasingly common shortcut for determining whether coexistence will occur. It assumes that coexistence is guaranteed when each consumer can increase from near-zero abundance in a subsystem in which the remaining consumers and resources have reached their limiting dynamics. Use of invasion analysis for determining coexistence has been strongly advocated by Siepielski and McPeek (2010), McPeek (2019a) and Grainger et al. (2019), among many others. Of the recent highly cited articles considered in the following chapter, only Barabás et al. (2018) raise concerns about using invasion analysis to determine whether consumer species will coexist in cases with a small number of species. The ability of all competitors to 'invade' in the sense defined above is not a good indicator of conditions allowing coexistence. It has long been known that this criterion is not appropriate for analysing coexistence in 3-species Lotka–Volterra systems that exhibit cycles (May and Leonard 1975; Gilpin 1979), as coexistence of some or all species pairs may be impossible. However, there are at least four problems with using the invasion-implies-coexistence criterion which are likely to apply more generally: (1) invasion from low densities does not ensure persistence; (2) invasions of species with Allee Effects can be successful if enough individuals are present; (3) systems exhibiting alternative outcomes depending on initial conditions are found in models with as few as two competitors, and become more common in larger systems (Gilpin and Case 1976); and (4) natural communities are seldom at their equilibrium (or on any non-stationary attractor) due to environmental variation, which can allow successful entry of new competitors even when equilibrium invasion is not possible. These possibilities are discussed in turn.

Initial increase in a system need not mean that a species continues to increase, even in a constant environment. Abrams and Shen (1989) provided an example of a 2-consumer–2-resource system in which initial invasion by the second consumer shifts the original 1-consumer–2-resource system to a different equilibrium, which then results in exclusion of the invader. Case (1995) gives examples of this phenomenon from non-resource-based competition models (Lotka–Volterra) with more than two competitors. Mylius and Diekmann (2001) provide examples from evolutionary biology and describe the general phenomenon as a 'resident strikes back' scenario. At present we cannot determine how frequently this is likely to occur in natural systems involving shared resources.

The second problem with invasion from near-zero densities is that it may not indicate the success or failure of invasion from a somewhat larger number of individuals. Barabás et al. (2018) note that adding Allee Effects to the LV model makes it possible to have coexistence without having invasibility from very low density. The frequency of occurrence of Allee Effects has been debated, but they are unlikely to be rare, and have probably been underestimated (Perälä and Kuparainen 2017). Evolutionary mechanisms for generating Allee Effects were discussed by Berec et al. (2018). Recently, Schreiber et al. (2019) have discussed the impacts of Allee Effects in competition models with fluctuating populations. Typically, the presence of an Allee Effect in the case of a single species population is not regarded as implying that the species in question cannot exist. It is therefore inconsistent to require invasibility of each single-species equilibrium as a prerequisite for coexistence of two species. This argument applies to multi-competitor systems as well.

The third problem with invasion analysis is that the set of species being invaded is unlikely to be at its equilibrium density, and often invasion and persistence can occur in such perturbed communities. Two-species Lotka–Volterra models with periodic variation in coefficients frequently have outcomes that depend on initial abundances, and can have alternative coexistence outcomes (see Chapter 8). The well-known rock–scissors–paper type of competitive interactions (3-species intransitive competition; Gyllenberg et al., (2006)) has no 2-species attractors, so examining invasion of 2-competitor subsystems is impossible. In addition, a given intransitive system may have a wide variety of different potential outcomes (Gyllenberg and Yan 2009). Even without intransitive pairwise abilities, it is common for alternative equilibria to exist in multi-species competition models (Gilpin and Case 1976; Case 1990, 1995), and it has long been known that most natural populations exhibit high levels of environmentally driven population fluctuations (Pimm 1991). This makes it quite possible for species to increase and fluctuate around a mean population size, without being able to increase from very low abundances into one or more other configurations of the system in which they are initially absent. Chapter 4 presents additional problems with invasion analysis that are related to the return of resources to a system after they are excluded by apparent competition.

The preponderance of competition theory assumes that species compete in a spatially homogeneous spatial domain that is cut off from immigration. However, an isolated system with no possibility of immigration will not have any species, and most populations occur in spatially heterogeneous domains. The raises the possibility (first noted by Levin (1974)) that, even with identical patches, two species that cannot coexist in one patch will be able to coexist in a 2-patch system if the competitive exclusion is based on a priority effect, and the initial relative numbers of the two species differ between patches. Metapopulations and metacommunities are the norm in nature (Holyoak et al. 2004), and their ultimate species composition depends on the timing and relative numbers of individuals in invasion events. Coexistence at the larger spatial scale cannot be determined by a single invasion event. More details on spatially structured competition are provided in Chapter 10.

Environmental variation can have particularly large impacts on competition, both within (Abrams 1997, 2009d) and between consumer species (Chesson 1994, 2000a, b, 2003; Abrams 2004b; Barabás et al. 2018). Large effects from temporal variation are particularly likely when consumer functional or numerical responses are nonlinear. In such cases, any analysis that assumes equilibrium (e.g., traditional isocline analysis, and most actual applications of invasion analysis) will be insufficient, regardless of the number of competing consumers. While variable systems will not be considered in detail at this point, it is important to have a framework in which the impacts of different types of environmental variation are not implicitly ruled out, and variability in different growth-related parameters can be compared. Resource-based models are needed to explore differences between the effects of variation that arises from the consumer–resource interaction, and the effects of environmental variation directly influencing resource growth. Abrams (2004b) showed that these two categories of effects differ significantly. Temporal variation is examined more closely in Chapters 8 and 9.

3.9 Summary

This brief review suggests that many of the consumer–resource interactions that underlie competition are quite different from those considered by MacArthur (1972), and most still have not been analysed in the context of competing consumers having homogeneous populations. Even the simplest nonlinear functional response forms proposed by Holling (1965) have seldom been incorporated into models of competition. Perhaps a better appreciation of the importance of functional and numerical response shape for competitive as well as predatory interactions would make the study of their shapes and dependencies a more popular research topic.

In most of the models considered in subsequent chapters I will restrict consideration to consumer populations in which individuals within a species are identical (or nearly identical) to each other in all their ecological traits. I also assume (in most but not all chapters) that there is no temporal environmental variation except that produced by the interaction of consumers and resources. These seemingly restrictive assumptions still encompass a wide range of models.

4

Competition theory

its present state

4.1 Introduction

The previous chapters suggested that a consumer–resource approach should always be taken in modelling competition, because competition is an indirect interaction that always involves resources. Nevertheless, resources have often been omitted, even in the most influential recent research. This probably reflects the undue influence of past literature based on Lotka–Volterra (LV) models. As Getz (1998, p. 541) pointed out, 'reason can be dominated by historical precedent in the development and elaboration of models in population ecology'. The original models by Lotka and Volterra did not contain resources, so many have simply followed this historical precedent.

Resources represent distinct dynamic entities that transmit the interaction, and all other indirect interactions of this sort (trophic cascades in 3-level systems, apparent competition, or mutualism via shared predators) are modelled with explicit dynamics of the transmitting entity. Thus, the first step in addressing the adequacy of recent theory on competition is to examine whether and how it accounts for resources. However, even in those models that incorporate resources, the range of cases examined in recent theoretical work has been limited; theory has retained its LV-era focus on systems with only two competing species, and models having equal numbers of consumers and resources. While reliable complete counts of the number of competing consumer species in natural guilds of competitors are rare, having only two is uncommon, even in species-poor groups (e.g., hermit crabs; Abrams, Nyblade, and Sheldon 1986; see discussion in Chapter 7). The number of resources consumed is almost always much greater than the number of consumers, if resources are distinguished properly (Haigh and Maynard Smith 1972; Abrams 1988b; see Chapter 3). Finally, the majority of influential studies that use a consumer–resource approach have used linear forms for the three basic components discussed in the previous chapter. Linearity is very unlikely and produces many unusual properties for the models. Thus, many of the findings may be misleading with respect to competition in natural communities.

Another metric for judging recent studies of competition is the extent to which they provide a description of the interaction that would allow prediction of the effects

Competition Theory in Ecology. Peter A. Abrams, Oxford University Press. © Peter A. Abrams (2022).
DOI: 10.1093/oso/9780192895523.003.0004

of various environmental changes on the abundances of the species. Environmental change is known to affect the several parameters of the simplest consumer–resource systems differently, and the same must be true of multi-consumer systems (Amarasekare 2015). Many studies in the past have either focused exclusively on the issue of when species can coexist, or have observed the consequences of addition or removal of one species on a small spatial scale. These experiments generally do not provide a basis for estimating responses to climate change, harvesting, or several other common types of environmental alterations.

4.2 Questions for assessing recent influential theory

Here is a list of questions that address the adequacy of theory to represent competitive interactions between species:

1. Does the model used include explicit resource dynamics (meaning separate dynamical equations describing temporal change in the consumer and resource populations)?
2. If so:
   a. What is the form of resource dynamics, and is more than one form explored?
   b. How many resources are present, and how does that compare with the number of consumers present?
   c. Do the resources interact directly with each other or via indirect pathways that do not include the consumers?
3. Does the model consider more than two competing species, and does it apply to systems with more species?
4. What are the characteristics of the functional and numerical responses of the consumers that are implied or assumed by the model?
5. Is the goal of the study to describe the interaction or only to predict whether the species will coexist?

These questions underlie my assessment of a set of highly cited articles listed in Table 4.1 below. I do not mean to argue that it is necessary that all the issues underlying these questions be addressed in every study. This list of questions is certainly not comprehensive in terms of the processes that could be included, the functional forms that could be used, and the questions that could be asked. Given our current rudimentary understanding of the interaction known as competition, it is impossible to give a full accounting of the requirements for a predictive theory. However, at least some of the required features of a better competition theory seem clear, and are listed below:

(1) A priori, it would seem almost impossible that resource properties would not affect competition, and many supportive examples are given throughout this book. Thus, a general theory regarding competition should incorporate resources, and be capable of dealing with different effects of consumed resources on the per capita rate of increase of the consumer. Even when resources can be assumed to quickly reach a quasi-equilibrium with respect to current consumer abundances, knowledge of resource dynamics and their demographic effect on consumers is required to calculate the relative impact of any environmentally driven change on the interaction between and abundances of consumers (Schoener 1974a, 1976; Abrams 1975).

(2) The 2-competitor case represents the minimum number possible for interspecific competition, and it is usually less than the number of species that share resources in natural communities (Yodzis 1988). Having equal numbers of consumer and resource species is also unlikely; given the roles of space and time in defining resources there are likely to be many more distinct resources than consumer species. Furthermore, equal numbers of consumers and resources imply that exclusion of at least one consumer is bound to follow upon elimination of one of the resources, provided the system has a stable equilibrium. Having equal numbers of consumer species and resource types also implies that, for a large class of models, a sufficient increase in the abundance of one competitor will eventually cause the extinction of all others, something that is not guaranteed when at least some consumers have exclusive use of one or more resources.

(3) An adequate theory should be able to predict how changed environmental conditions affecting one consumer alter the abundances of others, rather than just determining whether or not they can coexist. Most changes in abundance that are caused by competitors do not involve extinction, but do involve different effects on the resource-related traits of any given consumer. A useful description of the dynamics allows such changes to be understood and their effects on consumer abundances to be predicted.

(4) Finally, a general theory of interspecific competition should provide a framework that can be used to understand the co-occurring interaction of intraspecific competition and to understand the evolution of traits that determine both intra- and interspecific competition. These two areas of investigation involve the same types of resource-mediated effects as interspecific competition. The closely related interaction of 'apparent competition', defined by effects transmitted by shared predators, requires the same focus on the dynamics of the transmitting entity. Probably for historical reasons, theory on apparent competition has always included this feature.

Each of the above four requirements had been recognized by some investigators by the beginning of the 1980s, and at least the first two seemed to be widely accepted at that time. However, they do not seem to be a major concern for most of the recent articles on competition listed below.

4.3 Choosing articles to represent current competition theory

To what extent has competition theory embraced the use of consumer–resource models, included more than two competitors, addressed questions other than coexistence, and dealt with the other issues raised above? I examine a set of the most influential recent works on ecological competition to shed light on this question. To identify these articles, I focused on the top cited articles about competition among those published during the past four years at the time that the first draft of this chapter was written. This was based on a Web of Science search performed in December 2020 using two or three search terms. The combination (competition AND coexistence) was the one chosen; this produced more relevant articles than any other set of two or three of the following terms: 'theory', 'competition', 'ecology', and 'coexistence'. I included all articles within the top 15 of the most cited list that actually had some theory, dealt with ecological competition, and were not in published in a strictly mathematical journal (as opposed to mathematical biology journals). This produced five articles, all of which were published in 2017 or 2018. Citations are to some extent a function of publication date, so I added a more recent (2019) article on competition theory published in the journal (*Ecological Monographs*) having the largest number of papers (three of the five) in the highly cited set. This article was McPeek (2019a). While it is more consistent with the approach advocated here, McPeek (2019a) also has some of the limitations of the other articles, which will be detailed below. These five articles are listed in Table 4.1.

There have been many other prominent recent works published during the years since 2017. Most of these are discussed more briefly here or in later chapters. They include Germain et al. (2018), Hart et al. (2018), Chesson (2018), Letten and Stouffer (2019), Song et al. (2019), Broeckman et al. (2019), Ellner et al. (2019), Grainger et al. (2019), and Pásztor et al. (2020). It is important to note that a number of recent articles, not included in this list, have used approaches more in keeping with the arguments made here.

All the articles in Table 4.1 and almost all listed in the previous paragraph refer extensively to Peter Chesson's (2000a) review article. This is by far the most cited paper dealing with competition from the past 120 years. Chesson's paper received roughly twice the number of citations during the past year alone than the most cited recent article (Levine et al. 2017) received during the past four years. Treated in more detail later in the book (see summary in Chapter 12), this single work by Chesson seems to have been a major factor in the current focus on coexistence, a focus that has apparently come at the expense of studying the nature of the competitive interaction.

Five of the six articles in Table 4.1 have the word 'coexistence' in the title, but none mention 'competition' in the title. Mayfield and Stouffer (2017) do not have either word in their title, but their work is the only one that focuses more on the interaction (competition) than on the outcome (coexistence or exclusion). Several of the above, as well as many other recent articles, have proclaimed the existence of a 'modern' form of competition/coexistence theory, usually citing Chesson (2000a), who introduced

Table 4.1 Top cited recent papers from 2017 through 2020[*]

Rank	Citation
1	Beyond pairwise mechanisms of species coexistence in complex communities Levine, Jonathan M.; Bascompte, Jordi; Adler, Peter B.; et al. NATURE Volume: 546 Issue: 7656 Pages: 56–64 Published: JUN 1 2017
2	Higher-order interactions capture unexplained complexity in diverse communities Mayfield, Margaret M.; Stouffer, Daniel B. NATURE ECOLOGY & EVOLUTION Volume: 1 Issue: 3 Article Number: 0062 Published: MAR 2017
3	A structural approach for understanding multispecies coexistence Saavedra, Serguei; Rohr, Rudolf P.; Bascompte, Jordi; et al. ECOLOGICAL MONOGRAPHS Volume: 87 Issue: 3 Pages: 470–486 Published: AUG 2017
4	Linking modern coexistence theory and contemporary niche theory Letten, Andrew D.; Ke, Po-Ju; Fukami, Tadashi ECOLOGICAL MONOGRAPHS Volume:87 Issue: 2 Pages:161–177 Published: MAY 2017
5	Chesson's coexistence theory Barabás, Gyoergy; D'Andrea, Rafael; Stump, Simon MacCracken ECOLOGICAL MONOGRAPHS Volume: 88 Issue: 3 Pages: 277–303 Published: AUG 2018
approx. 50.	Mechanisms influencing the coexistence of multiple consumers and multiple resources; resource and apparent competition Mark A. McPeek ECOLOGICAL MONOGRAPHS Volume: 89 e01328 2019

[*] This is based on a Web of Science search in December 2020. Short summaries of some problematic features of each of these articles are provided in the chapter appendix (section 4.7). A more recent Web of Science search, just prior to submitting the final manuscript of this book in late February 2022 identified the same set of five top articles, although the relative ranks of the third through fifth had changed. McPeek (2019a) had only approximately 1/6 the number of citations of the currently lowest ranked of the top five in the later search. Holt and Bonsall (2017) was actually ranked second in the recent search, but would not have been included here because it deals almost exclusively with apparent competition.

the idea of separating coexistence into 'stabilizing' and 'equalizing' factors (but did not use the term 'modern'). Later articles by Chesson (2018, 2020b) updated this idea. Chesson (1994) had earlier introduced a three-category classification of 'coexistence mechanisms', (partitioning of resources and other limiting factors, relative nonlinearity, and the storage effect) which has since been almost universally adopted. The last two mechanisms require variable environments, but the primary subject of the majority of the articles in Table 4.1 is partitioning. The relative nonlinearity and storage effect mechanisms are discussed in more detail in Chapters 9 and 8 respectively. This subdivision and classification were collectively referred to as 'Modern Coexistence Theory' by Mayfield and Levine (2010) and HilleRisLambers et al. (2012), and this terminology (currently often denoted MCT) is now widely used; e.g., Germain et al. (2018), Barabás et al. (2018), Burson et al. (2019) and Song et al. (2019). The usefulness of this classification system has lately received some limited criticism in Barabás et al. (2018) and Song et al. (2019).

The large number of recent theoretical articles dealing with competition and the frequent claim that they are 'modern' seems to suggest major new advances in our understanding of competition, and particularly the theory underlying that understanding. However, this is contradicted by the fact that the approaches used and recommendations in the recent articles identified in Table 4.1 are inconsistent or contradictory regarding several important issues. One of these is the importance of explicit resource dynamics. Saavedra et al. (2017) base all, and both Levine et al. (2017) and Barabás et al. (2018) base most, of their treatments on models that lack explicit descriptions of resources. In a later influential article on quantifying competitive ability, Hart et al. (2018) use a Beverton–Holt model, which also leaves out any explicit resources. Mayfield and Stouffer (2017) argue for modelling competition as a direct effect, but one that includes both linear and quadratic terms in the expressions for per capita population growth rates. Unlike the other papers, this one is largely an empirical analysis of whether this quadratic model fits the data better than a linear (LV) model (it does). Of the five highly cited articles from Table 4.1, only one includes resources in all its models (Letten et al. 2017) and this only considers the very simplest forms of 2-consumer–2-resource models. Thus, of the five highly cited papers, two are based strictly, and two more are based largely, on models with no resource dynamics. The one paper of the six listed in Table 4.1 that explicitly argues in favour of resource-based models is McPeek (2019a), which is not in the highly cited category. However, most of its analysis is again of the 2-consumer–2-resource case, and its only 3-resource model also is characterized by equal numbers of consumers and resources. None of the five top cited articles that include equations with explicit functional forms for resource dynamics considers forms other than the two very simple ones considered by MacArthur (1972) or Tilman (1982). Although variable environments are given some coverage in Levine et al. (2017), Barabás et al. (2018) is the only article that gives a detailed treatment of the effect of variation on competition/coexistence.

Saavedra et al. (2017) is based on multi-species models, and both they and Levine et al. (2017) stress the need for more multi-species models. However, Saavedra et al. (2017) is strictly based on the LV model, and it is worth noting that a number of articles from the 1970s had analysed 3-or-more species LV models and stressed the need for more work on them (Gilpin 1975; Gilpin and Case 1976; Levine 1976; Lawlor 1979). Moreover, it had been shown that the case of logistic resources was unique among density-dependent growth models in predicting that the per capita impact of one consumer species on another was independent of the resource abundances (Abrams 1983a). Because addition of a third species changes abundances of at least some of the resources used by the original two, the results of Abrams (1983a) imply that the competitive effects between those first two will change in any nonlinear model.

Most of the multi-species works cited by Levine et al. (2017) are cases in which the indirect effects are transmitted via other competitors rather than via changes in resource abundance. The exception is the work of Huisman and Weissing (1999, 2001) who explored a set of conditions for a rock-scissors-paper type interaction to arise among three competing consumers, each requiring and using three abiotic resources.

The resulting chaotic dynamics can allow coexistence of even more consumers. How-ever, these unusual conditions for three-species coexistence arose only when the consumer species each had their most rapid intake of the resource they required in intermediate amounts (Huisman and Weissing 2001). This scenario seems unlike-ly on evolutionary grounds, since species would be expected to evolve the highest intake rates for the resources required in greatest amounts. Even if we ignore this, the parameter range required for multi-species coexistence is quite narrow (Schip-pers et al. 2001). The argument that interspecific effects between a pair of species are altered by a third species is valid for more probable reasons than the scenario described in Huisman and Weissing (1999).

The six works in Table 4.1 also differ on other issues. Letten et al. (2017) suggest that the division of forces promoting coexistence into 'stabilizing' and 'equalizing' factors proposed by Chesson (2000a) is a major advance, while Barabás et al. (2018) make the opposite point. The latter work (2018, p. 277) states that 'these concepts [equalizing and stabilizing factors as components of invasion growth rates] are use-ful when used judiciously, but have often been employed in an overly simplified way to justify false claims'. Barabás et al. (2018) point out limitations of invasion analy-sis in determining the outcome of competition, but other articles in the list make it a prerequisite for coexistence (Letten et al. 2017; McPeek 2019a; Levine et al. 2017). More recently Grainger et al. (2019) published a review in which they argue that inva-sion analysis is, 'a common currency for ecological research'. This book will point out numerous cases where invasion analysis cannot be used to assess coexistence. Both articles in Table 4.1 that focus on consumer–resource models, (Letten et al. (2017) and McPeek (2019a)), restrict their analysis to systems having equal numbers of con-sumers and resources. Graphical analysis of the dynamics of models was a central feature of Letten et al. (2017) and McPeek (2019a), but was (in my opinion, properly) ignored in other works (Saavedra et al. 2017; Barabás et al. 2018). Many of the features of isocline analysis, such as insight into the stability of equilibria cannot be extended to systems with more than two variables; this includes essentially all real competitive communities.

The second most cited paper in Table 4.1, Mayfield and Stouffer (2017, p. 1), states that studying communities 'often requires some simplification, such as the widespread assumption that direct additive competition captures the important details about how interactions between species impact community diversity'. How-ever, both this article and the related later work by Letten and Stouffer (2019) fail to provide a good reason why the simplification of direct effects (no resource dynamics) is or should be required.

The article by Barabás et al. (2018) is restricted to competition in temporal-ly variable systems, and understanding such systems is incompatible with some of the assumptions and methods advocated in other articles. Temporal varia-tion prohibits graphical analysis using isoclines, and it creates additional mech-anisms by which coexistence can occur when mutual invasion is impossible (see e.g., Chapters 8 and 9). While there is some reference to models with resources, the potentially large effect of the resource dynamics on the nature of variable competition is not explored. The focus of Barabás et al. (2018) is purely on coexistence, and they do note some limitations of using invasion analysis in this context. However, they refer to few of the previous works that had earlier made this case.

A surprising aspect of these differences in approach and outright disagreements is that some of the underlying issues had seemed to be close to resolution decades earlier. For example, the need to represent resource dynamics in competition theory was the subject of many works during the decade following MacArthur's seminal (1970) article (e.g., Schoener 1974a, c, 1976, 1978; Leon and Tumpson 1975; Levine 1976; Abrams 1975, 1977, 1980a; Vandermeer 1980; Tilman 1980). The importance of resources in understanding both the evolutionary and ecological consequences of competition was reinforced in the following decade (e.g., Lundberg and Stenseth 1985; Schoener 1986; Tilman 1987; Abrams 1986a, 1987f, g). Abrams (1987b, 1988a), and Abrams and Shen (1989) pointed out problematic features of the graphical analysis introduced by Leon and Tumpson (1975) and later popularized by Tilman (1982). The limitations of such graphical shortcuts were implicit, if not explicit, in all analyses of competition in variable environments (e.g., Chesson and Warner 1981; Abrams 1984a; Chesson 1994, 2000b, 2003). The fact that dynamics near an equilibrium point could not be extrapolated to competitive systems with sustained large amplitude fluctuations was the subject of several works on the limiting similarity of competing species during the 1970s (Abrams 1976; Turelli 1978). The fact that the effect on equilibrium population size of one 'competitor' species on a second could become positive when a third competitor was added was a central topic of Levine (1976), Gilpin and Case (1976), Lawlor (1979), May and Leonard (1975), and Gilpin (1975), among others. Three-competitor systems were central to MacArthur and Levins' (1967) introduction and analysis of the concept of the 'limiting similarity' of competitors. Thus, the need to extend competition theory beyond two species and/or Lotka–Volterra models seemed to be widely accepted well before 1980, and many of the consequences of that extension had already been documented at that point. One of these was the dependence of community composition on the order of invasion (Gilpin and Case 1976). This work implied that the ability of a focal species to coexist with others is not identical to its ability to invade from near-zero densities into an existing equilibrium community.

4.4 Forgotten results in 'modern competition theory'

Other issues raised in some of the recent articles on competition also have a long, but lately unacknowledged history. The supposedly modern result that competitors that are highly similar in resource use must also have small differences in overall competitive ability for coexistence to occur was known since the time of Lotka and Volterra, and had been quantified in several ways in early theory (MacArthur and Levins 1967; MacArthur 1972; May 1973, 1974; Armstrong 1976; Abrams 1975, 1977). May (1973, 1974) had several figures illustrating this point. Armstrong (1976) implicitly formalized this division in his concept of 'coexistence bandwidth' more than forty years ago. Despite this, none of these works suggested that dividing observed cases into stabilizing and equalizing factors was the key to understanding the dynamics of competitive systems. Resource-based models have shown that a change in the per capita consumption rate of a particular resource by one consumer making it more similar to a second consumer could either increase or decrease its effect on the population size on that second consumer (Abrams 2003; also see Chapter 6). And large fitness differences together with high overlap do not prohibit coexistence when each consumer has exclusive use of some resource that can sustain a population (Schoener 1976; Abrams 1977; this is discussed in greater detail in Chapter 6).

A second seemingly forgotten result is the importance of considering interactions between resources in studying consumer competition for those resources. Levine (1976) and Vandermeer (1980) established that between-resource competition could change the interaction between their consumers from competition to mutualism. However, this finding was not a major part of any of the analyses in the 2017–2019 articles mentioned above. (Steven Levine's 1976 work is referred to in several of the Table 4.1 articles, but its conclusions do not lead any of those modern articles to call for greater study of between-resource interactions.) Levine (1976) is discussed somewhat indirectly in another article that shares some authors with the top five considered here; Godoy et al. (2018). Of course, competition is not the only possible indirect interaction between resources, and, if that interaction is competition, its nature can be as diverse as that between consumers.

A third, related issue—the impact of consumer-caused resource exclusion on competitive interactions—was first raised by Hsu and Hubbell (1979), noted in Abrams (1980b), and was explored further in Abrams (1998, 2001a), Abrams and Nakajima (2007), Abrams, Rueffler, and Dinnage (2008), Abrams, Rueffler, and Kim (2008), and Abrams and Rueffler (2009). Its effect on a 2-consumer–2-resource MacArthur model was discussed in Chapter 3. Consumer-caused resource exclusion is the central focus of most studies of apparent competition (Holt 1977; Holt and Bonsall 2017). Nevertheless, resource exclusion in models of competition has received relatively little consideration in recent years, and that is true of most of the post-2016 articles mentioned above. McPeek (2019a) is the only exception among the six central articles addressed here, and it is a strongly qualified exception; this is discussed in Chapter 6.

This brief review suggests that our general theoretical approach for understanding of competition has in some respects been regressing rather than progressing over time. There have, of course, been advances in more specific questions about competition, such as the role of environmental variation in allowing coexistence (Chesson 1994, 2000b; 2003; Edwards et al. 2013; Miller and Klausmeier 2017; Kremer and Klausmeier 2017; Ellner et al. 2019; Schreiber et al. 2019; see review by Barabás et al. 2018).

Before ending this section, a few more points should be made regarding the current near-complete focus on exclusion and coexistence. This dichotomy leaves out a wide range of questions and phenomena that are tied to competition. Many potential changes in the environment that affect communities do not cause extinction, but simply alter the abundances of species. Understanding the meaning of such changes is essential for predicting extinction (Abrams 2002). If we want to avoid loss of species, it is important to understand the causes of changing abundance. Even if extinction and the number of species present were the only topics of interest, the factor producing extinction is certainly not always a competing consumer. Simple models with two trophic levels show that a given consumer species may be driven extinct due to changes that affect any species in the consumer–resource system. The much-studied 2-consumer–2-resource MacArthur system provides a good example in support of this claim. It is easy to show that each consumer species may be excluded by increasing the growth of the other consumer's preferred resource or decreasing the growth of its own preferred resource. Chapter 7 examines a model having two species on each of three trophic levels, to see which species have the greatest impact on the abundances of the mid-level consumer species; it is usually not the other mid-level consumer. Of course, exclusion may also be driven by negative changes in the focal consumer's growth parameters or positive changes in the other consumer's growth parameters. In models with three or more competitors or additional trophic levels, there are many other possible sources of a change producing extinction of a focal consumer.

The focus on coexistence has been justified because the question of coexistence or exclusion affects biodiversity (e.g., Levine et al. 2017). However, the presence of a large number of consumer species available to coexist depends on their past evolution, and competition has been cited as a primary driver of evolutionary diversification (Schluter 2000). Studying the role of interspecific competition in evolutionary processes also requires an understanding of the details of how traits influence resources and consumers, as well as their interactions. This topic will be explored in Chapter 11.

4.5 Why the Lotka–Volterra and MacArthur models are insufficient

The many references to consumer–resource models in previous chapters might seem to suggest that most investigators have accepted the resource-based approach and the nonlinearity that it usually entails. However, this is not the case. Beyond the articles in Table 4.1, there are numerous examples from the recent literature that lack explicit resources. Leimar et al. (2013), Barabás et al. (2016), and Hart et al. (2018), are just a few of these. Regarding coexistence, the top-cited article from Table 4.1 (Levine et al. 2017, p. 59) states, 'The assumption that the interaction between species is fundamentally pairwise is central to almost all coexistence theory. Yet from an empirical standpoint we have little idea of whether this assumption is correct.' The 'almost all' in the first sentence neglects scores of resource-based models, which were common in the 1970s and early 80s. It also neglects multi-competitor models, many of which were also published in that decade, as discussed below. The second sentence in the preceding quotation was not regarded as correct in the mid-1970s when Neill (1974) reviewed the earlier literature and presented the results of his investigation of this topic using aquatic systems in the laboratory. Theory based on consumer–resource models had long ago shown that virtually all systems must be characterized by impacts of a third competitor on the per capita effects of a focal pair of competitors on each other (Abrams 1980b, 1983a). Subsequent reviews of empirical studies (Jeschke et al. 2004; Sibly et al. 2005), and those models that included adaptive behaviour (Abrams 1989, 1990a, b), suggested that component functions of consumer–resource models should be nonlinear. Taken together these earlier findings mean that nonlinear competitive effects are almost certain to be prevalent, and that competitive effects between two consumers that are independent of the presence of other consumers are at best extremely rare.

Note also that linear density dependence (i.e., logistic intraspecific competition) is rare to non-existent in empirical studies of the growth of single species (Sibly et al. 2005). If this is the case, why should linearity prevail in interspecific competition? Levine et al. (2017) fail to cite Pomerantz et al. (1980, p. 311), who opened their article with the statement 'The linearity assumption in the logistic model of density dependence is violated for nearly all organisms.' This fact about resources implies that, even if all else in the consumer–resource model were linear: (1) a single consumer will have nonlinear responses to a press perturbation in any population growth parameter (Abrams 2009b); and (2) competitive effects of one consumer on another (measured by eq (3.2)) will be nonlinear, and additional consumers will change the equilibrium magnitude of those effects. Levine et al. (2017) do cite a later article by Pomerantz (1981), which argues that linearity *should* be assumed in multi-species competition models based on Occam's razor. However, later empirical studies of functional responses (Jeschke et al. 2004) and the large body of consumer–resource models reviewed here show that independence and linearity of interspecific effects is highly unlikely for the same reason that they are unlikely in single-species growth. Arguments against Pomerantz's (1981) justification had been presented in Abrams (1983a). Levine et al. (2017) cite Abrams (1983a) elsewhere in their article, but ignore the response to Pomerantz's 1981 work, and seem to argue that little is known about the linearity of competitive interactions (e.g., p. 61: 'few studies measure response variables that can be translated into dynamics through a competitive population dynamics model'). It is contradictory to admit that most intraspecific competition is nonlinear and reject nonlinearity for interspecific competition (as Pomerantz did). Nonlinearity in either inter- or intraspecific effects implies 'higher-order' interactions in a consumer-only model derived from a consumer–resource system by assuming quasi-equilibrium of the resources.

'Higher-order interaction' (often denoted HOI) is another term that has been used in various ways. HOIs are usually defined (Billick and Case 1994) as effects on per capita growth rate that cannot be represented by a sum of terms, each of which depends only on the abundance of a single other species or food web component. This means that the immediate effect of the population size of one species on the per capita growth rate of a second species is altered by the abundance of a third species. Nonlinear single-species effects are not covered by this definition, but have been included in HOI by some authors. Letten and Stouffer (2019) define HOI as non-additive effects of densities on per capita growth. In most multiple resource systems with saturating (type II, III, or IV) functional responses, the consumption rate of one resource by a given consumer will depend on the abundances of other resources, implying a higher-order interaction between the consumer and the resource. This is true for different nutritionally essential resources, even if they are encountered at a rate proportional to their abundance, provided there exists some mechanism for adaptively adjusting relative intake rates of different resources (Abrams 1987c). Thus, even the transmitting links in single-consumer–multiple-resource systems almost always involve higher-order interactions (Abrams 1980b, 1983a, 2001c, 2010a, b).

'HOI' for competitive interactions is actually an all-inclusive category in systems with nonlinear responses and adaptive foraging, as all must involve HOIs when resource populations are considered as dynamic entities. Even if resources are assumed to always be close to equilibrium with respect to current consumer density, which produces a model without explicit resource abundances, the effect of one consumer on a second consumer will be affected by the abundance of any additional consumer species in any model with nonlinear density dependence in the resources. More generally, the receiver species' per capita growth rate in an indirect interaction within a food web almost always includes terms that involve non-additive combinations of initiator and transmitter abundances (Abrams, 2001c). Thus, the basic conclusion of Letten and Stouffer's (2019) analysis of higher-order effects in competitive systems would not have come as a surprise to most ecologists familiar with the competition theory current in the mid-1980s.

4.6 Reasons for including resource dynamics

The majority of the articles reviewed in this chapter omit resource dynamics. Both of the articles in Table 4.1 that focus on consumer–resource interactions (Letten et al. 2017; McPeek 2019a) consider a very narrow segment of the spectrum of likely functional forms for both resource growth and the consumer's responses.

The absence of resource dynamics is problematic even for cases in which one or more consumers produce some of their effects on other consumers by direct contact or proximity. It is these types of systems in which resources are usually ignored; studies of competition between rooted plants are a good example of this. Here competition often entails some form of physical displacement; overgrowth of one plant by another, or removal from a required substrate, as in the case of some sessile invertebrates. These mechanisms have led many authors to continue to use models with direct interactions. As noted above, Hart et al. (2018) argue that the (nonlinear) multi-species Beverton–Holt model is a good representation of plant competition. However, even in this case, ignoring resources is problematic because overgrowth competition between rooted plants requires resources (light, water, and nutrients) to fuel that growth. The nutrients, like any other abiotic resource, have dynamics that may be crucial to the nature and outcome of competition. And stationary adult organisms need to disperse their gametes and offspring, which makes it possible to have coexistence based on pollinator, life history, or spatial differences. Haigh and Maynard Smith (1972) showed that localized interactions mean that a given substance (e.g., nitrogen) often constitutes many distinct, spatially localized resources, whose movement between patches influences the extent of competition between their consumers (e.g., rooted plants).

The argument for studying (and modelling) resource dynamics and consumer functional and numerical responses to understand competition is bolstered by the fact that no other indirect interaction has been modelled primarily by leaving out the dynamics of the entities that transmit the effects involved. I know of no study of apparent competition that omits a description of the dynamics of the shared predator(s). Similarly, there appear to be no theoretical analyses of top-down or bottom-up cascades in food chains that have omitted the dynamics of the intervening trophic level(s). Why should the indirect interaction of competition be treated differently? It is important to note that explicit modelling of the intervening trophic links in both apparent competition and trophic cascades has revealed that neither one is always characterized by the sign structure that it has in simple models with linear interaction terms. Shared predation can result in effects other than mutually negative ones (Holt 1977; Holt and Bonsall 2017; Abrams 1987d; Abrams and Matsuda 1996), and trophic cascades in 3-level systems can have effects other than (+,+) between the top and bottom species (Abrams 1992c, 1993, 1995; Abrams and Vos 2003).

Some consideration of consumer–resource interactions is required for useful descriptions of inter- and intraspecific competition. However, our understanding of consumer–resource interactions in general is also relatively limited. A large number of the theoretically possible functional forms that have been proposed over the last several decades have not been examined in empirical systems; relatively few have been incorporated into models of interspecific competition. Chapter 3 had a short review of this multitude of forms for the components of consumer–resource models, and it implies that much is left to learn about competition. The recent, highly cited articles reviewed in this chapter suggest that current theory is not moving in the direction of exploring these alternatives. The remainder of this book includes a wide variety of additional questions and examples, illustrating how a resource-based theory of competition differs from the more popular direct-interaction models.

4.7 Appendix: Problematic features in the focal articles

As a preface to this section, I should note that most of my own previous articles on competition suffer from one or more of the limitations I criticize here. Many of those articles deal with 2-consumer–2-resource systems, as do many articles by other authors from the 1970s and 1980s. I have frequently based analyses on systems in which the resources did not compete with each other, and have often modelled the species on the bottom (resource) trophic level using independent logistic growth.

1. Levine et al. (2017) is a broad review article covering a large body of work. It contains relatively little mathematical theory. The authors focus entirely on the question of coexistence. They argue that mutual invasion is sufficient to determine coexistence in a 2-competitor system, an idea that is refuted by Barabás et al. (2018) and here, based on previous work. They argue that almost nothing is known about 'higher-order interactions', so ecological theory should start studying these interactions by examining the effect of 'randomly assigned' values for them. In fact, higher-order interactions follow directly from most consumer–resource models, and these more appropriate models can provide a logical basis for what types of higher-order interaction terms make biological sense. In multi-species systems, Levine et al. (2017) argue for the 'structural stability' approach, also advocated by Saavedra et al. (2017). There is no reason to assume that most multi-species systems have stable equilibria (as required by the approach), and consumer–resource models are not expected to have the same stability properties as a simplified competition model of the same system derived by assuming resources are constantly at equilibrium with respect to current consumer abundance. Levine et al. (2017, p. 63) end by proposing that the sparseness of evidence of how coexistence comes about in multi-species models results from the 'intractability of empirically evaluating competition between many species.' The study of multi-species functional responses and relative magnitudes of consumption rates does not involve any problems that are not shared by almost all studies in community ecology. We know enough about the range of likely forms for these component functions that theoretical studies could provide a good indication of the differences that are likely in more realistic models having more resources than consumers and/or having more than two consumers.

2. Mayfield and Stouffer (2017) is more of an empirical study, so is considered separately here. They used a definition of 'Higher Order Interactions' as 'the quadratic density dependent effects on per capita fitness', which differs from the definition used here. This quadratic term limitation lacks a mechanistic justification from resource-based models and is incapable of being related to the biological details of resource consumption and conversion. It is based on observations of growth rates and neighbouring plants in an annual plant community. The abiotic nature of resources required for plant growth leads to nonlinear forms of interspecific effects, as was noted by MacArthur (1972). Unfortunately, mechanistic models of resource-dependent growth were not considered.

3. Saavedra et al. (2017, p. 471) seek to answer the question 'how much of this coexistence depends on mechanisms that require more than two species.' The reason for attempting this partition is unclear. Different competitors of a focal species will often have effects of different sign on the focal species, and an aggregate measure of dependence of coexistence on the entire set of other competitors has little meaning. Even if the question were well defined and of interest, a reasonable model of the interaction would be required to answer it. Instead, Saavedra et al. (2017) adopt the Lotka–Volterra equations. A more useful question about a multi-species

system would be whether there exists some negative or positive effect on one or more other competitors that could produce extinction of a focal species. Another possibility is the question discussed by Schoener (1993); i.e., whether interactions diminish in magnitude with greater food web distance (i.e., more intermediate links). The answer to these questions depends on which parameters of the model are altered, and is likely to depend on still other food web components (on trophic levels higher than the consumer or lower than the resources). Multi-species models also have the possibility that one or more of the 'other competitors' may go extinct as a result of the perturbation applied to the system, which rules out extrapolation from effects defined by the original equilibrium. Predicting any response to any perturbation will require some consideration of the appropriate functional forms of the consumer–resource interactions involved, most of which are incompatible with the linear effects in the LV model.

4. Letten et al. (2017) only analyse models in which resources are governed by chemostat dynamics, so do not consider biotic resources, which can be driven extinct by their consumers. Letten et al. (2017) also confine their analysis to 2-consumer–2-resource systems, and to consumers with linear functional responses. They do consider the possibility of nutritionally essential resources, but do not consider the effects of adaptive variation in relative consumption rates, something that is very likely in the case of such resources (Abrams 1987b; Abrams and Shen 1989). Thus, their models represent a very limited range of biologically plausible 2-consumer–2-resource systems. The inability of their favoured method of graphical analysis to provide generally applicable results in other cases (e.g., most cases with biotic resources) is not noted.

5. Barabás et al. (2018) is largely concerned with reviewing Chesson's body of theory on competition in temporally fluctuating environments. Resources are not explicit variables in most of their analysis. However, resources are implicit in the competitive effect parameter of their models, and do make an appearance in their supplemental material. On the other hand, Barabás et al. (2018) suggest that Chesson's general theory is most appropriate for systems with a single resource, which is to say, virtually no natural system. There are certainly many qualitative results in Chesson's large body of work on competition in variable environments that are valid in situations with many explicit resources. However, it is also true that the nature of the covariance of competition and the environment, which determines the possibility for coexistence, is strongly affected by resource dynamics. Consequently, models with explicit resources are usually required to determine the impact of sustained environmental variation on the interaction of consumers.

6. McPeek (2019a) argues strongly for a consumer–resource approach to competition, and considers some nonlinear functional responses, as well as the case with three consumers and three resources. However, he fails to note the effects of consumer or resource extinction on the competitive process and ignores (in some cases) or is incorrect (in other cases) about the effect of type II responses on the nonlinearity of competition, the probability of resource extinction, and the

relationship between overlap and competition. He also does not discuss a variety of other mechanisms that make resource-based competition inconsistent with the Lotka–Volterra model, and/or simple methods of analysis. McPeek (2019a) favours simplified graphical methods of analysis that are seldom sufficient for understanding multi-species consumer–resource models. Although this and several other details of McPeek's (2019a) article are criticized in this book, its stress on the need for consumer–resource models still makes it more likely to contribute to the future of competition theory than the other articles discussed in this chapter.

5

Understanding intraspecific and apparent competition

5.1 Introduction

The two phenomena mentioned in the title of this chapter are closely related to interspecific competition, but also differ from it in important ways. Intraspecific competition involves effects of the abundance of a consumer species on its own population growth, while apparent competition involves effects of one species on the population growth of others that are transmitted by one or more shared consumer species rather than shared resources. The reason for discussing models of both of these processes is that each usually co-occurs with interspecific competition in any competitive system having living resources. In keeping with Chapters 2 and 3, this chapter argues that understanding both of these phenomena requires an understanding of consumer–resource interactions.

Intraspecific competition always accompanies interspecific competition and influences its effect on population density. Apparent competition is not an absolutely necessary component of either intra- or interspecific competition, because it does not take place in systems having a single resource or in systems with abiotic resources. However, single-resource systems are extremely rare. Most real consumer species have two or more resources, and apparent competition occurs between the resources if they are living entities. In fact, apparent competition is a key determinant of the form of intraspecific competition in the single consumer, no matter how that form is measured. When intraspecific and apparent competition co-occur in a two-trophic-level system, the important distinguishing feature is the identity of the initiator and receiver of the effect. The two focal species occupy the lower trophic level in the case of apparent competition, and the focal species occupies the higher trophic level for intraspecific competition. 'Real' and apparent competition are closely connected with living resources because most consumers utilize two or more resources, and the nature of intraspecific competition in such species is always influenced by the indirect interaction of the resources via the consumer (apparent competition between the resources).

Intraspecific competition is closely related to the process of natural selection within a species. This is discussed in Chapter 11. In an asexual species, the presence of two genotypes with different ecological characteristics has dynamics that are almost

Competition Theory in Ecology. Peter A. Abrams, Oxford University Press. © Peter A. Abrams (2022).
DOI: 10.1093/oso/9780192895523.003.0005

identical to those of a system with interspecific competition. The one difference is the possibility that one genotype gives rise to individuals of the other type via mutation.

While arguing that apparent competition should be viewed as distinct from 'traditional' (resource-based) competition, I will also suggest that, like interspecific competition, it should be explored using a wider range of consumer–resource models than has generally been employed in the past. Models of apparent competition have always included the dynamics of the transmitting entity or entities (the consumer(s)). This makes the body of theory somewhat stronger than that for 'normal' competition. However, most of the literature on apparent competition has been restricted to the simplest possible system—two dynamically independent resource species with a single shared consumer species, usually assuming a stable equilibrium. This limitation seems to have been even more prevalent than the corresponding focus of models of interspecific competition on systems having just two consumers and at most two resource types. Apparent competition was unnamed and virtually unstudied before Holt's (1977) article, so it has had a shorter history. Perhaps as a consequence, models having some nonlinear functional components have been more common than in the literature on competition for resources.

5.2 Intraspecific competition

Chapter 2 noted that one of the earliest textbooks on population ecological theory (Wilson and Bossert 1971) had discussed within-species competition under the heading of 'density dependence'. This was standard usage for intraspecific competition, both before and after their book was published. The terminology is problematic in that it implicitly suggests a direct effect of abundance, rather than an effect that is mediated through interactions with resources. The terminology also supports the unfortunate view that intraspecific competition is fundamentally different from interspecific competition. Yet, both processes are likely to involve overlapping sets of resources. For both processes, the nature of the resource dynamics, including possible interactions between different resources, can greatly change the population-level consequences of environmental variables that act directly on a focal consumer species. Wilson and Bossert's (1971) textbook and much of the subsequent literature have treated the subjects of intra- and interspecific competition separately, and the questions asked about the two processes have also differed. This separation does not seem to be justified, and an understanding of intraspecific competition is needed for a full understanding of interspecific competition. Density dependence is actually a somewhat broader term, in that increased predation rates with higher prey density is usually considered to be a form of density dependence in the prey, but it is not an interaction via resources.

Intraspecific competition involves two or more resources in nearly all consumer species; true specialists are quite rare (Polis and Winemiller 1996). Thus, a basic understanding of intraspecific competition requires an understanding of the dynamics of systems with a single consumer of multiple resources. When the resources themselves are living species, this means that the process of apparent competition is inextricably linked to both intraspecific and interspecific competition.

5.2.1 The definition and mechanism of intraspecific competition

It is interesting that, like 'interspecific competition', the definitions of both 'intraspecific competition' and 'density dependence' have long been the topics of debates that are still not settled. Regarding density dependence, Herrando-Pérez et al. (2012) claim that most modern ecologists would agree on Murdoch and Walde's (1989) definition of, 'a dependence of per capita population growth rate on present and/or past population densities'. However, they go on to discuss a number of subsequent works by various authors who had different definitions. The definition of intraspecific competition used here is similar to the definition of interspecific competition proposed earlier; i.e., the effects of the abundance of a single species on its demographic rates that are caused by the associated changes in the consumption rates of shared resources. Intraspecific interference effects are included, as they generally modify consumption of some resources used by other individuals, as well as by the focal individual.

Most of the criticisms of the LV competition model covered in previous chapters also apply to the logistic and other continuous-time single-variable models of single-consumer population dynamics. In both types of models the instantaneous growth rate is determined by the current population size. Like the MacArthur consumer–resource model, the logistic model for single-species population growth can be derived from a standard continuous-time consumer–resource model in which the resources are nutritionally substitutable, all resources have logistic growth, and none of the resources have the potential to be driven extinct by the interaction. The derivation also requires that the consumer species have linear functional and numerical responses. These limitations mean that the use of the logistic model requires a rather unlikely set of assumptions to be an accurate representation, including the particularly troubling assumption that the logistic itself must apply to the growth of all of the resources. As one examines successively lower trophic levels in a food web, abiotic resources are likely to become more prevalent, and most plants use primarily abiotic resources. If all of the higher-level species had linear functional and numerical responses, this would result in their density dependence being concave (decreasing at a decreasing rate), rather than linear (Abrams 2009b, c). Even if conditions for linear density dependence apply at a stable equilibrium, the dynamics and mean densities as a function of the mean mortality rate are not well-described by the logistic model in fluctuating environments, a topic that is explored in Chapter 8. As noted in treatments of the MacArthur model in previous chapters, we know from empirical studies that the functional forms required by this model are relatively uncommon in systems where they have been quantified. Nonlinear forms of any of the three component functions in the MacArthur model imply that the simple logistic model will generally provide inaccurate predictions for the response of population size to any perturbation.

The connection of density dependence to resources and the underlying mechanisms of consumer–resource interactions has been recognized by many previous authors. For example, Begon et al.'s (2006, p. 411) textbook states: '[many studies] ... have been concerned to detect "density dependent" processes, as if density itself is the cause of changes in birth rates and death rates in a population. But this will rarely (if ever) be the case: organisms do not detect and respond to the density of their populations. They usually respond to a shortage of resources ...'. Even earlier, Krebs (1995, 2002) and Sibly and Hone (2002) had argued that a mechanistic approach should replace the concept of density dependence. One of Krebs' (2002) arguments against simple density dependence is that previous attempts to quantify growth rate as a function of density have found that the relationship changes greatly with time and with the location of the study. This is an indication of effects that depend on other food web components, as well as variable factors in the environment. Krebs (2002, p. 1218) ends his article as follows: 'My plea here is to concentrate our efforts on finding out in the short term why population growth rate is positive or negative. In doing this, we can abandon the worries about equilibrium ... and put more interesting experimental biology into population dynamics. By concentrating on what factors affect population growth rate, we can provide a science that will be useful to decision makers and managers of the diversity of populations on our planet.' The continuation of the opening quotation of this chapter, from Murdoch, Briggs, and Nisbet's book, 'Consumer–Resource Dynamics' (2003, p.1), states that, 'Virtually every species is part of a consumer–resource interaction ... Consumer resource interactions are, in addition, fundamentally prone to being unstable ... If we are to understand population regulation ... we therefore need to focus on consumer–resource interactions'.

The use of models of (consumer) population growth that include resource dynamics has a long history. A few of the many examples are Schoener (1973), Royama (1992), Rueffler et al. (2006b) and Johst et al. (2008). However, the relationship between consumers and resources is often ignored in applied areas of ecology, such as fishery regulation and conservation biology. It is common for fishery regulators to use simple models of single-species density-dependent growth to determine the abundance and harvest rate producing maximum sustainable yield, as well as the smallest population size that is consistent with minimal extinction risk. This can result in inappropriate regulations, as discussed in Matsuda and Abrams (2006, 2013).

Despite the opinions quoted above, many textbook treatments of intraspecific competition have continued to ignore the underlying consumer–resource interaction when discussing the topic of density-dependent growth. The rest of this section will examine what is needed for a mechanistic resource-based approach to intraspecific competition, and how the range of predictions of such an approach differ from those of simple direct-density effects that still dominate the literature.

5.2.2 Describing, measuring, and modelling intraspecific competition

The fundamental motivation for models of density dependence is to describe the relationship between population size and growth rate. However, this relationship is usually at least partly indirect, implying the presence of time lags. Adding or removing consumer individuals initiates a change in resource densities, but the consequences of such a change are not fully realized until resources respond and the system reaches a new equilibrium or dynamic attractor. By the time the resource response is measurable, the perturbed consumer population will likely have changed in size. Even if it were possible to maintain the consumer population at exactly its perturbed abundance by constant additions and removals, the effect of that population control often changes the pattern of change of the resource abundances to produce dynamics that would be impossible to achieve by altered values of neutral parameters. This is true, for example, in any system that would otherwise exhibit predator–prey cycles, as well as systems in which the flexibility of the manipulated species is required to maintain stability of the community in which it is embedded. As is true of interspecific competition, the changes in both immediate dynamics and ultimate population size caused by a changed property of the focal consumer species differ depending on the nature of the dynamics of the consumer's resources. The conclusion is that the dynamics of both consumer and resource must be modelled to describe density dependence.

Does the involvement of resources mean that the traditional view of a relationship between density and population growth must be discarded completely? The answer is no. Density dependence can be quantified in the same manner as interspecific competition; i.e., as a relationship between a neutral parameter and equilibrium or mean population size (Abrams 2009b, c). As was noted in Chapter 3, a potential experimental approach to quantifying the effect of one species on another is to impose a continuous but low addition or removal rate of individuals. Once the limiting dynamics have been attained following this perturbation, the change in the mean population size indicates the interaction strength (eqs 3.1 and 3.2). The same approach can be applied to intraspecific competition as well as interspecific. The measure for density dependence in species i is simply the denominator of eq. (3.2). As noted in Chapter 3, the effects of such a perturbation can be counter-intuitive, such that the perturbed population changes in the direction opposite to that of the perturbation (where the direction of the perturbation is defined by the effect of the perturbation on immediate per capita growth rate). A constant input of individuals can decrease the ultimate population size of the perturbed species; similarly, a newly imposed removal/death rate can increase the ultimate population size (Abrams and Matsuda 2005; Abrams 2009a). Abrams and Matsuda (2005) termed this counterintuitive change the 'hydra effect'. The scenarios exhibiting hydra effects include systems in which maintaining a constant population size of the focal species destabilizes its original equilibrium (Cortez and Abrams 2016), and, as shown in that article, such responses are by no means confined to predator/consumer populations. These possibilities have no analogue in models of direct density dependence. However, if constant mortality perturbations of a range of sizes were applied to a logistic growth model, the results would indicate a constant value for the decline in population size per unit change in mortality; namely, constant 'density dependence' over the range of densities.

While the addition or removal of individuals (i.e., neutral parameters) are useful for quantifying interaction strength, the effects of other perturbations are also important if we are to fully understand intraspecific competition. Environmental changes often affect the nature of the consumer–resource interaction directly, for example, by altering the rate of capture of resources, and/or the effects of those consumption processes on demographic rates. Climate change is likely to affect activity levels, and therefore resource uptake rates. This effect makes climate change a non-neutral perturbation, as it has an immediate effect on both consumer(s) and resource(s). An increase in its resource uptake rate can decrease a consumer population in a much wider (and different) range of circumstances than those leading to a decreased population due to decreased mortality (Abrams 2002, 2003, 2004a).

Decreasing the abundance of resources directly by consumption is not the only mechanism by which consumer individuals can affect the resource intake of other members of their population. As in the case of interspecific competition, individuals of a single species can directly interfere with other individuals of that species, decreasing their rate of food consumption or imposing additional mortality. Predators of the consumer species may have effects on the consumer's resource uptake, so a quantitative description of intraspecific competition may often require consideration of higher trophic levels.

The only major mechanism that differs between intra- and interspecific competition in animals is the (usually) negative impact of competition for mates and of mating itself, which are both (usually) interactions within a given species. These include energy expenditure and mortality on the part of females resisting mating, disease transmission, and direct negative effects of substances that are transferred by males in the process of successful mating (see Chapter 3 of Arnqvist and Rowe 2005). Males are also subject to mortality from other males while competing for females, and female-caused mortality due to cannibalism. Burke and Holwell (2021) describe an example of high female mortality as a result of violent male mating behaviour (which is, in part, a function of males trying to avoid cannibalism). This type of interaction could be a source of reduced population growth at higher densities. Mating-related mortality in males seems less likely to reduce population growth rates due to the abundance of mating opportunities for females of most species, once population size is simply moderate. However, the effects of male mortality are complicated, as reduced male numbers may also entail indirect positive effects on females due to reduced resource competition as a result of fewer males. For plants, higher population densities may attract more pollinators and/or reduce self-pollination, producing effects that may offset resource competition. The pollinator visits can be represented as resources, although their dynamics differ greatly from those of light and mineral nutrients. In any event, we know relatively little about the effects of such processes on population growth rates for most animals or plants, and distinguishing sexes is inconsistent with the homogeneous-population approach used in most of this book, as well as most previous competition theory. Thus, sexual conflict will not be included explicitly in the models discussed below, although it is one of many causes of intraspecific interference competition.

Even if the population-level consequences of sexual reproduction are ignored, the wide variety of potential resource dynamics and interference effects should have resulted in a comparably expansive variety of models of intraspecific competition.

Based on this, we should have expected a wide range of observed single-species population growth trajectories. The connection of consumer–resource models to density-dependent growth has been discussed occasionally over the last several decades (Schoener 1973; Schaffer 1981; Matessi and Gatto 1984; Abrams 2009b, c, d). Abrams (2009b, c, d) suggested that neither the logistic nor the somewhat more flexible theta-logistic (Pella and Tomlinson 1969; Gilpin and Ayala 1973; see eq. (5.1) below) is likely to provide a generally applicable model for the response of a consumer species to environmental change affecting population growth. Abrams (2009b) examined how the presence of multiple resources alters the form of intraspecific competition, as reflected in the response of a consumer species' equilibrium or mean population density to changes in a neutral parameter. Abrams (2009a, b, c, d) and Abrams and Matsuda (2005) illustrate various implications of type II consumer functional responses for the viability of a logistic growth model as an approximation to consumer dynamics. As in the 2-consumer case, nonlinear per capita growth of the resource, resource exclusion, and nonlinear consumer functional responses all produce nonlinear responses of population size to changes in neutral consumer parameters. They also produce the potential for discontinuous change in abundance with continuous change in a parameter (the latter being possible when at least some of the lower-level resources are self-reproducing).

Unlike traditional models of density dependence, resource-based models do not necessarily involve an immediate reduction in population growth rate following an increase in population size. The decrease happens after the additional consumers have significantly decreased the abundances of at least some of the resources. The one major set of exceptions to this generalization about delayed consequences is the category of effects produced by rapid (usually behavioural) change in either predators or prey in response to a change in predator abundance; prey may change activity to decrease their risk of predation in response to higher predator numbers. Predators may increase their foraging in the presence of others to ensure increased food intake before the local prey population is depleted. However, these behavioural effects are only one component of the feedback process by which current consumer abundance affects future consumer population growth. The feedback via altered abundance of resources always involves some delay. And because of predator–prey oscillations, even if they are damped and eventually disappear, there may be prolonged and oscillatory changes in the abundances of consumers and resources following a perturbation to their abundances (see Figure 2.1). Such oscillations are less common, but they also occur in some systems with abiotic resources. Chapter 8 shows that these lagged responses can have a major effect on the nature of consumer–resource systems in seasonal environments.

The transient dynamics of models of a single consumer with one or more resources, and their responses to parameter perturbations in non-equilibrium systems usually cannot be captured adequately with a model of direct density dependence (Reynolds and Brassil 2013; O'Dwyer 2018; see Chapter 8). It is nevertheless useful to determine how imposed mortality affects ultimate population size, as this reflects the action of the various processes that determine intraspecific competition.

It also provides a quantitative way to compare the responses of systems with two or more competing consumers to those with a single consumer species, as the mutual effects of altered mortality rates provide a standard measure of interspecific competition (Chapter 2).

For a consumer that uses biotic resources, analysing the interactions of those resource species with their own resources is often important for developing a model of the growth of the consumer population. The general need to consider lower-level entities may extend to two or more trophic levels below the focal species whose intraspecific competition is of interest. In any case, the prey's resources will display delayed responses to a change in prey abundance, contributing to a larger delay in the predator's response. This scenario calls for using a model with more trophic levels when the consumer species of interest is on the third or higher level. Having three or more dynamic entities opens up the possibility of more complex dynamics (Hastings and Powell 1991). Abrams and Roth (1994a, b) provide some examples of these dynamics in simple models where consumers on the top one or top two trophic levels have type II functional responses. The case when only the middle species has a type II response was considered in Abrams and Roth (1994a), who show that there are alternative attractors in the three-level system that have or lack cycles. With type II responses on both trophic levels, there are often alternative attractors (Abrams and Roth 1994b). Both types of three-level systems allow positive perturbations to growth of the bottom-level species to translate into decreases in the (mean) abundance of the top-level species, something that does not happen in simple continuous models of density dependence with a strictly food-limited top species. These systems can even exhibit extinction of the top-level species as a result of increasing the carrying capacity of the bottom-level species. Hydra effects are often possible for the top-level consumer; its abundance decreases with lower mortality due to reduced production at lower levels in the food web. Loreau (2010) has presented the case for including all levels down to the abiotic resources used by plants. Most of the models considered below only include two trophic levels, so they understate the likelihood and magnitude of departures from traditional models of single-species density dependence. These traditional models are considered in the following section.

5.2.3 Models of density dependence

It was recognized long ago that the logistic growth model of Verhulst (1838) could not represent all forms of density dependence. The logistic model describes the consequences of intraspecific competition as a linear and immediate negative effect of abundance on per capita growth rate. The main alternative for many years was the 'theta-logistic' model of Gilpin and Ayala (1973), which differs in allowing the effect of density to be a power function of abundance:

$$\frac{dN}{dt} = rN\left(1 - \left(\frac{N}{K}\right)^\theta\right) \tag{5.1}$$

Sibly et al. (2005) analysed existing lengthy time series of population sizes in an attempt to estimate the value of the positive exponent, θ. Their methods were widely criticized (e.g., Ross 2006; Doncaster 2006, 2008; Getz and Lloyd-Smith 2006), and, in many cases, we do not know whether competitors or predators of the focal species had some influence on the time course of population change. Their data were typically annual measurements of size, so it was actually a discrete version of eq. (5.1) that was used. Alternative nonlinear models of per capita growth were not considered. However, Sibly et al. (2006) argue for their original analysis, and their 2005 article remains the most comprehensive analysis of the form of single-species dynamics. The wide range of estimated exponents in that study is, at a minimum, strong evidence for great interspecific variation in the functional form of intraspecific competition. Figure 5.1 illustrates part of the range of shapes of the per capita growth rate function defined by eq. (5.1).

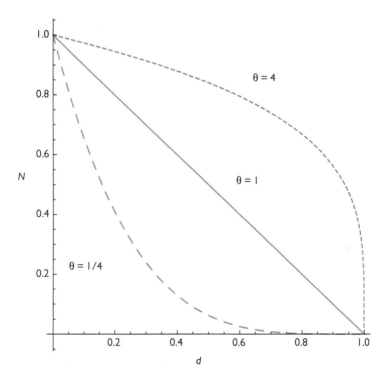

Fig. 5.1 The relationship between mortality and population size for the 'theta-logistic' model (eq. 5.1) under three different exponents.

In another review of empirical studies, Brook and Bradshaw (2006) analysed 1198 time series of abundances, and noted that clear evidence for any density dependence was lacking in approximately 25% of those studies. They attributed this primarily to large sampling errors and time series that were too short. However, it is quite possible that delayed compensatory dynamics were responsible for the apparent lack of evidence, as judged by short-term population responses. It would take a massive programme of exploring food web connections to know enough to understand the responses of populations to current environmental change. Given the rapid pace of environmental change, it is likely that some type of deductive approach will be required to make informed guesses about the nature of density dependence in species of management or conservation concern. The question addressed below is whether eq. (5.1) is sufficiently flexible to represent the range of departures from logistic density dependence.

The two basic characteristics of eq. (5.1) are that current abundance determines current per capita growth rate, and that the growth rate can be a linear, a uniformly concave, or a uniformly convex function of that abundance. The first of these possibilities is known to be inaccurate in a large array of cases, as reviewed above. The analysis in this section will examine the extent to which different exponents in eq. (5.1) can capture the shape of a relationship between external mortality and the equilibrium population size of a consumer species in a model with explicit resource dynamics. There are at least four major qualitative differences between the dynamics predicted by eq. (5.1), and those predicted by consumer–resource models. The latter (but not the former) can exhibit the following phenomena:

1. The relationship between the per capita mortality rate and population size often includes both concave and convex sections, and may exhibit discontinuous changes;
2. Hydra effects occur, in which greater mortality of the consumer increases its equilibrium population size;
3. A variety of non-stationary dynamics may result from the process of population regulation;
4. Alternative equilibria/attractors may occur.

These four possibilities are detailed below. In addition to exhibiting these qualitative differences, a consumer–resource perspective also provides a much more appropriate basis for understanding the evolution of the traits that define self-regulation, and therefore, of how the process is likely to change over longer time spans.

The first difference between eq. (5.1) and the set of consumer–resource models for systems of two competitors is that the predicted relationship between mortality rate and population size has a much wider variety of shapes in the consumer–resource models. A relatively general form for a consumer–resource (predator–prey) interaction with continuous dynamics and homogeneous populations was analysed in Abrams (2009c). It is:

$$\frac{dR}{dt} = Rf(R) - CRg(R, N)N$$

$$\frac{dN}{dt} = N\left(b\left(CRg(R, N)\right) - d\right)$$

(5.2a, b)

This has the structure of most of the consumer–resource models discussed in previous chapters, except that the functional forms of several components are not specified. The function f describes resource per capita population growth. The consumer functional response is a linear response, CR, multiplied by a function, g, which describes the alteration of the linear response form by factors such as handling time, behavioural modification of foraging, and predator (consumer) interference. The increasing function b describes the per capita birth (production) rate as a function of resource intake, and d is a per capita death rate of the consumer. At equilibrium, $b = d$; this equilibrium condition provides an implicit relationship between d and the value of N at equilibrium.

Abrams (2009c) illustrates how the shapes of the component functions of eqs (5.2) influence the relationship between consumer population size N and consumer per capita mortality rate, d. Some of the main conclusions of that analysis are as follows:

1. Linear forms for f and b, together with a constant value for g (i.e., linear resource density dependence, linear consumer numerical response, and linear consumer functional response) are required to obtain the linear N vs d relationship produced by the logistic model for consumer growth. Empirical studies suggest that this set of three conditions is seldom satisfied.

2. If the only exception to the conditions for a linear N vs d relationship is nonlinear resource density dependence, then the consumer 'inherits' the same form of density dependence as the resource. Thus, if eq. (5.1) described the resource density dependence in eq. (5.2a), the curvature of the resource per capita growth rate function would also describe the curvature of the consumer's N vs d relationship. If B is the slope of the numerical response, b, at the equilibrium point then,

$$\frac{\partial N_{eq}}{\partial d} = \frac{1}{BC^2} f'\left(\frac{d}{BC}\right)$$

$$\frac{\partial^2 N_{eq}}{\partial d^2} = \frac{1}{B^2 C^3} f''\left(\frac{d}{BC}\right)$$

(5.3a, b)

Equation 5.3b means that the curvature of density dependence at a particular mortality rate, d, has the same sign as, and is proportional to, the density dependence of the resource at this point. It is important to note that the relationships described by the above two formulas may have major differences from the power law of eq. (5.1). This is true when resource growth, or a large component of it, is based on an abiotic growth model similar to the chemostat model ($f = (I/R) - E$, with I being input rate, and E being per capita exit rate). This case is treated in Abrams (2009c; Figure 1, p. 324); a model characterized by a linear b, $g = 1$, and chemostat resource growth produces

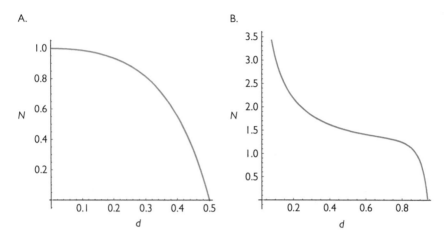

Fig. 5.2 Consumer population size vs mortality for 1-consumer–1-resource systems with a logistic resource and a consumer with a linear numerical response and a type II (panel A) or type III (panel B) functional response. The type II response is a disc equation of the form used for a 2-consumer model in eqs (5.5). The type III response has the same formula except that resource abundance, R is replaced by R^2. The parameter values in panel A are: $B = C = k = r = h = 1$. Panel B has identical parameters except that $k = 1/4$.

a uniformly concave N vs d relationship. However, this is not well approximated by any power-law relationship.

A wide variety of models can exhibit both concave and convex sections in their N vs d relationships. This is true of models with a logistic resource and a type III consumer functional response. It is also true of models that combine abiotic (or partially abiotic) growth with a consumer having a type II functional response. These two scenarios are illustrated in Figures 5.2 and 5.3 respectively.

The second major difference between density dependence as represented by eq. (5.1) versus consumer–resource models is that the latter often have a range of mortality rates over which the equilibrium consumer abundance rises with increases in its own mortality, i.e., a hydra effect. This was implicit in Rosenzweig and MacArthur's (1963) analysis of the possibility of cycles in predator–prey systems. They showed that the equilibrium predator abundance increased with predator mortality in unstable systems. However, they did not calculate mean densities in cycling systems, so it was not clear whether mean abundance would actually increase with increasing mortality over some range of mortalities. Abrams (2002) showed the potential for an increase in population size with mean density in a system having logistic resource growth and a type II consumer functional response, and Abrams (2009a) examined that relationship for a theta-logistic resource with resource immigration from an external source. The mean and equilibrium abundances can differ greatly, but there are still many cases in which the mean increases with increasing mortality (Figure 2 in Abrams 2009a, p. 466). The presence of population cycles makes

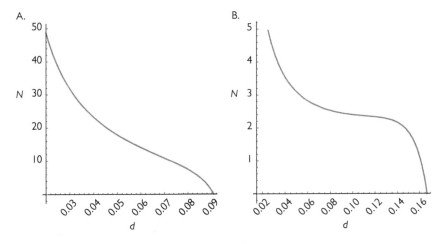

Fig. 5.3 Consumer population size vs mortality arising from 1-consumer–1-resource systems with abiotic or partially abiotic resource growth. Panel A describes a system with purely abiotic growth ($dR/dt = I - ER$), and a consumer with a linear numerical response and a type II functional response (as in eq. 5.4). It has parameters $I = E = 1$; $B = C = 1$; $h = 2$. Panel B shows consumer population size in a system with summed logistic and abiotic growth components and a type II consumer functional response. It has parameters $I = E = 0.1$; $r = k = 1$; $B = C = 1$; $h = 5$.

any model having the form of eq. (5.1) a poor description of dynamics. Cortez and Abrams (2016) showed that hydra effects can occur in other models for the predator or the prey species in stable predator–prey systems. If the factor causing predator mortality also reduces its capture rate of prey, a much wider range of models and conditions imply an increase in the ultimate population size with an increase in the consumer mortality rate (Abrams 1992b, 2002).

The third qualitatively different outcome of models with explicit resources is the possibility of sustained cycles. These are most often associated with saturating functional responses and self-reproducing resources. However, even within the framework of homogeneous populations with no time delays, cycles are not restricted to biotic resources. Abrams (1989) showed that limit cycle dynamics were possible in a system having an abiotic resource provided the consumer had a functional response that decreased over some range of resource abundances. Such a functional response was termed 'type IV' by Holling (1965; also Holling and Buckingham 1976), but most subsequent theoretical articles on functional responses have ignored it. Holling (1965) observed a type IV in experiments at very high prey abundances, and attributed it to disturbance of the predator's hunting and consumption behaviour being interrupted by the large number of encounters with prey. However, Jeschke et al. (2004) found a number of type IV responses in their review of experimental studies. Abrams (1989) showed that an adaptive functional response should decrease at high levels of foods that contained non-lethal toxins.

The response of the mean consumer population in a cycling population to changes in its mortality rate or other parameters can differ greatly from the corresponding response of the equilibrium point to the same mortality change (Abrams and Roth 1994a, b; Abrams et al. 1998; Abrams 2002, 2009a). The nature of responses of mean densities to consumer mortality in cycling systems has only been examined for a handful of models, mainly 1-consumer–1-resource systems using very simple functions for the three basic components. Cycling consumer–resource systems can be characterized by a relationship between the mean consumer abundance and its per capita mortality rate, but those must be determined numerically (as done in Figure 3.2 for a system with two consumers).

The last class of qualitative differences between the potential dynamics of the theta-logistic model (eq. (5.1)) and consumer–resource models is the presence of alternative attractors. Consumer–resource models with relatively simple components can have alternative equilibria or alternative attractors. Bazykin (1974; described in Bazykin 1998) long ago showed that models having a logistic resource and a predator with both a type II functional response and some direct density dependence in its mortality, could have a cyclic and a non-cyclic attractor. Two alternative stable equilibria are possible in this model. Alternative equilibria also can occur if immigration of the resource is added to a standard model in which the resource is logistic and the predator has a type II functional response (Reynolds and Brassil 2013). In these cases, the relationship between population density and mortality can have two (or more) distinct segments (Abrams 2009d, Reynolds and Brassil 2013).

Consumer–resource models should be used in modelling intraspecific competition because of their much wider range of dynamics and frequent inconsistency with single-species density dependence. However, in many cases little or nothing is known about the consumer–resource interaction(s) underlying density dependence. There are also cases in which the resources rapidly reach a quasi-equilibrium state with respect to consumer abundance, without complicated transient dynamics. In either of these circumstances a traditional model of density dependence may be unavoidable. This leads to the question of what function f should be used in $dN/dt = Nf(N)$? The most common approach has been to use the per capita growth rate in the logistic or theta-logistic model. If this is done in the future, it would be desirable to at least explore some of the implications of the more common forms of consumer–resource models for the dynamics of the system under consideration.

The consumer–resource models discussed above have largely used the same range of functional components that had been developed in the pre-MacArthur era. That pretty much reflects the range used in the recent models of competition that were reviewed in Chapter 4. Adaptive behaviour is probably the most consequential biological process omitted from these models. The need to consider adaptive behaviour in models of consumer–resource interactions was first suggested by William Murdoch in his work on switching behaviour in predators (Murdoch 1969). Incorporating foraging behaviour into models now encompasses a reasonably large body of literature. However, adaptive food choice has been more influential in behavioural ecology than in population or community ecology.

I incorporated the types of behaviourally influenced functional responses discussed in Chapter 3 into food chain models in a series of articles beginning in 1984 (including Abrams 1984b, 1990c, 1992c, 1995). Some of these treated 3- and 4-level systems. They all suggested that adaptive balancing of food intake rate and predation risk could produce very different models for the growth of the top species in a food chain than were present in typical 2-level consumer–resource systems. For example, Abrams (1995) examined a simple 3-level food chain model with linear functional responses, logistic growth of the basal resource, and nonlinear numerical responses for both consumer species. It had the potential for adaptive behavioural balancing of food intake and predation risk by the mid-level species. This allowed the abundance of the bottom-level species to have the same (negative) immediate effect on the per capita growth rate of the top species as did the abundance of that top-level species (eqs (9) in Abrams (1995)). Neither of these immediate effects existed in the model without behavioural flexibility of the mid-level species. The mid-level species' behaviour also allowed for a hydra effect in the top-level species. Increased abundance of the bottom-level species reduced its rate of mortality due to its consumer. All these behaviourally generated changes in the per capita growth rate are incompatible with simple models of density dependence having a form similar to eq. (5.1).

Since the development of basic theory about anti-predator behaviour in food chains in the 1990s, considerable empirical evidence for the type of adaptive foraging envisioned in those models has accumulated (Werner and Peacor 2003; Bolnick and Preisser 2005). However, the nature of the impact of these behaviours on the relationship between mortality and population productivity of a top-level consumer species has received relatively little attention since my 1995 article discussed above. Abrams (2009d) showed that adaptive defence was another possible mechanism that could produce changes in the curvature of the relationship between consumer per capita mortality and consumer population size, as well as alternative attractors (Figure 4 of Abrams 2009d, p. 300). Abrams and Vos (2003) had earlier demonstrated that several counter-intuitive responses of species to mortality (either their own or one of the other species) occurred in a food chain where each level could have direct negative intraspecific effects on immediate growth rate, and the middle level varied its traits adaptively to balance increased food intake with its associated increased predation risk.

5.2.4 One-consumer–multi-resource systems

As noted above, specialist consumer–resource systems are rare. However, the possibility of systematic differences in the shape of density dependence between specialist and generalist consumers has received very little attention, as noted in Abrams (2009b). Having two or more resources implies the existence of indirect interaction(s) between those resources, transmitted via the consumer (due to consumer satiation and effects on consumer abundance). In addition, there are often other interactions between the resources, including competition for their own foods when they are biotic. Finally, active resource choice by the consumer can greatly alter the indirect component of interaction between resources that is transmitted via the consumer.

This section will use the approach of the previous one in exploring the shape of the relationship between equilibrium or average population size and a constant additional per capita mortality. This is the best description of the effect of intraspecific competition in systems where resource dynamics imply that a single-species density dependent growth model cannot adequately describe the process of self-limitation. Many of the qualitatively different behaviours of models with multiple resources are exhibited by the simplest case, in which there are just two resources, so that is the only case considered here. Biotic and abiotic resources are considered, as are situations with and without interactions between the resources.

Having a consumer with two resources that do not interact directly differs from single-resource systems when the two resources differ from each other, either in their growth functions or their susceptibility or value to the consumer. A potential choice of resources means that there are a variety of adaptive behaviours that may affect the form of the consumer's functional responses. The consumer may be able to increase consumption of one resource at the expense of decreased consumption of the other. When the resources are nutritionally substitutable, this involves increased relative consumption of the resource that can be caught more rapidly, meaning that the attack rate parameter C_i increases with R_i (Abrams 1987c, Abrams and Matsuda 2004). When resources are nutritionally essential, a decrease in the attack rate, C, on the more abundant resource is likely, as the less abundant one is then likely to be more limiting (Abrams 1987c). If the original relationship between mortality and abundance is linear, choice will generally make it nonlinear. In addition to affecting consumer choice, the abundance of the second resource may change the optimal level of general foraging effort, producing other indirect interactions between the different resources.

I will first consider a system with two resources in which the consumer is unable to alter its 'attack rates', C_i, on the resources. If the two resources differ in their susceptibility (C) to the consumer, the N vs d relationship is changed from that in a comparable single-resource system. Given a difference in C-values, a smaller consumer death rate increases the relative contribution of the lower-C resource to the consumer's diet. A sufficiently low death rate results in exclusion of the more vulnerable resource when the consumer does not exhibit any adaptive switching behaviour. Figure 5.4 provides an example of a system in which the consumer is characterized by a multi-species type II response (see Box 2.1) and the two resources have logistic growth ($dR/dt = m + R(r - kR)$, where m is a very low rate of immigration). The system exhibits cycles over a broad range of consumer death rates. Resource 1 is captured at 1/5 the rate of resource 2, which means resource 2 is susceptible to extinction. The three panels show cases with different r-values for resource 1. All the panels show the equilibrium consumer population size as a function of its per capita mortality rate. A low enough consumer death rate causes extinction of the more vulnerable resource, and stable dynamics. As the death rate is raised, the second resource returns, which in turn causes a relatively abrupt change in system dynamics to cycles. The dynamics

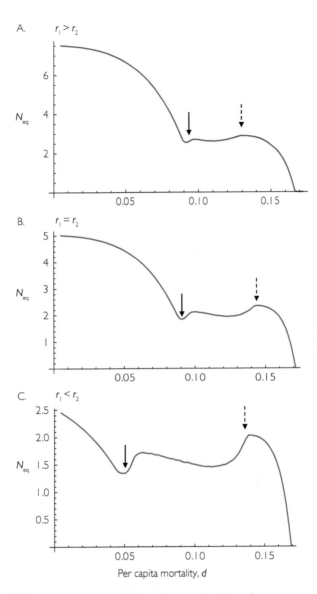

Fig. 5.4 Equilibrium consumer population size as a function of consumer per capita mortality rate, for systems with two logistic resources. The dynamics assume external resource immigration at a rate m for both resources. The parameters that are common to all panels are: $B = 1$; $m = 0.0001$; $k = 1$; $C_1 = 0.2$; $C_2 = 1$; $h = 5$; $r_2 = 1$. In panel A, $r_1 = 1.5$, in panel B, $r_1 = 1 (= r_2)$, and in panel C, $r_1 = 0.5$. The solid arrows give the point at which the more vulnerable resource (resource 2) is able to achieve a significant density (i.e., where it would persist without immigration), and the dashed arrows denote the mortality rate above which the system with both resources becomes stable.

switch again from cyclic to stable at sufficiently high consumer mortality rates. The overall shape of the N vs d relationship in all three panels is one with convex sections at low and high mortalities, separated by a relatively flat intermediate section over which there are population cycles. In this section, the mean consumer density is only slightly higher than the equilibrium shown here.

Abrams (2009b) illustrated the form of N vs d for a system with very many resources types that are characterized by a continuous range of different C-values. As consumer mortality is increased from very low values, extinct or nearly extinct resources can return to the system, resulting in a continuous increase in mean resource vulnerability, and a concave N vs d at low mortality rates, d. At high consumer mortality, all resources are present, and the consumer's type II functional response then leads to a convex relationship with a sharp and accelerating decline in equilibrium consumer abundance to zero as mortality approaches its maximum value. As Holt (1977) demonstrated, the most extinction-prone resources are generally those characterized by the greatest ratios of the consumer's attack rate to the resource's maximum per capita growth rate in the absence of the consumer. The loss of a resource species due to this apparent competitive exclusion produces a decrease in mean resource vulnerability to consumption and/or a decrease in total resource productivity as the death rate of the consumer is decreased. When combined with the fact that functional responses must saturate at a high enough resource abundance (high enough predator mortality), this also suggests that the most likely relationship between per capita mortality rate and population size is one that is concave (decelerating) at low mortality and convex (accelerating) at high mortality.

The models considered thus far lack adaptive foraging by consumers or defensive behaviour by resources. If the two resource types are caught using different foraging behaviours, or are found in different locations, the consumer can usually adapt quickly to increase its capture rate of the currently more rewarding resource. This usually entails a reduction in its capture of the less rewarding resource. The impacts of such dynamic switching behaviours were examined in Abrams and Matsuda (2003, 2004). McCann et al. (1998) and McCann (2012) represented switching using a response that is a special case of Abrams and Matsuda (2003). Subsequent works with more complex procedures for representing switching functional responses were later proposed by van Leeuwen et al. (2013) and Morosov and Petrovskii (2013). All varieties of switching responses tend to equalize the intake rates of alternative resources. As a result, extinction of one resource following a lowering of the consumer death rate occurs over a smaller range of mortality rates, or may not occur at all. However, the possibilities of unstable equilibria and of different potential forms for the dynamics of behavioural change again introduce complications. The impact of switching on the relationship between mortality and consumer population size does not appear to have been studied. Figure 5.5 presents an example using the very simple switching function that was used by Abrams and Matsuda (2003). With two or more prey (resource) types, switching multiplies the attack rate on resource i, C_i, by the function:

$$\omega \frac{(C_iR_i)^z}{\displaystyle\sum_{j=1}^{\omega}(C_jR_j)^z}, \tag{5.4}$$

where ω is the number of resource types, and z is an exponent that roughly affects the accuracy of the behavioural switching. (Larger values of z imply a greater 'preference' for the most abundant resource(s); see Abrams and Matsuda (2003).) With $z = 0$ the intake rates of all the resources are proportional to their attack rates, so there is no adaptive preference. Figure 5.5 shows the impact of incorporating switching on the mortality vs population size relationship in a 1-consumer–2-resource system. Each of the two panels compares systems with and without switching; panel A assumes linear functional responses, while panel B assumes type II responses. Switching prevents the extinction of the more rapidly caught prey that can occur at low predator mortality rates. Switching also expands the upper limit of the mortality rates allowing consumer existence. Switching maintains the generally concave shape of the relationship produced by models with linear functional responses and non-interacting resources, but alters the details of the shape. Although it is not shown in the figure, switching also changes the range of parameters that allows population cycles. In addition, switching significantly alters the nature of those cycles. Either switching or cycles alone would likely change any quantitative measure of the interaction.

In the models discussed above, the only linkage between the population dynamics of different resources has been via their shared consumer. There is also the possibility that resources (prey) compete for lower-level resources. The impact of competition between resources is illustrated by the simplest model of the 'diamond food web' (Leibold 1996; Noonburg and Abrams 2005), which embodies this food web structure. The diamond web involves a single basal resource (with population R), consumed by two mid-level prey species (N_1 and N_2), which are in turn consumed by a single predator (P). Coexistence of the prey in a stable system requires that the one that is better at resource exploitation (has a lower R^* in the absence of the predator) must be more susceptible to the predator. Without the predator, one prey (the one with the lower R^*) would exclude the other. If a low level of external immigration is added for both prey, the otherwise-excluded prey persists at a very low abundance.

The simplest (and most common) model of the diamond web assumes linear functional responses of all consumer species and logistic growth of the basal resource. If both prey species are present, mortality applied to the predator has no effect on its equilibrium abundance. An increase in the predator's mortality brings about a short-term decrease in its abundance, but the resulting increase in the relative abundance of the more vulnerable prey restores the predator's original abundance. If the predator were maintained at any fixed abundance, one of the two prey would be excluded, as this would imply two competing prey with a single limiting factor (the shared resource). There is no effect of an increased per capita mortality rate on the predator's population size. Of course, a sufficient change in the mortality rate of the predator will cause extinction of one of the prey; the predator-resistant prey will go extinct at high enough predator mortality rates, and the superior resource exploiter will go extinct at

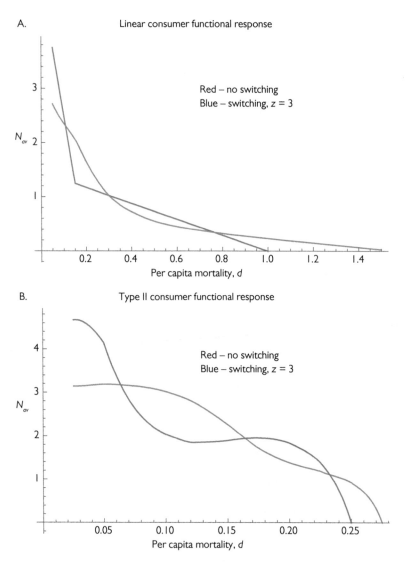

Fig. 5.5 The effect of mortality on the population size of the consumer in a system with two logistically growing resources and a single consumer species that may exhibit switching behaviour. In panel A, the consumer has a linear functional response, and in panel B, it has a type II functional response with a handling time, $h = 3$. In each panel, the red line represents the mean population vs mortality (d) relationship in the absence of switching, the blue line is a switching predator characterized by the exponent $z = 3$ in eq. (5.4). The other parameters are; $\omega = 2$; $r_i = 1$; $k_i = 1$, $C_1 = 0.8$, $C_2 = 0.2$; $B_i = 1$.

predator mortality rates below a lower threshold value. In either case, the slopes of the relationships between per capita mortality and predator population size and between per capita mortality and predator production (predator population growth rate) will change abruptly when the extinction occurs. The web and the resulting relationships between predator mortality and predator population size are shown in Figure 5.6A, while the resulting relationship between the productivity of the predator population and mortality is shown in Figure 5.6B. The latter has a minimum close to one-half of the mortality that would cause the top predator extinction. This contrasts sharply with the prediction arising from a logistic model of the predator's growth, in which mortality that was half of the maximum allowing existence would maximize the productivity of the population.

Giving the predator a type II (disc equation) functional response changes the dynamics of the diamond food web considerably. Here I assume a multi-species disc equation with equal nutritional values and handling times for the two prey species (Abrams and Matsuda 2005). In this case, increasing predator mortality often increases its abundance; i.e., higher average predator abundances are associated with larger (rather than smaller) per capita growth. Abrams and Matsuda (2005) proposed the general term 'hydra effect' to apply to this and any other scenario in which greater mortality of a given species led to a greater equilibrium population size. As was noted above, the hydra effect also occurs in the simple 2-species model with a type II predator and a logistically growing prey. Here, it has been known since 1963 (Rosenzweig and MacArthur 1963) that increasing predator mortality in an unstable system will increase the equilibrium predator abundance (see e.g., Abrams 2002, 2009a). The basic mechanism is that greater predator mortality increases the prey abundance, which leads to a larger amount of 'handling'. If the predator is initially overexploiting the prey (prey abundance less than the level that maximizes the population rate of increase), a lower effective capture rate (due to the increased total handling time) can increase the predator's abundance. This is somewhat complicated by the fact that mean abundance in cycling systems in this model can be much greater than the equilibrium abundance. However, this does not change the fact that mean predator abundance goes up with increases in its own mortality over some range of mortalities. When the system is at equilibrium, the imposed per capita mortality is again equal to the predator's per capita production rate, so higher mortalities are again associated with greater population production, contrary to standard density dependence.

Another scenario involving shared predation occurs when two prey species consume different biotic resources, and each prey adaptively adjusts its foraging time or rate based on the commonly observed trade-off between food intake and predation risk (Lima and Dill 1990). Abrams (1984b) analysed the single-prey version of this scenario, and Abrams (1987d) extended it to a predator with two prey, each with its own resource. In the two-prey system, adaptive prey behaviour can lead to an increase in the first prey's equilibrium density after adding the second prey. The basic mechanism is that the increase in the predator abundance after adding the second prey causes the first prey to exploit its resource at a lower rate, causing an increase in resource abundance that more than compensates for the lower capture rate.

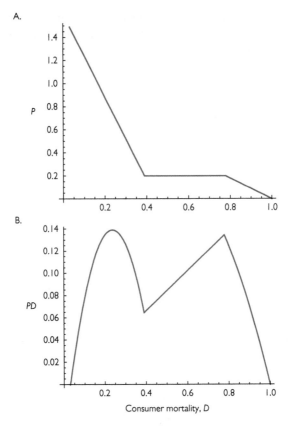

Fig. 5.6 Panel A illustrates the effect of the per capita mortality rate of the top consumer on its population density for the simplest diamond food web with linear functional components. This model has the form:

$$\frac{dP}{dt} = P(E_1 C_1 N_1 + E_2 C_2 N_2 - D)$$

$$\frac{dN_i}{dt} = N_i (B_i S_i R - d_i - C_i P), \quad i = 1, 2$$

$$\frac{dR}{dt} = R(1 - R - S_1 N_1 - S_2 N_2)$$

The minimum per capita mortality (*D*) of the predator is assumed to be 0.03; all additional mortality is assumed to be due to harvesting. Panel B shows the corresponding relationship between total death rate of the predator and amount harvested in this same food web. The parameter values are: $C_1 = 0.25$; $C_2 = 1$; $S_1 = 0.5$; $S_2 = 1$; $d_1 = d_2 = 0.1$; $E_1 = E_2 = 1$.

5.2.5 A more mechanistic approach to density dependence

Models of purely density-dependent growth continue to be common in the literature. Theoreticians use density-based models of growth to simplify multi-species models, which makes them easier to analyse. They are particularly likely to be used for species at the bottom trophic level of the food web when the focus of interest is a higher trophic level. I have certainly been guilty of this in many articles. Fisheries biologists (and others dealing with resource management) still frequently use models of single-species density-dependent growth for species at higher trophic levels. This is presumably because the relationships of population growth to resources and other factors are largely unknown for the species in question. Textbooks give detailed coverage to density-based models. This is perhaps due to historical inertia or the simplicity (and consequent pedagogical utility for a mathematically unsophisticated audience) of the models. All models require some simplifications, but it is vital to be aware of the phenomena and outcomes that may be rendered invisible by such simplifications. Loreau (2010, and elsewhere) has argued for including all lower trophic levels explicitly in models of population growth, and this approach should be used to verify whether assumptions of logistic resource growth affect the qualitative results of a model of competition.

The common view that consumer population density is the sole or main determinant of its per capita population growth rate has no doubt inhibited experimental approaches that could provide a better understanding of single-consumer population growth. The response of consumer population growth to a change in its own abundance is always affected by resources, which seldom, if ever, reach a new equilibrium instantaneously when the number of consumer individuals changes. Chapters 8 and 9 will illustrate some of the consequences of these delays in seasonally varying environments.

5.3 Apparent competition

Apparent competition is almost always present when one or more consumers exploit two or more resources. It is the interaction between the resources via those shared consumers (prey via shared predators). Apparent competition has already been mentioned in a variety of contexts in both this and preceding chapters. Many of the same models used to understand competition can be used to understand apparent competition. Apparent competition shares some of the definitional problems associated with interspecific competition. Robert Holt's (1977) article, which introduced the term 'apparent competition', noted that commonly used measures of the effect of one resource on another could be positive rather than negative, and most authors have subsequently referred to such cases as 'apparent mutualism' (e.g., Abrams et al. 1998; Holt and Bonsall 2017). Mutually positive effects seem particularly likely when consumer–resource systems undergo self-generated cycles (Abrams et al. 1998). This work also noted the possibility of (+,−) effects between resources that share a consumer. If 'competition' is used for all interactions involving shared resources, consistency argues for applying 'apparent competition' to all interactions via shared consumers (including parasites and diseases), regardless of the attendant set of changes in population size following neutral perturbations to each species. This is a question for the ecological community to decide.

Another diverse set of authors has argued in favour of dropping the 'apparent' and using the general term 'competition' to apply to both interactions via shared resources and shared natural enemies. Nicholson (1937) and MacArthur (1972) had mentioned shared enemies as a type of competition, and, more recently, Kuang and Chesson (2008) argued that these two indirect interactions were fundamentally the same in terms of coexistence. Kuang and Chesson's (2008) assumption of unrepresented resource species that would sustain the predators in the absence of the ones that are explicit in the model means that they treated the two trophic levels in a fundamentally different manner. As shown later in this section, there are some important differences between competition and apparent competition in a simple 2-consumer–2-resource case. While it is true that both competition and apparent competition can cause exclusion of one or more species on the focal trophic level, there are differences due to the directionality of energy flow, which goes upward from lower to higher trophic levels in food webs. The resource level can exist without their consumers, but the predators on those consumers cannot do so. This fundamental fact would argue for a distinction between the two processes, but there are several other important differences that are noted below. The effects of assuming type II rather than type I (linear) functional responses differ qualitatively depending on which trophic level is affected. The same is true of the role of increasing direct density dependence in the transmitting entity/entities. In addition, the fact that consumer behaviour changes in different ways in response to changes in the abundances of resources and predators is another reason for distinguishing between the two interactions.

As was noted above, theory regarding apparent competition has never displayed any tendency to leave out the dynamics of the predator of the indirectly interacting prey. Neither Lotka nor Volterra used what we now call apparent competition as an example of the phenomenon they were modelling. The importance of such factors as predator self-regulation, nonlinear predator functional responses, and unstable population dynamics have been noted repeatedly (Holt 1977, 1984; Holt and Kotler 1987; Abrams 1987d, Abrams et al., 1998), and were reviewed recently (Holt and Bonsall 2017). It has long been recognized that a shared predator can produce mutually positive effects between two prey species (Holt 1977; Holt and Lawton 1994; Holt and Bonsall 2017; Abrams 1987d; Abrams and Matsuda 1996). These articles show that type II or other nonlinear functional responses are an important mechanism by which such 'apparent mutualism' comes about in models of shared predation. This mutually positive outcome does not require consumer–resource cycles; it occurs whenever a type II predator also has some additional limiting factor other than its prey (e.g., good hunting sites). Nevertheless, population cycles often contribute to mutually positive effects via a shared predator.

A type II response in a 1-predator–1-prey system produces larger amplitude cycles when the predator can persist on a low abundance of prey, and when there is a relatively large handling time. The addition of a second prey with a similar handling time usually increases the amplitude of the cycles, and higher amplitude cycles, whatever their cause, tend to increase mean prey density. Thus, (+,+) interactions are relatively common when the interaction sign is determined by the difference in mean prey density between systems with one vs two prey species. However (+,−) effects may also occur when prey differ significantly in handling time or some other properties (Abrams et al. 1998).

Although theory regarding apparent competition has always incorporated explicit dynamics for the effect-transmitting species, it has suffered from some of the same deficiencies as theory on resource competition. Systems with more than one predator or more than two prey are seldom examined, and behavioural adaptations of the species involved are also rarely considered. The remainder of this section will provide a brief exploration of a few of the implications of these omissions.

A system with two consumers that are limited solely by two dynamically independent resources has been a staple of competition theory. As a result, one might think that it should have played a role in theory regarding apparent competition as well, since it involves both resources sharing both predators. However, this food web is virtually absent from the literature on apparent competition. This is because of a somewhat inconvenient truth. If the indirect interaction between the two resources is measured by the effect of a small change in the neutral parameter (r_i in eqs (5.5) in Box 5.1), one finds that there is no effect of either resource species on the other resource's equilibrium abundance at a stable equilibrium. (This is true provided the consumers have no limiting factors other than the two resources.) The equilibrium resource abundances are determined entirely by the consumer's parameters in this case (see Box 5.1, which deals with logistic resources and type II consumers). However, a sufficient increase in one of the resource intrinsic growth rates, r_i, will cause extinction of the predator which is more dependent upon the other resource. Similarly, a sufficient decrease in one r_i will cause extinction of the consumer that has a greater relative consumption rate of that resource. Extinction of one of the two consumer species leaves the system with a single consumer species, so it will display apparent competition with mutually negative effects under the normal circumstances for such a model (for example, all cases with a stable equilibrium in which both resources have positive densities). A sufficient reduction in r_i will cause extinction of resource i itself, leaving a simple 1-consumer–1-resource system, usually with a higher abundance of the remaining resource. Note that the independence of the equilibrium populations of the two resource populations (over some range of r_i) only depends on having a stable equilibrium and having consumer per capita growth rates that are solely dependent on the two resource abundances. Thus, adding consumer switching behaviour or changing to any other set of functional responses that are solely determined by resource abundances would not affect the lack of a between-resource interaction.

Box 5.1 Interactions in a potentially unstable 2-consumer–2-resource system

The model of two consumers and two resources used here is:

$$\frac{dR_1}{dt} = R_1\left(r_1 - k_1 R_1\right) - \frac{C_{11} R_1 N_1}{1 + C_{11} h_{11} R_1 + C_{12} h_{12} R_2} - \frac{C_{21} R_1 N_2}{1 + C_{21} h_{21} R_1 + C_{22} h_{22} R_2}$$

$$\frac{dR_2}{dt} = R_2\left(r_2 - k_2 R_2\right) - \frac{C_{12} R_2 N_1}{1 + C_{11} h_{11} R_1 + C_{12} h_{12} R_2} - \frac{C_{22} R_2 N_2}{1 + C_{21} h_{21} R_1 + C_{22} h_{22} R_2}$$

$$\frac{dN_1}{dt} = N_1\left(\frac{B_{11} C_{11} R_1 + B_{12} C_{12} R_2}{1 + C_{11} h_{11} R_1 + C_{12} h_{12} R_2} - d_1\right)$$

$$\frac{dN_2}{dt} = N_2\left(\frac{B_{21} C_{21} R_1 + B_{22} C_{22} R_2}{1 + C_{21} h_{21} R_1 + C_{22} h_{22} R_2} - d_2\right)$$

$$(5.5a, b,c, d)$$

In the simplified example, the two resources have equal growth parameters, all of the B_{ij} are identical to each other, as are the h_{ij}; in addition, $d_1 = d_2$ in the system before any perturbation. The consumers are assumed to have mirror image consumption rates, with $C_{11} = C_{22} = C$, and $C_{12} = C_{21} = 11 - C$, where $1/2 < C < 1$. The equilibrium point of these equations can easily be found, but the formulas for the consumer densities are still quite lengthy. However, the resource densities at equilibrium are determined entirely by the zero per capita growth conditions for the consumers from eqs (5.5c, d). These conditions reduce to two linear equations in two variables. The solution is independent of any parameter that appears solely in the resource equations, if that parameter value does not imply extinction of one or more of the four species. The parameter simplifications adopted above eliminate the possibility that consuming a resource has a negative effect on the consumer by preventing consumption of a more nutritious resource (because of a relatively low B or high h). Cases where optimal diet choice implies that the lower-value resource should be ignored at high abundances of the preferred resource require a modification of this model (see Fryxell and Lundberg 1994; Ma et al. 2003; Abrams et al. 2007). However, this possibility does not change the fact that equilibrium resource densities are independent of the resource growth parameters. This independence, which prevents apparent competition, only depends on the absence of terms involving consumer abundance in the expressions for consumer per capita growth rate. If consumer densities directly affect their own functional responses, this independence will no longer hold.

The consequences of mortality (reduced r) of one resource species for the population size of the other are more complicated if the system exhibits sustained cycles before any perturbation. In this case, we can again get some general insights by examining a symmetric system having 'mirror image' competitors using the same two resources that initially have identical growth parameters, but differ by having opposite susceptibilities to being consumed by the two consumers. The interaction between the resources is again measured by a change in their mean densities following a change in the neutral parameter r_i of each on the other. Mortality applied to either resource (lowering r_i) reduces the amplitude of the population cycles.

Lower amplitude cycles in this system are detrimental to both resources because the mean consumption rate increases when the consumers spend less time handling resources. As a result, both resources decline in mean abundance in response to a lower r of a single resource for many parameter sets. This phenomenon is influenced by the degree of difference between the two consumers in their relative consumption rates of the two resources. It can also be affected by the presence of switching predation. Hydra effects in the resource (increases in the mean density of a resource with a decreased r) occur in some cases, and the between-resource interaction has a complicated, nonlinear form. As noted above, a sufficient decrease in r (e.g., increase in the mortality) in one resource will cause the extinction of the consumer that is more specialized on that resource. Once that happens, the interaction (now in a 2-resource–1-consumer system) often reverts to apparent competition. However, if the cycles are still relatively large in amplitude at this point, apparent mutualism is still possible (Abrams et al. 1998). Further increases in the mortality of one resource will eventually cause its extinction, leaving a 1-consumer–1-resource system.

Figure 5.7 illustrates the indirect interaction between the two resources in a potentially unstable version of the 4-species web described in Box 5.1. The figure illustrates the mean abundances of the two resources as a function of the intrinsic growth rate of one of them (r_1). Because of the symmetrical structure of the community, the effects of changing r_2 on the two resource abundances are identical (with resource identities reversed). Figure 5.7A has a moderate degree of resource partitioning between consumers ($C = 0.7$), while panel B of the figure assumes a higher level of resource partitioning ($C = 0.8$). In both panels, the baseline case assumes equal intrinsic growth rates of the two resources ($r_1 = r_2 = 1$). The main message from the figure is that, except for values of r_1 that are low enough to cause extinction of consumer species 1, the interaction of the two resources via the consumers often cannot be described as mutually negative. In addition, the presence of cycling implies that there is normally some effect of a change in the resource per capita growth rate on the mean abundances of both resources. (Recall that models of the form of eqs (5.5) with a stable equilibrium exhibit no effect of a changed neutral parameter of one resource species on the equilibrium abundance of the other.) An increase in the r-value for resource 1 increases the mean abundances of both resources over the part of the parameter range illustrated in Figure 5.7A in which all four species are present. The expected lack of effect on the equilibrium R_2 does not occur when r_1 is increased because the higher amplitude of consumer–resource cycles increases the mean level of consumer satiation, and therefore decreases the consumers' net effect on both resources. The symmetry of the original system guarantees that a reduction in r_2 would have the same relative effects as reducing r_1 by the same amount; resource 2 would increase in population size and resource 1 would increase to a smaller extent. Although abundances for $r_1 > 1.2$ are not shown, both resources continue to increase with a larger r_1 throughout the range of r_1 values that allow predator 2 to exist (it becomes extinct for r_1 slightly above 2.20).

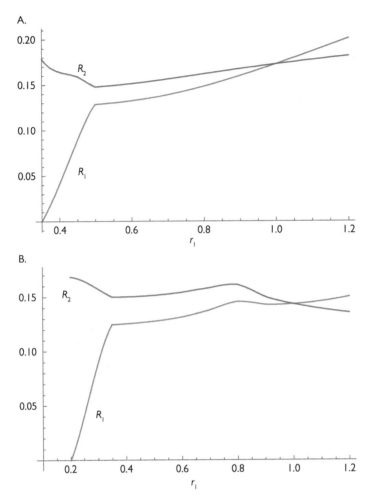

Fig. 5.7 Apparent competition in 2-consumer–2-resource systems with type II consumer functional responses. The model (eqs 5.5) is given in Box 5.1. Each panel shows the effect of changing the maximum per capita growth rate of resource 1, r_1. The two resources have identical r's, and therefore identical equilibrium and mean abundances when $r_1 = 1$. The parameters in the system illustrated here are: $r_2 = 1$, $k_1 = k_2 = 1$; $h_{ij} = 2$; $B_{ij} = 1$; $d_1 = d_2 = 0.1$. Panel A assumes that $C_{ii} = 0.7$ and $C_{ij} = 0.3$; panel B assumes greater resource partitioning by the consumers; $C_{ii} = 0.8$ and $C_{ij} = 0.2$. In panel A, the abrupt change at r_1 values slightly below 0.5 corresponds to loss of consumer 1 at that point. Resource 1 is extinct if $r_1 \leq 0.35$. Both resource species increase in mean abundance as r_1 is increased above 0.5. In panel B, resource 1 is extinct for $r_1 < 0.2$, and consumer 1 is extinct for r_1 values below approximately 0.35. The system exhibits cycles for the entire range of parameters shown.

The greater level of resource partitioning assumed in panel B of Figure 5.7 produces a somewhat different range of interspecific effects between the resources, but there is again some interaction over the entire range of r_1 shown (and for larger values that are not shown in the figure). The apparent competition at low r_1 reflects the fact that consumer 1 is absent. Following the entry of consumer 1 to the system, further increases in r_1 produce relatively weak positive effects on both species. In Holt's (e.g., Holt and Bonsall 2017) terminology this would be categorized as apparent mutualism. Still higher values of r_1 (> 0.8) are characterized by mutually negative changes in resource abundances following a small increase in r_1. The negative effect of r_1 on the abundance of resource 1 constitutes a hydra effect, so the interaction with the other resource could be considered positive or negative depending on whether the perturbation to resource 1 is regarded as positive (a higher per capita growth rate) or negative (a lower mean population size). For $r_1 > 0.9$, greater r_1 again increases the mean R_1 and decreases the mean R_2, so the interaction is again apparent competition. Although it is not shown in the figure, the effect of greater r_1 becomes positive for both resource abundances for r_1 slightly above 1.4.

The qualitative pattern of change in the two resource densities with altered per capita growth (larger or smaller r) of one of them, shown in Figure 5.7, differs for other levels of resource partitioning between the consumers (different values of C), which are not illustrated. If $C = 0.5$, the two consumers are identical generalists, so they are unlikely to coexist. This reduces the system to the 1-consumer version of eq. (5.4), which was discussed in Abrams et al. (1998). The 1-consumer version does have some parameters producing mutually positive effects, although they are less prevalent than in the 2-consumer case with cyclical dynamics illustrated here. Relatively low levels of resource partitioning (e.g., $C = 0.6$) produce mutualistic interactions between the two prey, but a relatively modest reduction in r_1 leads to the exclusion of predator 1 and stabilization of the system, so further reductions in r_1 will increase the equilibrium R_2. Very large values of C (close to unity) imply near-specialized predators, so there is very little indirect interaction between them.

There is much to be learned about interactions between two or more prey that share two or more predators. However, the common assumption of mutually negative effects is clearly far from universal. One topic that has received relatively little attention is the interaction between biotic resources in systems where saturating functional responses may produce cycles and the consumers also display adaptive switching. This is of interest because switching can increase the range of parameters that allow persistence of all of the species; i.e., it greatly reduces resource exclusion caused by apparent competition. The effects of switching can also be like those of adaptive evolution of the consumer's resource-capture abilities, when there is a trade-off between greater consumption abilities of different resources (as in Abrams 2006b). Preliminary numerical exploration of eqs (5.5) with the addition of the switching function (eq. 5.4) suggest that in cycling systems the positive effects of an increase in the r value of one resource on the abundance of a second resource is usually significantly larger than in a comparable model without switching. The switching model also has positive effects over a larger range of parameter values. This may be in part because of the increased synchrony in the dynamics of the two species on a single level, which tends to increase the amplitude of population cycles. Further exploration of this case, and other models of adaptive predation, should help clarify the role of adaptive processes in shaping apparent competition between two resources sharing two consumers.

Cases of shared predation (apparent competition) with both more than two prey and two or more predator species are also understudied. Holt's original (1977) article considered an arbitrary number of prey with a single predator, but did not examine cycling systems. Some cases with two-or-more predators and three-or-more resources are looked at to address specific questions in Schoener (1993), Kamran-Disfani and Golubski (2013), and Rossberg (2013).

In ending this chapter, it is interesting to look back at the question of whether apparent competition should just be considered as another type of competition. As noted previously, Kuang and Chesson (2008) and Sommers and Chesson (2019) have revived an old argument for considering both apparent competition and resource competition as similar feedback processes. It is certainly true that both predators and resources can limit the ability of a species on an intermediate trophic level to exist, but they have many differences in their mechanisms and evolutionary consequences (e.g. Abrams 2000a, Abrams and Chen 2002a; see Chapter 11). The lack of any interaction between resources when stable systems have equal numbers of consumer and resource species (and when there is no direct effect of consumer abundance on consumer per capita growth rate) represents another important difference between competition and apparent competition. All of this argues for the traditional practice of regarding resource-mediated and consumer-mediated indirect effects as different interactions, despite some obvious analogies in their mechanisms.

6

The negativity, constancy, and continuity of competitive effects

6.1 Introduction

Previous chapters have suggested that both the Lotka–Volterra (LV) and the MacArthur consumer–resource model have many unique properties that do not characterize even slightly more realistic models. In describing an entirely different field, Richard Lewontin (1974, p. 8) stated, 'We cannot go out and describe the world in any old way we please and then sit back and demand that an explanatory and predictive theory be built on that description.' Descriptions must incorporate enough details in an accurate or flexible enough way if they are to provide a good basis for predictive theory. Consider the properties of the Lotka–Volterra model. Its defining feature is that the per capita effects of the abundance of one species on the per capita growth rate of a competing species are negative and are independent of the abundances of all additional species that compete with either member of this pair. This linearity also applies to the intraspecific effect of the abundance of the species on its own per capita growth rate. In two-species LV models, this means that the ratio of inter- to intraspecific effects is independent of the magnitude of the parameter change or the starting abundances. When a parameter that directly affects only one species (e.g., its own per capita mortality) is changed in a system with two coexisting species, the two equilibrium abundances change linearly and continuously in opposite directions until one species goes extinct. Under the LV model, the only case with alternative outcomes is the possibility of alternative exclusion, based roughly on the initially more abundant species excluding the less abundant one. (The word 'roughly' is needed because, if the species have different maximal per capita growth rates, the initially less abundant one can achieve numerical superiority in the early stages of growth.) It is difficult to conclusively show exact linearity, but the first study that examined this issue in a laboratory system more than half a century ago (Ayala 1969; see also Ayala et al. 1973) found obvious and pronounced nonlinearity.

The MacArthur model is still employed as a justification for continued use of the LV model, at least for assessing effects on equilibrium abundances (e.g., Kuang and Chesson 2008). In MacArthur's (1970, 1972) analysis, the model appears to exhibit the same range of outcomes and the same sorts of effects of consumer populations on each other's abundance as the LV model. However, important differences between

Competition Theory in Ecology. Peter A. Abrams, Oxford University Press. © Peter A. Abrams (2022). DOI: 10.1093/oso/9780192895523.003.0006

these two models had been pointed out in the 1970s. As noted in previous chapters, the 2-consumer–2-resource MacArthur model was shown by Levine (1976) to predict mutually positive effects of consumers on each other's abundance when there was competition between resources. In some 2-consumer–2-resource MacArthur models with resource competition added, a neutral parameter change in species i that increases its per capita growth rate may actually decrease its own density (Abrams and Cortez 2015a; Cortez and Abrams 2016). Whether eq. (3.1) or (3.2) is used to classify the interaction between two consumers, all potential sign structures ((+,+), (+,–), or (–,–)) are possible. A second departure from LV model behaviour under the 2-consumer–2-resource MacArthur model is that one or more of the resources may go extinct with a large enough change in a neutral parameter affecting either consumer. This always produces extinction of one consumer and may produce discontinuous change in the population of the remaining consumer. However, the extent of the differences between the LV and MacArthur models has yet to be widely acknowledged.

This chapter will begin by showing that the 2-consumer–2-resource version of the MacArthur model produces a wider range of qualitatively different outcomes than does the 2-species Lotka–Volterra model. The primary reason for this is the possibility of resource exclusion via apparent competition. Section 6.2 will also explore some of the consequences of incorporating more than two resources in the MacArthur model. These suggest that the model provides much less of a justification for the Lotka–Volterra model than is commonly believed.

Sections 6.3 and 6.4 examine a range of consumer–resource models that are only slightly more realistic than MacArthur's model, in that they change the linear form of one of the component functions, without changing the basic structure of substitutable resources exploited by consumers lacking adaptive behaviour. Section 6.3 examines consequences of the form of resource dynamics, while Section 6.4 reviews work on nonlinear functional responses. Both sections argue that it is highly unlikely for the per capita effects of one consumer on another to be independent of their abundances. Such independence requires linear density dependence in every resource (Abrams 1980b, 1983a) and linear functional responses (Abrams 1980a), as well as continued persistence of all resources (Abrams 1998; Abrams et al. 2008a). The linear density dependence of resources assumed by the MacArthur model is now known to be uncommon, based both on field studies of density dependence (Sibly et al. 2005) and on consumer–resource models with a single consumer species (Abrams 2009b, c, d; see also Chapter 5). Jeschke et al.'s (2004) review of empirical functional response studies found mainly nonlinear responses; they provide some examples described as type I, but their article uses an unusual definition of type I that includes a partially linear response with a fixed maximum consumption rate at high resource abundance. This is effectively type II. An example of some of the consequences of type II functional responses in a 2-consumer–3-resource model was provided in Chapter 3 (Figure 3.2). These consequences will be examined in somewhat greater detail for a range of models in Section 6.4. The wide variety of novel functional response shapes reviewed in Chapter 3 has thus far been largely ignored by competition theory.

Section 6.5 is a short consideration of multi-consumer models. These are usually characterized by dependence of the magnitude of competitive effects between any pair of species on the abundances of other consumers. As in the case of Lotka–Volterra models, it is likely that a wide variety of dynamics are possible, and there are often cases with multiple attractors for a given system. Negativity of effects of a neutral parameter on the abundances of competitors need not characterize interactions involving three or more competitors in the Lotka–Volterra framework, and that is also true of resource-based models with three or more competitors.

The chapter ends with a brief overview (Section 6.6) of some of the other elements of consumer–resource interactions that are missing from most theory dealing with interspecific competition. Changes in negativity, constancy, and continuity are also likely to arise in models that incorporate these elements.

6.2 Resource extinction and quasi-extinction in MacArthur's model

Beginning with MacArthur (1970, 1972), biotic resources in consumer–resource models of competition have usually been assumed to exist at positive abundances regardless of their level of consumption. This assumption underlies the justification for the common practice of computing LV competition coefficients based on overlap in two consumer's spectra of resource capture rates (e.g., Vincent and Brown 2005; Leimar et al. 2013), without considering the possibility that one or more resources could be excluded by increased consumer abundance. High consumer efficiency was shown to produce exclusion of one biotic resource in 1-consumer-2-resource models by Holt (1977), who referred to even earlier articles suggesting this possibility. Exclusion of more than one resource may occur via apparent competition when three or more resources are initially present. It is unclear why MacArthur ignored these cases.

Complete resource exclusion can be avoided by a low rate of immigration of resources from outside the system being modelled. However, the distinction between systems with low levels of resource immigration and those that are totally closed to immigration is missing from most recent works on competition. This is due to the lack of a metapopulation perspective in non-spatial models of competition. Resources are usually more widely distributed than their consumers because resources can exist in some places where their equilibrium level is too low to support a consumer population. This is likely to cause some input from these resource-only patches into other locations that have consumers. If an analysis of competition only considers locations containing consumers (which is normally the case for studies of competing consumers), such input is most often simply assumed to be absent.

Most natural populations also experience some level of input of consumers, as documented by studies of island biogeography; e.g., Warren et al. (2015). Having a very low level of immigration of competitors (i.e., consumers) in a LV model only has slight effects on abundances. For example, it changes outcomes of exclusion to presence at very low abundance. Minimal abundance of one species is effectively equivalent to exclusion in its effects on the population levels of other competitors. In cycling systems with heteroclinic (ever expanding) cycles (possible for LV models with three or more species), adding low-level consumer immigration can be essential for persistence (e.g. May and Leonard 1975), but such cases seem to be rare, even in theoretical models. In any event, continued low-level abundance of a competing consumer population that is maintained by immigration should be (and usually is) classified as functional or quasi-exclusion. Low-level immigration of the resources used by those competing consumers often has more profound effects than does consumer immigration. Including a low rate of external resource input can avoid some of the alternative outcomes predicted by spatially homogeneous models that lack this input. The impact of immigration of both consumers and resources is considered here, but the focus will be on the latter.

Researchers have explored 2-consumer–2-resource models with biotic resources and low-level resource immigration to examine the effects of quasi-extinction of resources (e.g., Abrams 1998, 2001a, Abrams and Nakajima 2007). However, published work has not compared these to analogous systems without any resource immigration. The models with immigration have shown that greater overlap in resource use may lead to smaller effects of species addition or removal on equilibrium population size over a wide range of consumer overlap values (Abrams 1998). This is contrary to the usual assumption of theory based on the MacArthur model. Figure 6.1A shows how the proportional decrease in population size caused by a competitor can change as the similarity of the two competitors increases. The unimodal change shown at low mortality rates is due to the fact that, for such efficient consumers, a relatively low consumption rate maximizes population size in a 1-consumer–1-resource system with logistic resource growth. Given an efficient consumer, the resource that is caught less rapidly can support a high enough population of consumer to exclude (via apparent competition) the more rapidly caught resource. If there is resource immigration, this is actually 'quasi-exclusion', but the very low density of the almost excluded resource has a minimal effect on the population sizes of consumers. Sufficiently similar values of the two capture rates, c_{i1} and c_{i2}, do not produce resource exclusion in the single consumer system. The range of values of C and d that produces resource exclusion in a 1-consumer system is shown in Figure 6.1B.

McPeek's (2019a) reanalysis of the 2-consumer–2 resource version of MacArthur's (1970, 1972) model assumed no resource immigration and also assumed that each consumer must come to equilibrium with its resources before the other consumer is introduced, to judge whether coexistence will occur. Such an analysis may be appropriate in a laboratory system, where complete exclusion is possible in a reasonable time frame (due to small population sizes) and immigration can often be prevented. However, when resource exclusion is a possibility, ruling out immigration drastically changes the nature of competitive relationships and coexistence. Furthermore, the underlying assumption of zero immigration is improbable for most natural systems, because without immigration, any empty patch would remain empty. Movement in

A. Sympatric to allopatric populations density vs.1 - C in a system
with $d = 0.05$ (solid line) and $d = 0.25$ (dashed line)

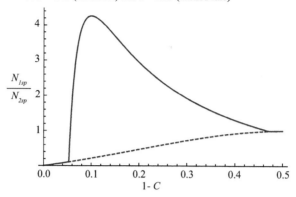

B. Range of parameters producing exclusion

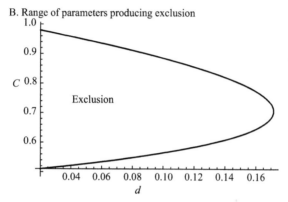

Fig. 6.1 Competitive effects in a 2-consumer–2-resource MacArthur model with the parameters given in the text. These are based on eqs (6.1a), with 'mirror image' consumers, where C is the capture rate of the 'preferred' resource and $1 - C$ is that of the non-preferred resource. Panel A shows the ratio of the consumer population size when present alone to that when it occurs with the 'mirror image' competitor. The solid line assumes $d = 0.05$, and the dashed line assumes $d = 0.25$. The smaller d produces apparent competitive exclusion over most of the range of C, while the larger d never does so.Panel B shows the maximum and minimum capture rates, C, of the preferred resource that allow coexistence as a function of the mortality rate, d. The 'exclusion' zone is characterized by elimination of the earlier arriving consumer by the later arriving competitor when there is no resource immigration and the first consumer comes to its equilibrium density before the second is introduced. The other parameters are $r = k = 1$, and $d_1 = d_2 = d$. Values of $d > 0.17157$ never result in exclusion. The figure shows that for $d < 0.12772$, more than half of the possible values of C produce exclusion.

some cases may be sufficiently rare that complete resource exclusion is possible, so it is worth exploring the outcomes that are produced in that case. However, McPeek (2019a) did not consider the final outcome of the process of sequential consumer invasion, except to note that it would not be coexistence. The actual result is an outcome that has apparently never been observed, as explained below.

The basic model analysed in this section was introduced in Chapter 3, but immigration of resources is added as a possibility and some changes in notation are also made here. The 2-resource version of this model is given by the following which was used in fig. 6.1 above.

$$\frac{dN_1}{dt} = N_1 \left(b_{11}c_{11}R_1 + b_{12}c_{12}R_2 - d_1\right)$$

$$\frac{dN_2}{dt} = N_2 \left(b_{21}c_{21}R_1 + b_{22}c_{22}R_2 - d_2\right)$$

$$\frac{dR_1}{dt} = I_1 + R_1 \left(r_1 - k_1 R_1\right) - c_{11}R_1 N_1 - c_{21}R_1 N_2$$

$$\frac{dR_2}{dt} = I_2 + R_2 \left(r_2 - k_2 R_2\right) - c_{12}R_2 N_1 - c_{22}R_2 N_2 \qquad (6.1a)$$

Equations (6.1b), below, represent the 3-resource extension of eqs (6.1a), in which each consumer has an exclusive and a shared resource, the latter being resource 2.

$$\frac{dN_1}{dt} = N_1 \left(b_{11}c_{11}R_1 + b_{12}c_{12}R_2 - d_1\right)$$

$$\frac{dN_2}{dt} = N_2 \left(b_{22}c_{22}R_2 + b_{23}c_{23}R_3 - d_2\right)$$

$$\frac{dR_1}{dt} = I_1 + R_1 \left(r_1 - k_1 R_1\right) - c_{11}R_1 N_1$$

$$\frac{dR_2}{dt} = I_2 + R_2 \left(r_2 - k_2 R_2\right) - c_{12}R_2 N_1 - c_{22}R_2 N_2$$

$$\frac{dR_3}{dt} = I_3 + R_3 \left(r_3 - k_3 R_3\right) - c_{23}R_3 N_2 \qquad (6.1b)$$

As in previous chapters, resource populations are R, and consumer populations are N. The maximum per capita growth rate of resource i is r_i, and the per capita reduction in that rate with greater resource abundance is given by k_i. The per capita consumption rate of consumer i on resource j is c_{ij}, and b_{ij} is the corresponding resource conversion efficiency. Resource immigration is given by the subscripted I parameters. Comparable models with abiotic resource growth are discussed in Section 6.3. Abrams (1998, 2001a) explored eqs (6.1) but did not consider cases without resource immigration. Other than the presence of resource immigration, the 2-resource version of the model is equivalent to the 2-resource linear functional response model analysed by McPeek (2019a).

I will first address the 2-resource case, eqs (6.1a). This provides a good illustration of the expanded range of outcomes arising from full consideration of resource dynamics. To simplify the analysis, assume that both consumers have the same initial mortality, d, and equal conversion efficiencies (all $b_{ij} = 1$, for consumer i and resource j), as well as equal mean consumption abilities. Differences between consumers in their ratios of b_{ij}/b_{ik} can affect the outcome of competition (Schoener 1974a; Chase and Leibold 2003). Such differences can promote or inhibit coexistence, but we unfortunately have very few empirical studies of resource conversion efficiencies.

First consider a case in which the two competitors have symmetric partitioning of two resources. The two resources are assumed to have the same growth parameters. The two consumers' resource consumption rates are mirror images; $c_{11} = c_{22}$, and $c_{12} = c_{21}$. Without loss of generality, I assume that $c_{i1} + c_{i2} = 1$ and $c_{ii} > c_{ij}$. This allows c_{ii} to be replaced by C with $1/2 < C < 1$. The condition for apparent competitive exclusion of the resource consumed more rapidly by a single consumer is that the consumer would achieve a high enough equilibrium abundance on its low-capture-rate resource alone, to exclude the more rapidly caught resource. This requires that C not be too close to either $1/2$ or 1. If C is close to $1/2$, the resources suffer roughly equal consumption rates, so neither goes extinct. If C is too close to 1, then the consumption rate of the lower-c resource $(1 - C)$ is close to zero, so the consumer population could not achieve a high abundance on that resource alone unless the consumer's mortality was very close to zero. An intermediate C, slightly above 0.707107 is most likely to lead to resource extinction, as shown in Figure 6.1B.

The outcome with resource immigration is coexistence as shown in Figure 6.1A. However, without resource immigration, and assuming complete equilibration of the 1-consumer system before introduction of the second consumer, there may be a different outcome. This is true in those cases where the first consumer causes exclusion of the resource it captures more rapidly. If consumer 1 arrives first and has a low enough mortality, resource 1 will be excluded. When consumer 2 is then introduced, it excludes consumer 1 because of consumer 2's greater consumption rate of the only resource present (resource 2). Consumer 1 cannot reinvade this system, and the 'no resource immigration' assumption means that resource 1 never does so. This outcome is a posteriority effect (i.e., later arriving consumer species always excludes the earlier arriving one). Under the same assumption of no resource immigration, but both resources initially present, simultaneous introduction of both consumers would allow them to attain the 2-consumer outcome. Thus, three outcomes are possible, depending on the timing of consumer introductions.

A second simple scenario of competition under this model (eqs 6.1a) has one generalist (species 1) and one semi-specialized consumer (species 2) that share two resources. The generalist is assumed to have $c = 1/2$ for each resource, and is assumed to have a slightly larger mortality, d (a cost of generalization). In this case, the single-consumer system with only the generalist consumer always has both resources present. As a result, regardless of resource immigration, the semi-specialist can invade if it has parameters that allow it to increase. Coexistence is relatively easy to achieve when both consumers are introduced while both resources are present; it occurs over a broad range of relative values of d_i for the two consumers. However, if the semi-specialized consumer 2 is introduced first and allowed to achieve equilibrium with no resource immigration, it will cause extinction of its 'preferred' (higher c)

resource. Following this, introduction of the generalist will result in exclusion of the specialist, because the generalist has a much greater c for the one remaining resource. Thus, the outcomes in this sequential introduction case (with no resource immigration) are exclusion of the specialist, if it is introduced first, and coexistence, if the generalist is introduced first or both are introduced while both resources are still present. Resource immigration eliminates the resource extinction that keeps the specialist from returning in the scenario where the specialist is the first consumer present.

Resource exclusion in the no-immigration case becomes more likely (and occurs over a wider range of consumer efficiencies) if the resources themselves compete (Abrams and Nakajima 2007). Resource exclusion also becomes more likely in at least one of the single-consumer systems if the two resources are characterized by unequal r-values. It is necessary that each consumer have a death rate that is low relative to the maximum that allows it to persist in order to have resource exclusion in each of the single-consumer subsystems. Resource exclusion also occurs over a wider range of consumer efficiencies (i.e., d values) if the resources experience relatively little density dependence until they are close to their carrying capacities—e.g., θ-logistic growth with $\theta > 1$ (Abrams 1998, 2001a). While the extent of consumer-driven resource exclusion in natural communities is unknown, such exclusion is unlikely to be rare, given the high predator effects shown by experiments on trophic cascades (Shurin et al. 2002; Terborgh 2015), as well as the results of studies that have shown strong apparent competitive effects in systems with a single predator species (reviewed in Holt and Lawton 1994; Holt and Bonsall 2017). In the symmetric 2-resource model considered here, an intermediate level of consumer specialization on the higher-c resource leads to the greatest apparent competitive effect of the lower-c resource on the higher-c resource (Abrams 1998; Figure 6.1B above), and each consumer must have such an intermediate level of specialization for alternative exclusion outcomes to occur in the four-species system.

In a metapopulation with many patches experiencing different environmental perturbations, such variation (or underlying spatial heterogeneity) would increase the probability that resource species arrive via migration, and would increase the probability that different patches would have species occupying different dynamic attractors (assuming they exist). There may be some cases in which the system of interest has extremely low levels of resource immigration, or is totally isolated. In these cases, the alternative outcomes described for the no-immigration cases above may be more likely. It remains to be determined whether any such cases occur in natural systems.

Immigration also plays a key role in determining the outcome of the 3-resource system given by eqs (6.1b), in which each consumer uses one shared and one exclusive resource. This food web (with different resource dynamics) has a long history in competition theory (Schoener 1974a, c, 1976, 1978; Abrams 1975, 1977), but has been ignored in many recent studies (including all of the six focal articles identified in Table 4.1). The articles by Schoener and by Abrams from the 1970s concentrated on abiotic resources, where resource exclusion is impossible, and resource input is indistinguishable from immigration. However, Abrams (1998) extended this shared-exclusive scenario to biotic resources with low external immigration of all three resources. Consumer extinction in the 2-consumer–3-resource model is impossible when a consumer can maintain a population on its exclusive resource alone. This means that it is possible to have very high overlap in resource use ability (i.e., high relative values of C for the shared resource) and large differences in neutral consumer parameters, but no possibility of consumer exclusion.

The outcomes for the 3-biotic-resource system described in the preceding paragraphs are altered if there is no immigration of resources. If one or both consumers produce apparent competitive exclusion of one resource in the 1-consumer system, that resource is lost permanently. In most of the discussion below, I make the same simplifications as in the above discussion of the 2-resource model; the resources have identical r_j, and identical k_j, and the consumers initially have identical d_i. The B_{ij} are again set to unity. The resource utilization rates of the two consumers are again mirror images; $c_{11} = c_{23} = C$ (for the exclusive resources, 1 and 3) and $c_{12} = c_{22} = 1 - C$ (for the shared resource, 2). In this case, if $C < 1/2$, the shared resource is the only resource that could be excluded in a single-consumer system, and this outcome occurs under the same conditions as in the comparable 1-consumer–2-resource model discussed above (Figure 6.1b). Exclusion of the shared resource means that there will be no competition between the consumers after the second consumer invades and the two species will coexist if that second species can persist on its exclusive resource alone.

If each consumer eats its exclusive resource at a greater rate ($C > 1/2$), the exclusive resource in each single-consumer subsystem becomes extinct if d is sufficiently low (and C is not too close to the maximum of 1). When extinction of its exclusive resource occurs with consumer 1 present, the introduction of consumer 2 will always cause extinction of consumer 1. (There can be a neutrally stable line of equilibria after consumer 2's exclusive resource goes extinct, but population fluctuations would likely result in loss of the initially present consumer in most realistic scenarios having environmental variation that affects the two species differently.) Consumer 2's exclusive resource will eventually go extinct in any event, and the final state will be consumer 2 persisting on the original shared resource, which it captures at a rate $1 - C$, which is $< 1/2$. An important feature of this 3-resource system with no resource immigration is that the qualitative outcome is changed by even a small difference from equality in the mortality rates (or a difference between consumers in their consumption rate of the shared resource). In these cases, the alternative outcomes are exclusion of the initially present species when this consumer has a higher R^* ($= d/C$) for the shared resource, or coexistence when the initially present consumer has a lower R^* for the shared resource. In both cases, the exclusive resource of the initial consumer goes extinct.

The symmetric 3-resource system with resource immigration was analysed in Abrams (1998). This scenario always allows coexistence of both consumers when each consumer is able to exist on its exclusive resource alone. However, as in the 2-resource

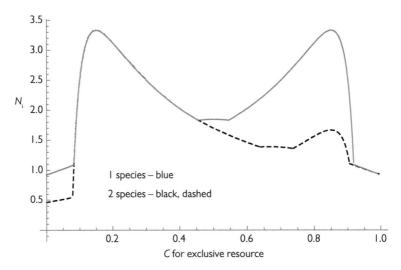

Fig. 6.2 Population sizes in 1- and 2-consumer systems as a function of the attack rate C on the exclusively used resource in a 3-resource system with shared and exclusive resources. The shared resource is consumed at a rate of $1 - C$. The death rate of both species is $d = 0.075$, with $b = r = k = 1$. The lines are identical when the exclusive resources are extinct.

system, the possibility of quasi-exclusion of resources (one resource maintained by immigration) changes the relationship between similarity and the proportional reduction in population size caused by the competitor. This reduction is shown as a function of consumer similarity in Figure 6.2. As in the 2-resource case, maximal competitive reduction in population size occurs for low levels of similarity, in which each species has a much higher capture rate of its exclusive resource than of the shared resource.

The symmetric 3-resource case just discussed is not the only one in which the outcome of an invasion analysis differs depending on the presence or absence of resource immigration. Consider competition (again using eqs (6.1b)) between a semi-specialized consumer 1 ($1 > c_{11} > 1/2$) and a perfect generalist consumer 2 ($c_{22} = c_{23} = 1/2$). Also assume d_2 is low enough that the generalist can exist when only resource 3 is present; i.e., $d_2 < r/(2k)$. If consumer 1 is the initial resident, it will exclude resource 1 if c_{11} and d_{11} lie within the range illustrated in Figure 6.1B. In this case, arrival of consumer 2 excludes consumer 1 provided $d_2 < d_1/C_{12}$. If, for example, consumer 1 is characterized by $c_{12} = 1/4$ and $c_{11} = 3/4$, then a four-fold or greater death rate disadvantage of the complete generalist consumer 2 is required for consumer 1 to persist after arrival of consumer 2. If the ratio of death rates is smaller than four, then exclusion of a resident consumer 1 occurs following introduction of consumer 2, and coexistence of both occurs when consumer 2 is the initial resident. Resource immigration ensures coexistence of both consumer species regardless of invasion order, as resource 1 will immediately increase following the introduction of consumer 2.

A final scenario in this 3-resource model (with the same resource parameter values and $c_{ij} + c_{ji} = 1$) assumes that the utilization patterns of the consumers are identical but displaced along the resource axis; here $c_{11} = c_{22}$, and $c_{12} = c_{23}$. Here, I assume that species 1 has a higher attack rate on its exclusive resource, while consumer 2 has a higher attack rate on the shared resource. In a low-d scenario where exclusion of the more rapidly consumed resource occurs with a single consumer, the species having the higher attack rate on its exclusive resource (consumer 1) is easily excluded by consumer 2, when consumer 1 is the initial resident. When consumer 2, with a higher consumption rate of shared resource, is initially present alone, the shared resource is excluded (given a low d), resulting in coexistence with no competition after the first consumer invades. If the scenario is changed to have low-level immigration of both resources, the final state is similar to the final state under no resource immigration and consumer 2 arriving first (i.e., coexistence at unequal abundances). The only difference is a very low abundance of the shared resource due to immigration, but this has little impact on either consumer's population size.

Coexistence is not the only outcome affected by resource exclusion in the two models described by eqs (6.1a, b). Another feature that is potentially affected is the phenomenon under which continuous directional change in a neutral parameter of one consumer species can produce a discontinuous change in the equilibrium densities of both consumers. (This is impossible in the Lotka–Volterra model with two species.) A discontinuous jump occurs when a resource either goes extinct or regrows from a near-zero density. The cause of this discontinuity is apparent competition, mediated by the consumers. For the systems considered above (eqs (6.1a, b)), a gradual change in the mortality of one species frequently produces an abrupt change in the equilibrium densities of both species. Figure 3.2 illustrated the 2-resource system (eqs (6.1a)) as a function of d_1 in an example with mirror image competitors having c values of 3/4 and 1/4 (i.e., $C = 3/4$). The discontinuous drop in abundance to zero for consumer 2 is accompanied by the extinction of resource 1 at the critical d_1 ($= 1/30$). Note that this disappearance occurs when the population of N_2 is only approximately 11% lower than its equilibrium when both species have identical death rates and abundances. The loss of consumer 2 is accompanied by a more than threefold jump in the abundance of consumer species 1. The more general condition for discontinuous changes in equilibrium abundances of consumer species as the mortality rate of one competitor is reduced, is $d < 2C - 1$ in the symmetrical 2-consumer–2 resource model considered above. For example, if $C = 3/4$, a discontinuous change in the equilibrium of both consumers at a sufficiently low d_1 occurs when the initial mortalities of both species are less than 1/2, which is coincidentally half the value above which no consumer can persist. Note that this qualitative result is not altered by a low rate of resource immigration, although immigration changes the N vs d relationship very close to the point of discontinuity. There is a second discontinuity in Figure 3.2, when d_1 exceeds 3/10. This is characterized by the loss of both resource 2 and consumer 1, and by a discontinuous jump in the equilibrium density of consumer 2, which more than doubles its population size.

Another lesson from the system described by eqs (6.1a, b) is relevant to 'modern coexistence theory', which interprets coexistence in terms of equalization and stabilization. The outcomes described above are not consistent with the view that coexistence always occurs when general competitive abilities are equal and there are differences in resource utilization rates (as is implied by at least some versions of 'modern coexistence theory'). With no resource immigration and temporally spaced introductions of consumers, it is often more difficult for moderately different species to coexist than it is for more similar species. If they do coexist because of resource immigration it is often true that pairs of very different competitors produce greater reductions in each other's population size in sympatry than do pairs of consumers with more similar resource capture rates (Abrams 1998). Increasing the ability of consumer species 2 to obtain its resources, reproduce, or survive (i.e., increasing inequality of the consumers without changing overlap in resource utilization abilities) cannot exclude consumer 1 if the latter can persist on its own exclusive resource(s) alone.

These simple examples also show that ecologists have still not properly understood the role of resource extinction and resource immigration in determining the nature of competitive relationships between consumers. McPeek (2019a, b) calls for more attention to apparent competition, but, by ignoring resource immigration and focusing exclusively on consumer coexistence, his analyses failed to reveal either the posteriority effect or the possibility of coexistence in systems having different single-consumer subsystems, each characterized by the apparent competitive exclusion of one resource. The earlier analysis (Abrams 1998), that included resource immigration, found that a competitor's effect on equilibrium population was greatest for relatively low levels of similarity in resource attack rates, a possibility not mentioned in McPeek (2019a). Results for the 3-resource model (eqs 6.1b) with no resource immigration are also inconsistent with the widespread view that greater similarity always makes coexistence more difficult and greater differences make it easier. Chesson (2018, supplemental material) defends a Lotka–Volterra approximation for understanding consumer–resource models in spite of the existence of resource extinction with the following statement: '(A56) [an LV equation derived from MacArthur's consumer–resource equations] would still represent valid Lotka–Volterra dynamics, but their interpretation in terms of specific resources would be compromised by resource extinction'. This (correct) claim is problematic because almost all the conclusions that may be of ecological interest (including coexistence or not) change abruptly when one or more resources go extinct.

Both the LV model and MacArthur's flawed analysis of his corresponding consumer–resource model have led to the conclusion that competitive effects are decreased and the ability to coexist is increased by having a greater difference in resource attack parameters (i.e., reduced overlap in resource utilization, often termed 'niche overlap'). The preceding results show that neither of these generalizations is valid in the case of two very efficient competitors that each uses two resources. Models having more consumers and/or more resources also fail to support these generalizations (Abrams et al. 2008a; Abrams and Rueffler 2009).

Apparent competition is the cause of the resource extinctions noted above. However, in a 2-consumer–2-resource system with no direct consumer density dependence, apparent competition (i.e., shared predation) is functionally absent over a wide range of parameter perturbations. The Yodzis (1988) method of measuring indirect effects (embodied in eqs (3.1) and (3.2)) implies that the total apparent competitive effect of one resource on the other can be measured by the population response of resource j to a change in the neutral parameter, r_i of resource i in eqs (6.1a). However, this effect is zero in the 2-resource system when there is no self-limitation in either consumer species. Self-limitation is assumed to be absent in eqs (6.1a) and most published analyses of consumer–resource models of competition and apparent competition. In its absence, apparent competition does not produce any effect on equilibrium abundance provided that all resources and consumers are present. Small to moderate change in the r_i of one resource species will not affect its own abundance or the abundance of the other resource—both resource abundances (assuming they are positive) are entirely determined by the equilibrium requirements of the two consumers. However, a sufficiently large parameter change affecting one resource in this 4-species system will produce extinction of one of the consumers. This will often be followed by extinction of one resource, depending on the efficiency of the remaining consumer. Just as in the case of between-consumer interactions in this 2-consumer–2-resource system, apparent competitive effects become discontinuous, with the equilibrium resource population jumping from a significant density to zero with a very small (growth-increasing) change in the parameter. Other implications of apparent competition in subsystems of this simple model, and in models with more resources are treated in Abrams (1998, 2001a, 2009b), Abrams and Rueffler (2009), and Abrams et al. (2008a).

Although it does not involve resource extinction, it is worth examining one of the possible outcomes of having a third consumer in the 3-resource system described by eqs (6.1b). The third species is a specialist on the shared resource (R_2), which means that its presence fixes the abundance of this resource at $d_{32}/(b_{32}c_{32})$. For consumer species 3 to exist, this resource requirement must be lower than the equilibrium R_2 that exists in the absence of consumer 3. Once it is present, species 3 eliminates any competition between consumers 1 and 2, as neither of these consumers affects the equilibrium abundance of the only resource they share. Thus, even without resource exclusion, the 3-consumer–3-resource MacArthur system shows that competition coefficients are not independent of the presence or abundance of additional consumer species, even when those species do not bring about resource extinction. In other words, the multi-species MacArthur model is again inconsistent with the Lotka–Volterra model.

The limitations of many previous analyses of the MacArthur model (eqs (6.1a, b), and examples having more species on one or both levels) go beyond the failure to take resource extinction into account. Many conclusions arising from the model depend on the exact linearity of its component per capita growth rate functions. Nonlinearity in resource density dependence results in very different competitive interactions, as does competition between the resources themselves. These possibilities are discussed below.

6.3 Consequences of non-logistic resource growth

Abrams (1980b) discussed the impact of non-logistic resource growth on competition. That article assumed the theta-logistic model (see Figure 5.1), in which the per capita growth rate declines at a rate proportional to the resource abundance (R) raised to the power θ. Consumer competition coefficients that are independent of resource abundance only occur for the case of $\theta = 1$ for all resources (i.e., logistic growth). Changes in either relative or absolute resource abundances alter the competition coefficients when $\theta \neq 1$ for one or more resources. Figure 6.3 provides some examples of the response of consumer abundance to changes in one consumer's mortality for simple 2-resource models with biotic resources, and with symmetric resource use by the two consumers (both have attack rate C on their 'preferred' resource and $(1 - C)$ on the less preferred). In these cases, the contribution of a given resource to the competition coefficient (eq. (3.2)) changes depending on its abundance. The contribution increases with greater abundance when $\theta < 1$ and decreases with greater abundance when $\theta > 1$ (Abrams 1980b). Figure 6.3 spans the full range of d_1 values allowing coexistence for the given d_2. Because $C > 1/2$, a low mortality of consumer 1 (d_1) means that resource 1 has low relative abundance, and therefore a high weighting in calculating the competitive effect in Figure 6.3A. Because consumer 1 has a three-fold advantage in consumption rate on resource 1, its relative effect on consumer 2 (α_{21}) is approximately three times that of consumer 1 on itself when resource 1 is rare. Conversely, when resource 2 is very rare, the effect on consumer 2 approaches $1/3$. Thus, with $\theta = 2$, as in Figure 6.3A, a low d_2 means that consumer 2's 'preferred' (higher C) resource is rare, so its impact on consumer 1 is largely determined by this resource. The reverse is true with $\theta = 1/2$, as in Figure 6.3B; here a resource has a higher weighting when it is common; thus the lines giving the competition coefficients have opposite slopes in the two panels of Figure 6.3. It is important to remember that the θ-logistic is only one of many possible models of nonlinear density dependence (Chapter 5), and it is only being employed here because it is mathematically simple and is the most widely used nonlinear model.

The potential for discontinuous change in the equilibrium abundance of a consumer in response to continuous change in its own or another consumer's mortality rate is driven by extinction (or addition) of a resource. The mortality rate at which resource extinction (or addition) occurs is solely determined by the consumer growth equations in the models considered here, which lack any direct effect of consumer abundance on consumer per capita growth rate. This means that the mortality d_i required for resource extinction is not affected by the exponent describing resource density dependence. However, the question of whether extinction of one consumer occurs before the d_i of the manipulated species reaches the point of resource extinction is affected by θ. The lowest value of C (with $C > 1/2$) that can produce discontinuous change is smaller for larger values of θ. For the example of a MacArthur system ($\theta = 1$) shown in Figure 3.2, if C is decreased, the discontinuity at the lower d_1 disappears when $C < 0.589$, while for $\theta = 2$, discontinuous change in consumer 2 at the lower d_1 occurs until $C < 0.5341$. This means that discontinuous change occurs for

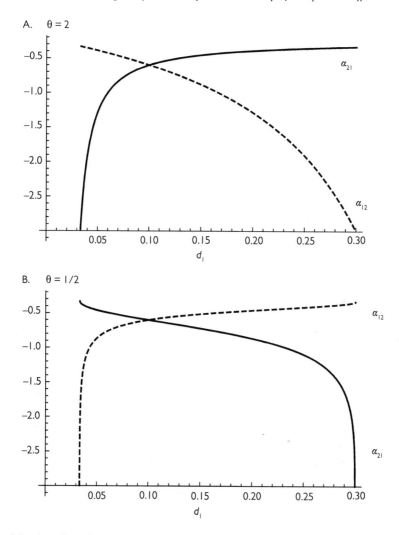

Fig. 6.3 The effect of nonlinear resource density dependence ($\theta \neq 1$) on the shape of consumer competition as measured by eq. (3.2), for a range of different mortality rates of consumer species 1, d_1, when $d_2 = 0.1$. The other parameters are; $r = k = 1$; and $C = 3/4$. In panel A, $\theta = 2$, and in panel B, $\theta = 1/2$. In both panels, the dashed line is α_{21}, and solid is α_{12}. The value of both competition coefficients using MacArthur's formula is −0.6. The actual coefficients shown below only have that value when death rates are equal, ($d_i = 0.1$).

a wider range of consumer similarities for resources having $\theta = 2$ than for otherwise similar resources with $\theta = 1$. More generally, the range of C producing discontinuous change in equilibrium consumer abundance in response to a continuous change in mortality expands with larger θ regardless of the initial value of this (positive) exponent.

Levine (1976) and Vandermeer (1980) showed that competition between resources in a 2-consumer system could produce mutually positive effects of each consumer species on the other. Both studies assumed the MacArthur model with the addition of direct and instantaneous linear competitive effects between resources. However, Levine did not consider cases in which resource species became extinct, and Vandermeer (1980) only examined equilibria, so did not examine the phenomenon of discontinuous change in abundance with continuous parameter change. This was extended to systems with type II consumer functional responses, including the possibility of large magnitude perturbations and resource exclusion by Vandermeer (2004, 2006), Abrams and Matsuda (2005), and Abrams and Nakajima (2007). The results in these works showed that the sign of the between-consumer effects often changed with the magnitude of the perturbation to the consumer. Abrams and Nakajima's (2007) results also confirmed the finding of Abrams (1998); i.e., the largest competitive effects on abundance frequently occur in cases with relatively low consumer similarity in resource uptake rates.

A second commonly used representation of resource growth is the chemostat model, which is routinely used to describe non-reproducing (abiotic) resources. The consequences of having abiotic resources were known to be nonlinear competitive effects between two consumers as early as MacArthur (1972). MacArthur briefly considered resources that were externally supplied at a constant rate and only left the system by consumption (i.e., a chemostat with no loss of unconsumed resources). This model produces lower weighting of resources whose abundances are lower when calculating a competition coefficient. However, unlike biotic resources, externally supplied (abiotic) resources cannot be driven to extinction. Unfortunately, MacArthur (1972) did not treat this case in detail, and he instead stressed that competitive effects could be approximated by a linear model for abundances very close to the equilibrium point. Unfortunately, the effects of very small changes (in densities or neutral parameters) near an equilibrium point are seldom measurable, and are only of ecological significance if they reliably indicate the responses to larger magnitude changes. Schoener (1974c) showed that MacArthur's abiotic resource model, with resources assumed to be at equilibrium led to extremely nonlinear competitive effects across the full range of competitor densities in a system where each consumer had one or more resource(s) that were not shared with the other consumer(s). Abrams (1975) demonstrated that such models produced quite different magnitudes of the competition coefficient at equilibrium than those produced by an otherwise similar MacArthur model. Systems with abiotic resources that obey the full chemostat equation (which includes resource loss due to causes other than consumption) were later explored by Abrams (1977) and Schoener (1978); this again led to nonlinear competition when resources were assumed to reach a quasi-equilibrium with consumers. In all of these cases, the relative weighting of a particular resource in determining the competition coefficient is smaller when the resource is less abundant. Lowered mortality of a given consumer differentially depletes the resources it consumes at a higher rate, so these contribute less to its competitive effect(s) on the other consumer(s).

The probability of resource exclusion increases with the number of resources. Analyses of 2- and 3-consumer systems with very many resources (Abrams et al. 2008a and Abrams and Rueffler 2009) showed that competition coefficients frequently increase with greater consumer differences in resource utilization. These studies used models with independent logistically growing resources having identical growth parameters. Abrams and Rueffler (2009) examined a model with three consumer species, and found that coexistence was impossible for some cases where the middle consumer had a resource utilization curve that was not exactly intermediate between the two other consumers. This result again required high enough consumer efficiency (e.g., low d) that resources used at a high enough rate by one or both species were driven extinct.

Pronounced nonlinearity of competitive effects was observed in some of the early experimental work on two-species competition (Ayala 1969; Gilpin and Justice 1972; Gilpin and Ayala 1973). Such nonlinearity was shown to be true of single-species growth (intraspecific competition) in many species by Pomerantz et al. (1980). These authors began their paper (p. 311) with the following statement: 'The linearity assumption of the logistic model of population growth is violated for nearly all organisms'. The type of nonlinearity that arises in abiotic growth models usually leads to lower weights associated with rarer resource types when calculating expressions (3.1) or (3.2) from Chapter 3. (Recall that measure (3.2) is equivalent to the competition coefficient at equilibrium.) For other types of growth models, the nonlinearity in single resource growth should translate into nonlinearity of competitive effects between consumers of those resources. This makes it particularly puzzling that competition theory has retained its focus on the linear case.

6.4 Consequences of nonlinear functional responses

The model used in most of the preceding sections assumes linear functional responses of both consumer species, which was previously argued to be rare in nature and to have special properties. While the empirically more common (Jeschke et al. 2004) type II responses were shown to produce nonlinear competitive effects with logistic resources nearly half a century ago (Armstrong and McGehee 1976a,b, 1980; Abrams 1980a), they have still not often been included in models of competition. McPeek (2019a) recently revived the argument for using such responses with a consumer–resource approach, but failed to cite the results of most of the previous studies that had done so. The only differences in outcomes for type II responses noted by McPeek (2019a) were: (1) a lower possibility of resource exclusion by apparent competition; and (2) the possibility for coexistence of more consumer species than resource species in cycling systems (documented by Armstrong and McGehee (1976a,b, 1980)). In fact, the nature of competition is changed significantly by type II responses in systems with two or more resources, even without cycling or resource exclusion

(Abrams 1980a). The competitive effect—measured using either eq. (3.1) or eq. (3.2)—is quite different when the MacArthur consumer–resource system is modified by adopting multi-species type II responses. Furthermore, there is actually a higher probability of resource exclusion in stable systems with type II responses (contrary to McPeek (2019a)), as shown in this section. Finally, the presence of cycles due to type II responses does not only affect consumer coexistence; it leads to a variety of inter-specific effects because perturbations to one consumer alter the nature and amplitude of cycles of all species. These effects were noted by Armstrong and McGehee (1980) and documented in more detail in Abrams and Holt (2002), Abrams et al. (2003), and a number of subsequent works. Another possibility is a hydra effect (an increase in the equilibrium or mean density of a species with an increase in its own mortality), which can occur even in models with linear functional responses having competi-tion between resources (Cortez and Abrams 2016). Hydra effects are more likely with type II responses, as they can then occur even in systems with a single consumer species (Abrams 2002). Hydra effects in 2-consumer systems change the sign of the interaction as measured by eq. (3.2).

Consequences of type II responses in stable systems are discussed in the first subsection 6.4.1 below, while the consequences of cycles are explored in 6.4.2.

6.4.1 Effects of nonlinear functional responses on consumer competition in systems with stable equilibria

A type II functional response alters the relationship between overlap in resource use and competition coefficients; it also changes most other possible measures of the strength of competition, even in systems that have stable equilibria Abrams (1980a). Type II responses also have a major effect on the nature of apparent competition between resources (Abrams et al. 1998).

The presence of a type II response in a system that is otherwise identical to the 2-consumer–2-resource MacArthur model of eqs (6.1a) significantly changes the relationship between similarity in resource use and the competition coefficient (eq. (3.2)). Figure 6.4 assumes a symmetrical 2-consumer–2 resource system with two independent logistic resources, linear consumer numerical responses, and type II functional responses. Consumer 1 has attack rates of C and $(1 - C)$ on resources 1 and 2 respectively, while consumer 2 has attack rates of $(1 - C)$ and C. Thus overlap in utilization is 0 when $C = 1$ and overlap is 1 when $C = 1/2$. For any value of $h > 1$, a small enough d makes the equilibrium point unstable. The competition coefficient (eq. (3.2), when $d_1 = d_2 = d$) is:

$$\alpha_{12} = \alpha_{21} = \frac{-2C(1-C)}{1-(1-d)h+dh^2-2C(1-2(1-d)h+2dh^2)+C^2(2-4(1-d)h+4dh^2)}$$
(6.2)

Equation (6.2) becomes independent of C and equals -1 when $d = (h-1)/(h(1+h))$. This is also the threshold mortality, d, below which the equilibrium is unsta-ble. (For values of $h \leq 1$, the equilibrium is stable for all values of d.) Thus, formula (6.2) implies that systems with $h \leq 1$ have little dependence of competition on similarity when the system is stable, but close to the stability threshold. Larger h values in a stable system always imply an increase in expression (6.2). Figure 6.4 shows the competition coefficient as a function of C at several different mortality rates for systems with two different handling times. In each case, mortality rates that are closer to the stability threshold yield values of the competition coefficient close to −1 over a wider range of consumer similarities. Even values of C close to 1, implying very low overlap in resource use, produce nearly equal inter- and intraspecific competition. Note that, in all cases, competition is stronger than it would be in a system with no handling time, although it approaches the

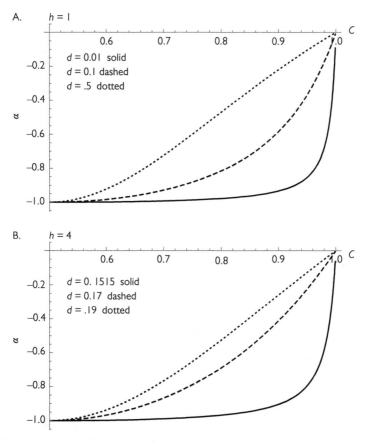

Fig. 6.4 The relationship between overlap, C, and the competition coefficient (α) in a symmetric 2-consumer–2-resource systems with type II functional responses having equal handling times, h, for all consumer–resource pairs (and $r_1 = r_2 = k = 1$), and attack rates of C and 1 − C for its more- and less-preferred resources. Thus, over the range shown, larger C implies less overlap in resource use. In panel A, $h = 1$, which gives a stable system for all mortality rates, d, and a maximum d of 1. In panel B, $h = 4$, which yields a stable equilibrium for $d > 0.15$, and no positive equilibrium for $d > 0.2$. In both cases, lower death rates in stable systems result in a competition coefficient closer to unity over a wider range of overlaps.

no-handling-time interaction strength when the mortality rate is close to the maximum that allows consumer persistence in the absence of competition. Note also that, if $h = 0$, expression (6.2) reduces to the MacArthur–Levins (1967) formula, $\alpha = -2C(1-C)/(C^2(1-C)^2)$, which changes from -1 to 0 in a sigmoid fashion as C increases from 0.5 to 1 (overlap going from 100% to 0). The MacArthur–Levins formula is nearly identical to the line for $d = 0.5$ in Figure 6.4A. If there is competition between the two resources, greater handling time increases the parameter range characterized by mutually positive effects on abundance in the 2-resource model considered here (Abrams and Matsuda 2005; Abrams and Nakajima 2007). Note that MacArthur and Levins (1967), and most subsequent work, use the unsigned (positive) formula for the competition coefficient.

Another consequence of type II functional responses is that the competition coefficients generally differ from each other in this system with perfectly symmetrical resource use parameters when the mortality rates of the two consumers differ (i.e., when their equilibrium abundances are unequal). It is possible to get a closed form expression analogous to eq. (6.2) for cases when $d_1 \neq d_2$. The two competition coefficients for the simple 2-resource system assumed in Figure 6.4A ($h = 1$) are plotted as a function of the mortality rate of consumer 1 in Figure 6.5, with low overlap in panel A, and moderate overlap in panel B. In both panels, the lines cross at the d_1 value that is equal to the (fixed) d_2; the competition coefficient at this point is always larger in magnitude than the value for a corresponding system with linear consumer functional responses (given by the blue line). The x-axis spans the full range of d_1 values that allow coexistence; consumer 2 is extinct for d_1 values lower than the range shown, and consumer species 1 is extinct for d_1 above that range. The (negative) per-individual impact of species 1 on 2 (i.e., α_{21}) is smaller in magnitude when species 2 is relatively common due to a high death rate of consumer 1; if the death rate of species 1 is decreased, its negative impact of species 2 (per unit change in d_1) becomes larger in relative magnitude, and even exceeds the magnitude of the change in the abundance of species 1 on itself ($\alpha_{21} < -1$). The reverse patterns occur for the impact of species 2 on species 1 (i.e. α_{12}). Panel B shows the same patterns for a system with greater overlap in resource use (and hence a narrower range of d_1 allowing coexistence). Note that for both panels, the competition coefficients are usually significantly greater in magnitude than would be the case in a system with linear functional responses (given by the blue line). This figure assumes a low enough handling time that all systems have a stable equilibrium. Some examples of the impacts of press perturbations to mortality in cycling systems are provided in Section 6.4.2 below.

The type II functional response is only one of a wide range of nonlinear multi-species functional response forms, as noted in Chapter 3. Even within the category of Holling type II responses, a multi-resource system may have multiple independent single-resource functional responses if the resources are present at different times, so that handling one resource type does not interfere with capturing others. Adaptive choice by the consumer will also change the relationship. Thus, the possibilities illustrated here represent a small fraction of what is likely to occur in natural systems.

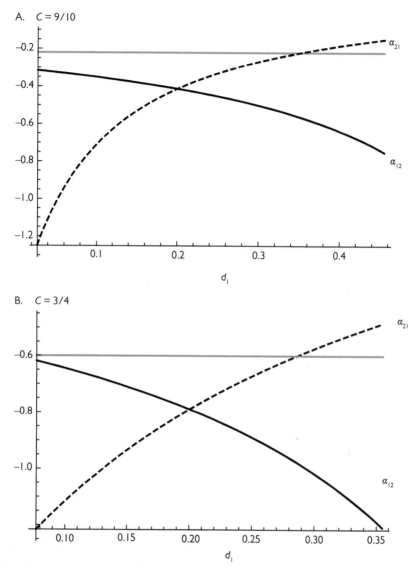

Fig. 6.5 The competition coefficients, α_{12}, and α_{21} (eq. (3.2)) as a function of d_1 for $d_2 = 0.2$ in a 2-consumer–2-resource model with logistic resource growth and a type II consumer functional response with a handling time $h = 1$ for all consumer–resource pairs, and mirror image C values as in Figure 6.4. The blue line is the MacArthur formula for the product of the two α for linear functional responses; the dashed black line is α_{21}, and the solid black line is α_{12}. As d_1 increases, the competition coefficient of consumer 2 on 1 decreases, and that of consumer 1 on 2 increases. The two panels illustrate two levels of overlap in resource use.

The only one of the six focal articles discussed in Chapter 4 that gave explicit consideration to functional response shape was McPeek (2019a). However, even here, the claimed potential consequences do not include some that had been demonstrated many years earlier. The one consequence of type II responses in stable systems claimed by McPeek (2019a) is a reduction in apparent competition in systems with a single consumer. McPeek (2019a, p. 1) states: 'Saturating functional responses do not qualitatively alter the conditions for multiple consumers and resources to coexist at a stable point equilibrium, but do increase the range of apparent competitive abilities for resources that can invade and coexist'. Conditions for resource coexistence in single-consumer systems having type II functional responses were studied by Abrams et al. (1998) and Abrams (2009b, d). All of these works showed significant differences from the case of linear functional responses. However, they did not show a decreased likelihood of apparent competitive exclusion. As shown in more detail below, exclusion of resources in stable single-consumer systems is actually more likely in systems having type II functional responses than in otherwise comparable systems with linear responses.

In determining how functional response shape changes the range of parameter values allowing all resources to persist with one consumer, it is important to compare systems with equivalent consumer efficiencies (measured by the reduction in resource densities they bring about). Equivalent consumer efficiency is important because higher efficiencies strengthen apparent competition by increasing the consumer effects on all resources (Holt 1977). If type II functional responses (or higher handling times) implied lower efficiencies, species with such responses would be outcompeted. In general, the range of 'apparent competitive ability' of one resource species allowing coexistence with another is most appropriately measured by assessing the fraction of the total range of a neutral resource parameter of the focal species that allows coexistence with and without the competitor. The parameter r_i is the only neutral parameter in eqs (6.1a, b). Under a linear (MacArthur) model, an increase in r for one prey leads to a reduction in the equilibrium abundance of the other prey species in the presence of the single predator species. The range of relative magnitudes of r_1 and r_2 that allows resource coexistence in a system where $C_1 = C_2$ provides a measure of the breadth of conditions for apparent competitive coexistence. Another measure is provided by the fraction of the potential range of consumer mortality rate that allows coexistence of both resources when $C_1 \neq C_2$. Both of these potential range measures are examined below for 1-predator–2-prey systems and for 2-predator–2-prey systems, to determine how predator handling time affects the breadth of conditions under which prey coexist. It is shown that type II responses, and particularly those with greater handling time, should more often lead to apparent competitive exclusion of one of the two resources than in the corresponding linear response systems, at least when the systems are stable.

First consider the range of relative r-values allowing resource coexistence in a 1-consumer–2-resource system. This is a measure of coexistence bandwidth

(Armstrong 1976) for resources, as r is a neutral parameter in the single-consumer version of eqs (6.1a), modified to have a handling time, h. Consider a simple case with $r_1 = 1$; $C_1 = C_2 = 1/2$; $k_1 = k_2 = 1$; and $B_1 = B_2 = 1$. The value of r_2 that is just large enough to cause extinction of resource 1 is given by

$$r_2 = \frac{1 + 2d - dh}{1 - dh} \tag{6.3}$$

It is easily confirmed that this expression increases with h, seemingly suggesting that apparent competition is reduced. However, a comparison between systems with and without handling time should be characterized by similar effects of consumer presence on the resources in both systems; i.e., equal equilibrium resource densities in the baseline case where $r_1 = r_2$. This can be achieved if $d_{t2} = d_{t1}/(1 + hd_{t1})$, where the t1, t2 subscripts denote type I (linear) and type II (saturating) functional responses. Substituting the formula for d_{t2} into expression (6.3) yields $1 + 2d_{t1}$, which is the value of expression (6.3) for $h = 0$. Thus, 1-consumer–2-resource systems with type I and type II responses that have identical values of all parameters other than handling time are characterized by the same range of r values permitting coexistence of resources.

In the present analysis (and in McPeek, 2019a), the focus is on interactions between consumer species, so looking at the range of consumer mortalities allowing resource coexistence is more relevant than the range of r values of resources. Here I assume that the resources are characterized by equal r-values but unequal vulnerabilities (C) to the consumer. The question is then, what fraction of the range of potential mortality rates (d) of the consumer results in exclusion of the higher-C resource. Using a 1-consumer–2-resource subsystem consistent with the previous 2-resource example ($k_i = r_i = B_{ij} = 1$; $C_{j1} + C_{j2} = 1$), the maximum death rate that a predator with a linear functional response can sustain is $d = 1$. The maximum mortality for a predator with a handling time h is $d = 1/(1+h)$. For example, if $h = 1$, $d_{max} = 1/2$, which is 1/2 the maximum for a linear response. I examine the d at which apparent competitive exclusion occurs with consumer 1 present when $C_{11} = C > 1/2$ and $C_{12} = 1 - C$. Exclusion of resource 1 occurs when

$$d < \frac{(1 - 3C + 2C^2)}{-C + h - 3Ch + 2C^2h}. \tag{6.4}$$

If $C = 3/4$, for example, resource exclusion occurs for $d \leq 1/6$ when $h = 0$, i.e., one-sixth of the maximum d (= 1) allowing consumer persistence. For the case of $h = 1$, exclusion (of the exclusive resource 1) occurs for $d \leq 1/7$. This represents 2/7 of the maximum d of 1/2. Thus, apparent competitive exclusion is a more likely occurrence (occurs over a larger fraction of admissible d-parameter space; 2/7 rather than 1/6) when comparably efficient predators have saturating functional responses, rather than linear responses. Figure 6.6 compares the fraction of d parameter space

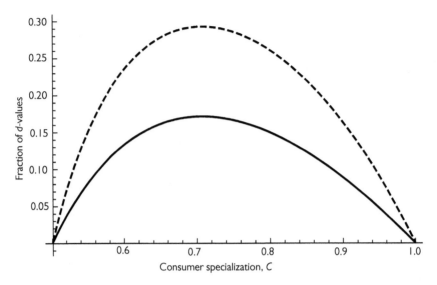

Fig. 6.6 The lines give the proportion of mortality (d) parameter space that yields extinction of the more vulnerable resource via apparent competition in a 1-consumer–2-resource model based on eqs (6.1a). The x-axis is the C-value of the more vulnerable resource ($C_1 + C_2 = 1$). Logistic resource growth is assumed with $r = k = 1$. The solid line assumes a type I functional response ($h = 0$), while the dashed line assumes $h = 1$. Handling times > 1 produce a larger fraction of admissible parameter space characterized by exclusion than does $h = 1$, but also allow for population cycles, which requires numerical analysis.

over which apparent competitive exclusion occurs for $h = 0$ and $h = 1$, for a system in which there are equal growth parameters of the two resources and $C_1 + C_2 = 1$. Within the range of handling times where all equilibria are stable (both lines in Figure 6.6), apparent competitive exclusion becomes more likely the greater the handling time. In the system examined here, the highest potential values of d in the parameter space are likely to imply extinction of a consumer species in a fluctuating environment. Thus, the total realistic range of mortalities is likely to be smaller than implied by the deterministic model. As a result, the effective fraction of consumer mortality rates leading to resource exclusion is even larger than indicated in the figure. In systems that may exhibit cycles ($h > 1$ in this case), the exclusion conditions would have to be determined numerically when d is sufficiently low because of population cycles. This is illustrated in the following subsection.

6.4.2 Interactions in unstable systems with type II responses

It is useful at this point to illustrate the types of interactions that arise in models with biotic resources when nonlinear functional responses are considered in conjunction with resource exclusion and/or unstable equilibria. The results are again based on eqs (6.1b) above after modification by substituting multi-resource type II functional responses for the type I responses. Here, the analysis must be numerical. The nonlinearity of the interaction can be illustrated by examining population sizes following different magnitudes of a press perturbation to a neutral parameter (per capita mortality, d will again be used below). The results provide examples of how population densities change with the mortality rate of one consumer, and how the resulting competition coefficient (eq. (3.2)) changes. A full exploration of parameter space for the model is beyond the scope of this work, but the examples presented are not exceptional. They show a very complicated nonlinear response to a range of magnitudes of press perturbations applied to one of the competitors.

In these examples, the consumers each have a shared and an exclusive resource, and all resources have a very low rate of external immigration. Here the conversion efficiencies (B_{ij}) are again set to unity, and the linear functional responses have been changed to type II responses with equal handling times ($h = 2$) for all consumer–resource pairs. Other parameters are given in the figure legends. In the two systems illustrated in Figure 6.7 and Figure 6.8 both consumers spend the majority of their time 'handling' resources when all resources are near carrying capacity. (Recall that 'handling' includes digestion.) The resource immigration rate keeps the resources from going extinct, but it is low enough to have minimal impact on equilibrium abundances.

Figure 6.7 examines competition in an example of the 3-resource system described in the previous paragraph. For both consumers the shared resource is characterized by $C = 1/3$ and the exclusive resource by $C = 2/3$. The baseline mortality rate ($d_1 = d_2 = 1/4$) is low enough to produce unstable equilibria for most of the systems being plotted. (Other parameters are: $r_1 = r_2 = r_3 = 2$; $K_1 = K_3 = 1$; $K_2 = 2$; $b_{ij} = 1$; and $d_2 = 0.25$.) Panel A (which is identical to Figure 3.2 in Chapter 3) examines the competition coefficient α_{21} (eq. 3.2) as a function of the mortality rate of consumer 1 (d_1). The line corresponds to an analysis based on the equilibrium points, while the dots are based on mean populations determined before and after a small (1%) change in d_1. The two points characterized by the largest death rates in panel A are stable systems, which is why the point values describing the completion coefficient lie on the equilibrium line. Discontinuities in the equilibrium formula for the competition coefficient occur at four different points. Two of these (at $d_1 = 1/8$ and $d_1 = 11/28$) correspond to points at which one resource goes extinct, while the other two (at $d_1 = 1/20$ and $d_1 = 13/44$) are values for which a small increase in d_1 has no effect on the equilibrium population size of consumer 1 making the denominator of eq. (3.2) zero. The corresponding MacArthur system (equal C_{ij} values, no handling time) yields $\alpha_{ij} = -1/3$ for all cases. Panel B shows both the equilibrium (lines) and numerically determined mean densities (dots) of the consumers over the range of d_1 that allows persistence of consumer 1. Panel B also shows equilibrium resource abundances. There are many cases where the effect of a change in d_1 on mean consumer densities has a sign opposite to that predicted by the equilibrium analysis, as well as cases where the effects on the mean and equilibrium population sizes differ greatly in magnitude.

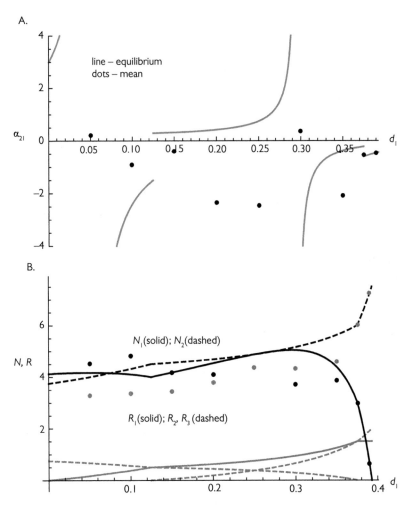

Fig. 6.7 (Note: Panel A is identical to Figure 3.2.)
A. The competition coefficient based on equilibrium densities (line segments) and mean densities (dots) for eqs (6.1b) modified by a type II response. The integrals determining the mean were evaluated for a selection of points spanning the range that allowed coexistence of the two species. Positive values of the point location mean that the two consumer species change in the same direction in response to increased death rate of species 1; they change in opposite directions for negative values. The parameter values are: $r_1 = r_2 = r_3 = 2$; $k_1 = k_3 = 1$; $k_2 = 0.5$; $C_{11} = 2/3$; $C_{12} = 1/3$; $C_{22} = 1/3$; $C_{23} = 2/3$; $B_{ij} = 1$; $I_j = 0.00001$; and $d_2 = 0.25$. All handling times are $h_{ij} = 2$. This graph has four discontinuities corresponding to where a resource becomes extinct ($d_1 = 0.125$ and $d_1 = 0.3928$), and where greater mortality of consumer 1 does not alter its density, at $d_1 = 0.05$ and $d_1 = 0.2954$. The system exhibits cycles at all of the black points except for the two highest values.
B. The population densities of consumers and resources corresponding to panel A. The two black lines are the equilibrium densities of consumer 1 (solid) and consumer 2 (dashed). The mean densities of each consumer species corresponding to these lines are given by the dots for selected values of d_1 (grey dots for consumer 2 and black for consumer 1). The (lower) grey lines are the abundances of the three resources. The dashed lines are for resources 1 and 3, with the decreasing line corresponding to resource 3. The solid grey line is the density of resource 2.

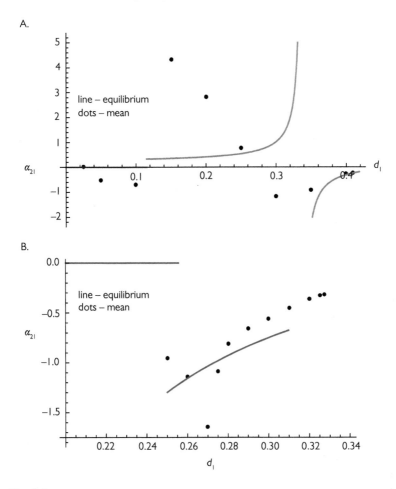

Fig. 6.8

A. The value of the competition coefficient (eq. (3.2)) based on both equilibrium densities (line) and actual mean densities (black dots), for a 2-consumer–3-resource system described by eqs (6.1b) modified to have multi-species type II responses. The parameter values for eqs (6.1b) are: $r_1 = r_2 = r_3 = 2$; $k_1 = k_3 = 1$; $k_2 = 0.5$; $c_{11} = 0.9$; $c_{12} = 0.6$; $c_{22} = 0.6$; $c_{23} = 0.9$; and $B_{ij} = 1$; The handling times (not shown in eqs (6.1b)) are $h_{ij} = 2$. Immigration of resource and baseline mortality are $I_j = 0.00001$ and $d_2 = 0.3$. Note that the abundance of consumer 1 is greater than that of consumer 2 at the value of $d_1 = 0.3$, where $d_1 = d_2$, even though the system is perfectly symmetrical for these values. There are two cyclic attractors here having asynchronous dynamics of the two species but reversed relative abundances; only one is represented here.

B. Competition in another shared-exclusive resource system; this one is identical to that illustrated in Figure 6.7 except that the attack rates of the shared and exclusive resources have been reversed for both consumers; $c_{11} = 1/3$, $c_{12} = 2/3$, $c_{22} = 2/3$, and $c_{23} = 1/3$. The baseline mortalities are 0.25 for both species; at this point the shared resource is extinct in the absence of immigration (and only present at very low levels given the resource immigration rate assumed here (as in Figure 6.7). This means that there is no competition. Competition occurs at d_2 values significantly above 0.25, and cycles characterize this range. The cycles allow both consumer species to persist above the value ($d_1 = 0.30986$) where species 2 would be predicted to go extinct assuming a stable equilibrium.

Figure 6.8 illustrates the competition coefficient α_{21}, as a function of d_1 for 3-resource models that differ slightly from those used in Figure 6.7. In Figure 6.8A, the relative use of the shared resource is higher for both competitors than in Figure 6.7, and the absolute capture rates of all resources are also somewhat higher; $C_{\text{exclusive}} = 0.9$ and $C_{\text{shared}} = 0.6$. The figure also assumes somewhat higher baseline mortalities for both consumer species ($d_i = 0.3$). All other parameters are identical to those in Figure 6.7. The result for the corresponding MacArthur model without handling times would be a competition coefficient (eq. (3.2)) of -0.5455, regardless of the mortality rate, provided that the mortality rate did not cause resource exclusion. The results in Figure 6.8A again show a complicated pattern of change in α_{21} with altered d_1. The lower discontinuity in the blue line corresponds to the lowest d_1 having a positive equilibrium for the shared resource. However, the shared resource is actually present for a range of d_1 values where the equilibrium is zero; this is due to population cycles. The black dots at mortalities of 0.05 and 0.1 indicate strong competition at these low mortalities. The discontinuity in the blue line just below $d_1 = 0.11538$ corresponds to the equilibrium density of the shared resource becoming positive; above this mortality rate, consumer 1 exhibits a hydra effect, so greater d_1 increases the mean densities of both consumers. However the actual competition coefficient (taking cyclic dynamics into account) is much larger than that for the equilibrium analysis until mortality d_1 approaches 0.3, the same value as d_2. When both consumers have identical death rates of $d_j = 0.3$, there are two alternative attractors, each with asymmetrical cycles in which one consumer has a greater mean abundance than the other (only one of the two attractors is represented on the figure). The discontinuity in the blue line at the higher d_1 occurs where the hydra effect disappears (and d_1 therefore has no effect on the abundance of consumer 1 at this precise density). The system becomes stable when d_1 is slightly below the value at which consumer 1 goes extinct (given by the right-most value shown on the graph). This stable equilibrium makes the mean population correspond to the equilibrium value given by the blue line.

Figure 6.8B has a considerably different pattern of competitive effects at different mortality rates than Figure 6.8A. The parameters are identical to those of Figure 6.7 except that the attack rates (C values) of the exclusive and shared resources have been reversed ($C_{11} = C_{23} = 1/3$; $C_{12} = C_{22} = 2/3$), giving a system with significantly higher overlap. The competition coefficient based on MacArthur's formula is -0.8889. The shared resource is extinct for a wide range of lower d_1 values, resulting in a competition coefficient of zero. Cyclic dynamics only occur for relatively high d_1 values (0.27 and greater in the figure). The effective competition coefficient is again significantly different from what the equilibrium formula, eq. (3.2), would suggest. The actual system can persist for d_1 values greater than what the equilibrium analysis suggests because of the presence of cycles.

In summary, it is clear that the presence of saturating functional responses greatly alters both apparent competition between resources and the interaction between their consumers. The conclusions of McPeek (2019a), that type II responses in stable systems reduce apparent competition and otherwise do not alter the interaction, are incorrect. Neglecting the effects of cycling means neglecting many of the most significant effects of type II functional responses on the interaction between consumers. While McPeek (2019a) is the only one of the six recent articles discussed in Chapter 4 to strongly advocate a resource-based approach, some of his specific assumptions and analyses fail to reflect the main differences between models with and without explicit resources. This is no doubt in part a reflection of how little attention has been given in other recent articles on competition to the many older results on resource-based phenomena.

Restricting studies of competition to the special case of two competing species has been criticized repeatedly (including earlier in this book). The next part of this chapter will briefly examine how the presence of one or more additional consumer species alters the interaction of a given pair of consumer species.

6.5 Interdependence of competitive effects with more consumers

Another phenomenon first noted in the 1970s was that the 3-species Lotka–Volterra model allowed a wide range of outcomes that had not been observed in similar 2-species models. Two of the first studies of unusual dynamics in 3-species competition systems were May and Leonard (1975) and Gilpin (1975), who noted cyclic outcomes. An earlier analysis by Strobeck (1973) had wrongly suggested that cases with unstable equilibria having positive abundances of all three species necessarily implied competitive exclusion. Additional outcomes of 3-species LV competition were explored by Hallam et al. (1979). Such systems were later shown to be capable of exhibiting several different limit cycles for a given set of parameters (Gyllenberg et al. 2006, Gyllenberg and Yan 2009). However, these and most subsequent studies of multi-species systems treated the interaction of a given pair as being independent of the presence of other competitors, and have used the LV model, thus ignoring resources. This tendency to ignore resources when there are three or more competitors has continued through the past decade (e.g., Saavedra et al. 2017; Levine et al. 2017; Hart et al. 2018; and many others).

The type of intransitive ordering of competitive abilities that is best known for producing cyclical dynamics in 3-species systems (Laird and Schamp 2006, 2015) requires a large interference component in the pairwise interactions. The pairwise exclusion that underlies the competitive ordering requires some interference competition, but most models of this phenomenon have not included a separate exploitative component. Exploitation is usually required for interference to be evolutionarily favoured. Huisman and Weissing (1999, 2001) is an exception to this, and they provide an additional example where having three competitor species and three nutritionally essential resources allows a wide range of non-equilibrium dynamics. No interference competition is required for this outcome. However the complex unstable outcomes require that every consumer species be best at consuming an essential nutrient that is needed in intermediate amounts, something that is evolutionarily unlikely.

A resource-based approach suggests that in almost all cases, adding a third consumer will alter the interaction of the original pair. In general, both the scaled (eq. (3.2)) and unscaled (eq. (3.1)) responses of one consumer species to change in a neutral parameter of the second species are changed by the presence of additional consumer species. In most cases this dependence will occur even if the added consumer only affects resources used by just one of the original consumer species. This comes about because a change in the abundance of one of the original consumers implies altered resource densities, which, as detailed above, usually changes the per capita competitive effect transmitted via each of the resources. This interdependence of competition coefficients has been shown to affect arguments about the limiting similarity in sets of three competitors that have exclusively exploitative competition, even in the MacArthur model (Abrams and Rueffler 2009). The dynamics of multi-species consumer–resource systems with both exploitative and interference effects remain largely unexplored.

Golubski and Abrams (2011) examined the effect of multiple species in changing pairwise interactions in a variety of food webs. Recent work by Letten and Stouffer (2019) also supports the presence of effects from other consumers on the competition coefficients between a given pair. Even cases with nutritionally substitutable, non-interacting, logistically growing resources that are consumed with linear functional responses by all consumer species can have pairwise interaction coefficients that are affected by the presence or absence of an additional consumer species.

A simple way to determine whether a third consumer species could alter the interaction between two other consumers in the models presented above is to examine the effects of different fixed abundances of the third consumer. (Fixed abundance could come about because of limitation of that third consumer by a specialist food-limited predator, combined with consumption of other substitutable resources not explicit in the model.) In the 3-resource model used here, the effect of such a competitor at a stable equilibrium is equivalent to reductions in the intrinsic growth rates, r, of the resource(s) it consumes. It is easy to show that, if such a species consumes the shared resource in the system considered above, it would lead to exclusion of the resource over a much wider range of other parameter values; this eliminates competition between the original two species. Of course, elimination of the resource is not necessary to produce large changes in the competition coefficients that characterize the original two consumers when the resources have non-logistic growth.

6.6 Other neglected aspects of consumer–resource models

The simple models considered here lack many near-universal features of natural systems. A seldom-noted feature of the MacArthur model is its assumption that resources are substitutable; i.e. the per capita growth rate is an increasing function of a weighted sum of the individual consumption of rates of each resource. Consuming two or more resources has repeatedly been shown to imply a nonlinear effect of the consumption rate of one resource on the consumer's growth rate (Leon and Tumpson 1975, Tilman 1982). The addition of adaptive adjustment of intake rates (which is particularly likely for non-substitutable resources) further changes the nature of the interaction between consumers in such system (Abrams and Shen 1989). For the most part, competition theory that deals with interdependence of resource effects on consumer fitness remains focused on the simple 2-resource models considered by Leon and Tumpson (1975) and Tilman (1980, 1982). These are characterized by functional responses that are only affected by the abundance of one resource. Even if the resources are nutritionally substitutable, either adaptive switching behaviour (Abrams and Matsuda 2003, 2004) or digestive constraints (Abrams 1990a, b) are sufficient to cause the abundance of one resource to affect the consumption rate of another, and thus, to affect competitive interactions involving either resource. Properly accounting for these scenarios requires models of the dynamics of behaviours that determine resource choice. This is another largely neglected topic in competition theory (see Chapter 3 and Abrams et al. 2007).

Another omission from most recent competition theory is that age and stage structure are lacking in most models, whether or not they contain resources. De Roos and Persson (2013) summarize the case for inclusion of population structure in models. Determining the nature and outcome of competition becomes more complicated in these cases, and there is a possibility of alternative stable equilibria in many models, even with a single consumer species (Abrams and Quince 2005; Schreiber and Rudolf 2008; de Roos and Persson 2013). Population structure makes the representation of competition more complicated, even in the absence of explicit resource dynamics. As a result, the many potential impacts of different types of population structure have received little attention. The presence of between-individual variation in resource use, even if this is not related to size or age, also alters competitive interactions.

Most traditional models of competition also neglect temporal variation in the parameters of the model. The majority of the theory regarding competition in variable environments has not been based on traditional differential equation models of seasonal environments; Li and Chesson (2016) is one of the few exceptions to this. Chesson has produced a large body of important work on competition in temporally variable environments (Chesson and Warner 1981; Chesson 1994, 2000b, 2003, 2018), but most of this lies outside of the ordinary differential equation framework that has dominated the rest of competition theory. (The theory developed by Chesson has recently been reviewed by Barabás et al. (2018), Barabás and D'Andrea (2020), and Chesson (2020a, b).) Chapters 8 and 9 will take a closer look at the impacts of seasonal variation in parameters in the context of simple consumer–resource theory.

A final neglected element of competition theory is the impact of changes in non-neutral parameters—those that affect the per capita growth rate of more than one species. Once a consumer–resource framework is adopted, there are usually a number of such parameters. In MacArthur's very simple consumer–resource model, every per capita resource capture rate parameter (denoted c_{ij}, or C_{ij} in the models considered here) is non-neutral. A beneficial (i.e., fitness-increasing) change in any of these parameters need not produce the same direction of change in population size as a beneficial change in a neutral parameter. (The same is of course true of detrimental changes.) Some of the effects of changes in capture rates in competitive systems were discussed in Abrams (2002, 2003, 2004a, 2009d) and Abrams et al. (2003). One of the most important features of a predator's (consumer's) capture rate is that adverse changes can increase the predator's population size; this happens when the resource is biotic and is 'overexploited'. The latter term simply means that the resource abundance is low enough that the reduced resource abundance (R) from increasing capture (C) is sufficient to reduce the product, CR. In multi-resource systems, the meaning of overexploitation is that increases in one or more capture rates will result in a lower total intake at equilibrium. In some competitive systems, this means that a beneficial change to one consumer can increase the population size of its competitor, or that a detrimental change to the first can decrease the population size of its competitor.

The 2-consumer–2-resource MacArthur model (eqs 6.1a) displays these counterintuitive changes when the equilibrium resource densities are below half their carrying capacity in the original system (Abrams 2002, 2003). Figure 6.9 provides an example. Each consumer in the system illustrated initially has a fourfold greater consumption rate of its preferred resource, but the consumers are otherwise equivalent. This yields a competition coefficient at equilibrium of 0.4706, which is significantly

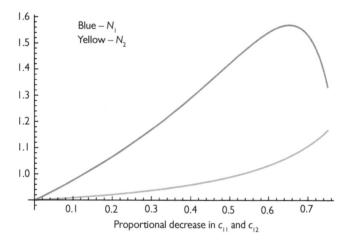

Fig. 6.9 The changes in the abundances (N_1 and N_2) of two consumer species in a MacArthur model (eqs (6.1a) with two consumers and two resources. The initial parameters are: $r_1 = r_2 = k_1 = k_2 = b_1 = b_1 = 1$, $c_{11} = c_{22} = 0.8$, $c_{12} = c_{21} = 0.2$; $d_1 = d_2 = 0.1$. The x-axis gives the fractional reduction in consumer 1's capture rates, c_{11} and c_{12}. The graph ends at the x-value (0.75) where consumer 1's equilibrium drops to zero, as explained in the text.

greater than that of the traditional case of multiple resources and Gaussian utilization curves separated by one standard deviation of the curves (see Figure 7.1). Figure 6.9 shows how the two equilibrium consumer abundances change when consumer 1 experiences equal proportional declines in its consumption rates of both resources. The fractional decline is given on the x-axis. Consumer 1 actually increases in abundance with proportional decreases in its consumption rates, and increases much more than its competitor, whose consumption rates remain constant. Consumer 1 benefits from reduced overexploitation of its preferred resource. Consumer 1 begins to decline once its proportional decrease in consumption rates exceeds approximately 65.3%. The graph ends at a proportional decline of 75%, because, at exactly that point, several discontinuous changes occur. Consumer 1 goes extinct, consumer 2 more than doubles in abundance to a density of 2.5, and consumer 2's preferred resource goes extinct. (Note that both species had initial abundances of 0.9.) The exact pattern of change depends on the degree of initial resource partitioning and the efficiency of the consumers, as measured by the extent to which they reduce the resource population sizes. Abrams (2002) provides some additional examples.

The example in Figure 6.9 is a good illustration of Lewontin's statement from the first paragraph of this chapter. The pattern of changes in abundance shown here would be totally incomprehensible using a Lotka–Volterra model, or MacArthur's model with resource extinction assumed not to occur.

7

Resource use and the strength of interspecific competition

7.1 Theory regarding the strength of competition

What is the level of competition in natural communities? Do most species experience large decreases in their own population size following large increases in the population sizes or population growth traits of other species with which they share resources? Are most species susceptible to complete elimination by a competitor with a small negative change in their own fitness or a small increase in the fitness of the competing species? More generally, how is similarity in resource use related to these population-level effects? One might think that these broad-brush questions about competition should already have been addressed in empirical studies. While competition has been the focus of many field studies, they have not answered these questions, in part because of the limited duration and spatial extent of experimental treatments. In a relatively recent assessment, Adler et al. (2010, p. 1019) admit that, 'Despite decades of research documenting niche differences between species, we lack a quantitative understanding of their effect on coexistence in natural communities'. One reason for this was stated many years earlier by a theoretical ecologist who chose the laboratory over the field for testing theory: 'the direct study of . . . communities outdoors labors under crippling disadvantages . . . Environmental parameters vary uncontrollably in time and in space, inordinate numbers of . . . species are present, the individual life histories of most of these species and their interactions in pairs or in larger sets are poorly known, and experimental manipulations may be infeasible or doomed to yield ambiguous results' (Gilpin et al. 1986, p. 23).

The answers to the questions raised in the previous paragraph can be approached theoretically as well as empirically. However, some level of theoretical understanding is a prerequisite for being able to refine the questions in a way that allows a meaningful approach by observation or experiment. How should similarity be quantified? Which aspects of similarity have the greatest impact on competitive effects? How broad a spatial and/or temporal scale must be considered to obtain a meaningful estimate of effect size? These questions need to be understood from a theoretical perspective to determine what observations or experiments would be informative in measuring competitive effects. For example, is the presence of additional competing species likely to significantly change the per individual effect of one

Competition Theory in Ecology. Peter A. Abrams, Oxford University Press. © Peter A. Abrams (2022).
DOI: 10.1093/oso/9780192895523.003.0007

consumer species on a second? The answer affects the applicability of any empirical measurement of competition between the two species. More generally, theory can suggest what types of organisms might be likely to experience stronger or weaker competition. And, because there are many potential measures of competitive strength, different perturbations to a system may change the various measures of competition in opposite directions. This first section deals with theory that discusses the relationships between measures of the similarity in resource use among co-occurring species and the population level interactions between them.

7.1.1 Theory from the early 1970s

Early natural history observations by Hutchinson (described in his 1965 book) suggested that significant differences between species were needed for them to coexist, and that competition between species might lead to a constant ratio of sizes in sets of three or more competitors. Much earlier, Volterra (1931) had argued that, if there were several competitors for a single limiting factor, the one capable of surviving and reproducing at the lowest level of that factor would displace all others. Hutchinson's later observation suggested that a different sort of limitation might be needed for coexistence of species that consumed many resources (as most consumers do). From 1967 through 1974 Robert MacArthur, Richard Levins, and Robert May, inspired by these and other early observational results, established what is still a widely accepted framework for thinking about the relationship between similarity in resource use and the strength of interspecific competition, and the consequences of that relationship for coexistence.

The initial analysis by MacArthur and Levins (1967) was based on a Lotka–Volterra model with additional assumptions about how the difference between two consumer species in their mean values of a consumption-related resource characteristic translated into a competitive effect. These assumptions were also part of MacArthur's (1970, 1972) consumer–resource model. All three of these analyses assumed a constant homogeneous environment. All argued that coexistence was very unlikely (although not strictly impossible) for species that had a high enough overlap in resource use, but one that was significantly below complete similarity. The requirement for substantial differences in resource use was later suggested to be more stringent by May (May and MacArthur 1972; May 1973, 1974), using Lotka–Volterra models that included environmental variation in consumer per capita growth rates. In these works, a level of similarity well below complete identity always led to exclusion. This result was the basis of a large amount of later theoretical work on the idea of a 'limiting similarity' of competitors.

All these initial studies contained a core set of assumptions that were not based on any supporting evidence. First, the resource use of each species could be characterized by plotting their maximum resource capture rates as a function of a single variable (often a measure of the size of the resource item). Secondly, this function was Gaussian. These two assumptions seemed to make it possible to obtain a simple analytical formula for the amount of competition between species just by measuring the distance between the peaks of the 'utilization curves' of the two competitors. Figure 7.1 illustrates the basic idea. Here, the trait distinguishing resources is given on the x-axis, and the maximum per capita consumption rates of three evenly spaced consumers is given by the curves, $C_i(x)$ for consumer species i. In the figure, these are Gaussian curves with a standard deviation of 1, and with mean values also separated by 1 unit on the resource axis. The competition coefficient (formula (3.2)) proposed by MacArthur and Levins (1967), and still widely used today is:

$$\alpha_{ij} = \frac{\int C_i(x)C_j(x)}{\int (C_i(x))^2} \tag{7.1}$$

This expression has a maximum value of 1 when the two utilization curves are identical. The assumption of Gaussian utilization curves having a common mean and a common standard deviation then makes eq. (7.1) equivalent to $\exp(-D^2/W^2)$, where D is the distance between the curves of the two species, and W is the (common) standard deviation of the curves. In the situation illustrated (with $D = W = 1$), the competition coefficient between adjacent species is approximately 0.368. This allows coexistence for a reasonably broad range of other parameters when there are only two competitors, but that range is greatly reduced in the case of three or more species as in Figure 7.1. Even coexistence of two species rapidly becomes less likely when the D/W declines below unity; $\alpha \cong 0.779$ when $D/W = 0.5$.

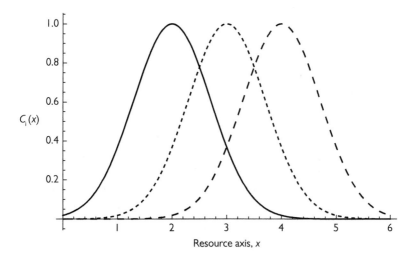

Fig. 7.1 A plot of standard resource utilization curves for three competitors as envisioned by MacArthur. The curve plots the value of the consumption rate of resources having the quality x (e.g., size) given on the x-axis. In the figure the standard deviation of these Gaussian curves is unity, as is the spacing between the curves. This spacing was MacArthur and Levins' (1967) 'limiting similarity'.

The above formula from MacArthur and Levins' (1967) Lotka–Volterra model turned out to be identical to the Lotka–Volterra approximation to MacArthur's (1970, 1972) consumer–resource model under the assumptions of an infinite number of substitutable resources having identical growth parameters and identical nutritional values to the consumer. This is the extension of eqs (3.4) to a large number of resources. The effects of unequal values of the parameters on the competition coefficients are shown in Chapter 3 immediately below those equations. MacArthur also used the further (unstated) assumption that extinction of one or more resources did not occur. Neither uniform growth and nutritional value nor absence of resource extinction is likely to apply generally, as has been noted in preceding chapters. The derivation of the formula for competition coefficients required assumptions about independent logistic resources of equal value to the consumer that were consumed by linear functional responses, and whose consumption rates combined additively to determine consumer fitness. All of this was required to reduce the model to Lotka–Volterra form (Schoener 1974a; Abrams 1983b).

A key feature of extensions of the 2-species Lotka–Volterra model to multiple competitors was that the interaction of any pair of species was independent of the presence of all the other consumer species. The three-or-more consumer version of the MacArthur consumer–resource model suggested that this was indeed the case. However, it should be noted that this is again dependent on the linear per capita growth rates of that model and on the continued existence of all the resources. A third consumer species often increases the total consumption rate of some resources, which has the potential to cause their extinction.

For the case of similar Gaussian utilization curves for all consumers, eq. (7.1) means that the range of a neutral consumer parameter (e.g., its per capita death rate) that allows coexistence is narrow when the distance between the utilization curves is significantly less than the standard deviation of the curves, particularly with three or more species. Early work (May and MacArthur 1972, May 1973, 1974) added stochastic variation to the consumer per capita growth rates and argued that this would ensure extinction of one competitor in an array of three or more, if the species had utilization curves separated by slightly less than one standard deviation of the curves. The stochastic forms of the models assumed that conditions near an equilibrium point also applied when one or more of the competing species was close to extinction (Abrams 1976, 1983b; Turelli 1978).

The evolution of traits related to resource use is also expected to influence observed levels of competition. However, in the early 1970s the evolutionary response to competition had not been analysed rigorously. Brown and Wilson (1956) argued that competitors would diverge in characteristics related to resource use. Grant (1972) suggested the possibility of convergence, but did not present a mechanism based on resource use. MacArthur did not analyse the evolutionary aspects of competition, and mathematical theory for those aspects was not developed until later in the decade (Lawlor and Maynard Smith 1976; Slatkin 1980). The evolution of competitors will be discussed in Chapter 11. However, early evolutionary models (Slatkin 1980; Taper and Case 1985; Abrams 1986a) did not predict that species would evolve to the limiting similarity predicted by MacArthur and Levins (1967). Regardless of the direction of evolutionary responses, the end result of those responses must be consistent with any ecological restrictions on the coexistence of very similar species. MacArthur's ecological theory seems to imply that species could only achieve their 'limiting similarity' by convergent evolution of initially more different forms, but there was no mechanism suggested for such a change. If two species initially violated the 'limiting similarity', it is unlikely that they would have diverged evolutionarily at a rapid enough rate to achieve coexistence prior to exclusion. This problem was not noted in the literature from the 1970s. Even if divergence was rapid enough to prevent extinction, there is no theoretical reason why it should stop at the exact amount that was sufficient to produce long-term coexistence.

7.1.2 Early questioning of MacArthur's limiting similarity

It did not take long for the above early results based on MacArthur's approach to be challenged. Theory from the decade following MacArthur's works showed clearly that there were many possible relationships between 'similarity' (however it was measured) and any chosen measure of the 'strength' of competition. Various works (Schoener 1974a, 1976, 1978; Abrams 1975, 1977, 1980a) showed that similarity–competition relationships were affected by: (1) the nature of resource growth, and differences between resources in their population growth functions; (2) the intake rates of each resource as a function of all variables affecting that rate (consumer functional responses); and (3) the effects of those intakes on consumer demographic rates (numerical responses).

The first factor (resource growth function) was acknowledged in MacArthur's original (1970, 1972) work, but was hidden by his decision to concentrate on the simplest case where all resources had identical logistic growth functions. MacArthur briefly noted that abiotic resources produced a different formula for the competition coefficient, but he chose to emphasize that the expression (7.1) was approximately correct when abundances were very close to their equilibrium values. Unfortunately, questions about coexistence depend on what happens far from equilibrium. MacArthur also failed to consider the possibility that extinction of some resources would be caused by efficient consumers. However, Hsu and Hubbell (1979) showed that that competition was changed by such extinctions. A final problem was that the most commonly assumed differentiating factor was the size of the resource items. However this usually leads to unidirectional change in the weighting factors; for example, larger items have a larger nutritional value (a larger b in eqs (3.4) or (6.1)). The values of the resource growth parameters are also known to change systematically with size, and it is highly unlikely that all of these unidirectional changes would exactly cancel each other. Figure 6.3 showed the effects of nonlinear resource density dependence on the competition coefficient.

The second factor limiting the applicability of expression (7.1) was the form of the functional response. The problem with the underlying assumption of type I functional responses was explored in Abrams (1980a), who showed that type II responses made competition coefficients change greatly with consumer population sizes (see Figure 6.4). Most subsequent models continued to assume linear functional responses or to use MacArthur's formula without noting its dependence on this assumption.

The third factor affecting the strength of competition was the nature of the numerical response. Leon and Tumpson (1975) initiated theoretical work on one of the most important aspects of numerical responses; i.e. how different resource intake rates combined to determine per capita growth rates. Leon and Tumpson considered the effects of non-additive contributions of different resources to per capita population growth; Tilman (1982) popularized their approach—like the original work by Leon and Tumpson (1975) he restricted his modelling to the case of exactly two limiting resources. Both works also shared assumptions of independent functional responses to each resource, and linear numerical responses to the intake rate of the limiting resource(s). A broader exploration of the range of numerical responses or their effects on competition has yet to appear.

In any case, studies from the 10 years following MacArthur's (1972) book demonstrated that the effects of the abundances of competing species on each other's growth rate were usually nonlinear. A sustained perturbation to the per capita growth rate of one consumer species that did not affect any consumer–resource interaction directly would have a different effect on its competitors' per capita growth rates depending on the initial abundances of the perturbed species and the competitor(s), and depending on the magnitude of the perturbation. This was one of the main points of the theoretical articles from the 1970s by Schoener and by Abrams. It means that any single measure of the strength of competition may only be a good approximation over a narrow range of initial abundances.

The theoretical arguments for a limiting similarity were also shown to be weak in the 1970s (Schoener 1976, 1978; Abrams 1975, 1977; Turelli 1977, 1978). The results depended on assumptions about population dynamics of consumers and resources that were seldom satisfied, and the stochastic versions of the theory had been improperly analysed. The work suggesting an absolute limit to similarity was based on extrapolating very limited simulation results (May 1973, 1974). This extrapolation turned out to be invalid (Turelli 1977, 1978, 1981). The original May and MacArthur results also assumed uncorrelated environmental effects that were independent of similarity, an assumption that was both unlikely and could reverse the qualitative results (Abrams 1976). Finally, the limiting similarity results with or without stochastic variation assumed the underlying linear functional response/logistic resource growth model of MacArthur, as well as a single resource at each position on the resource axis. Relaxing either of these assumptions could allow coexistence of similar species over a much wider range of relative mortality rates of the consumers (Abrams 1975, 1983b). However, some other potential assumptions made coexistence more difficult than in the MacArthur model.

The idea of simple Gaussian resource utilization functions was also far from a good approximation in the real-world systems where it had been closely examined (Wilson 1975). Thus, although a requirement for coexistence does limit the similarity of species that differ in mean fitness-related parameters, there was no generally applicable rule for the nature of these limits (Abrams 1983b). The difficulty of achieving coexistence for any given measure of similarity depended strongly on the details of the consumer–resource interactions involved. A general problem with using the Lotka–Volterra model for sets of three or more species was that it assumed that pairwise interactions were independent of the presence and abundance of additional competing species. This was shown to almost never be true for simple models based on consumer–resource interactions of homogeneous populations (Abrams 1983a). The presence of these 'higher-order' interactions meant that the relationship between similarity and any measure of competition for any pair of species was almost always changed by the presence of additional competing species.

If most organisms were close to a limiting similarity with their competitors, extinctions should be very frequent, as new mutations or environmental changes with species-specific effects would likely provide a slight advantage to one of the species. There is no general reason why being reduced in population size should increase selection for traits that would save the population. In almost all cases, a trait that provided an individual advantage would have been at least equally advantageous before the process of exclusion began. Mutations with such a trait would become less likely as population size decreased.

Whether or not consumers are close to a limiting similarity, it is important to try to understand the causes of any differences in the strength of competition that are observed for different taxonomic groups or in different ecological circumstances. All of this depends on having a quantitative description of the interaction. Here again, theory from the previous millennium provided a variety of reasons why the assumption of linear competitive effects (from both the Lotka–Volterra and MacArthur consumer–resource models) would often be misleading.

7.1.3 Recent and potential future theory on overlap and competition

The preceding section concentrated on theory published between 1970 and the early 1980s. Some later work has continued to explore the relationship between similarity in resource utilization and competition. In consumer–resource models with living resources, the departures from Lotka–Volterra dynamics and MacArthur's competition coefficients are particularly large when one or more resources may become extinct because of a perturbation to one or more of the consumer species. Abrams (1998) showed that the increasing relationship between similarity in resource use and magnitude of the competitive effect on equilibrium abundance (assumed to be universally true) did not hold for competition between predators in 2-predator–2-prey models when the predators were able to survive on resource abundances that were sufficiently low relative to their carrying capacities. This efficiency of the predators could produce apparent competitive exclusion of one resource in a system with a single predator type, if that predator was moderately specialized. Exceptions to the supposed rule of competition increasing with similarity were even more common in models with competing prey species (Abrams and Nakajima 2007) and in models with an infinite array of independent prey species (Abrams et al. 2008a).

The resource extinctions in these models all represent cases of exclusion due to apparent competition, which is itself a widely accepted phenomenon (Holt and Bonsall 2017). So far as I know, however, the implications of recognized cases of apparent competitive exclusion of resources for competition between two or more consumers have yet to receive any empirical study. This is at least potentially due to the almost complete focus on scenarios with a single predator species when apparent competition is discussed (Holt and Bonsall 2017). The presence of a second predator in a system with two prey does, in fact, reduce or eliminate the apparent competitive effect between the prey, if apparent competition is measured by the effect of a small change in a neutral parameter of one prey species on the abundance of its 'apparent competitor' (see below or Chapter 5). It is possible that apparent competitive exclusion is rare in cases with multiple predator and prey species; if so, determining why this is the case would represent a major advance in our understanding of the nature of both competition and apparent competition in natural communities. However, apparent competitive exclusion in situations with two or three consumers and three or more resources is common in simple models (Abrams et al. 2008a, Abrams and Rueffler 2009). Returning to cases with a single predator, one possibility is that consumers in natural communities seldom decrease their biotic resources by the relatively large amounts required for resource extinction. However, Abrams et al. (2008a) pointed out that empirical work on the effects of predators on prey included many examples with prey reductions of a factor of 10 or more (Shurin et al. 2002).

Apparent competitive resource exclusion is not the only factor that limits the applicability of MacArthur's competition–similarity relationship. Another problem is the idea that the sets of maximal per capita resource capture rates of two consumers are sufficient to determine the competitive effects between them. Almost all functional forms and parameter values in a standard consumer–resource model affect any measure of the strength of competition. The determinants of competitive effects include the form of and parameters characterizing the consumers' functional and numerical response functions, and the resource growth rate functions. The latter include any interactions between different resources. The presence of resource choice behaviour on the part of the consumers also affects the shapes of their functional responses (Abrams 2010a, b) and the amount of competition between them (Abrams 2006a, b). These works primarily discuss behavioural effects on competition in the context of two resources. The theoretical effects of adaptive consumer foraging in the context of a continuous array of resources have not been studied. This is probably because of the lack of a simple closed-form solution for any reasonably widely applicable sets of functional forms for the behavioural model components in these systems. In any case, the lack of uniform applicability for any of the potential functional forms, and the relatively small number of resources per consumer species identified in food web models (Cohen et al. 1990) both suggest that exploring cases with a relatively small number of resource types is likely the most useful approach.

MacArthur's consumer–resource model showed that if there were different nutritional values for different foods, a formula for the competition coefficient in the Lotka–Volterra approximation to that model would entail higher weighting for the more valuable foods. However, most of his analysis of similarity and competition assumed all foods were equally valuable, which was already known to be inaccurate in almost all systems. Schoener (1974a) later re-emphasized the importance of the weighting terms in the full version of MacArthur's formula. Nevertheless, they continued to be ignored. In part this was due to a lack of experimental study of their values at that time. However, the variable that was most commonly assumed to differentiate the use patterns of competitors was resource-item size (e.g. seed size in seed-eating birds). Caloric or nutrient content (which would likely be the most important property influencing effects on demographic rates) are often roughly proportional to weight. However, the handling times usually are not. Thus, the lack of terms modifying the products of attack rate ($C_i(x)C_j(x)$) in eq. (7.1) is a serious deficiency. The weighting terms also include resource growth parameters, which are also likely to differ between different resource types.

Including weighting terms for different resources in a model of competitive dynamics is even more important when the resources are not linearly substitutable in their effects on fitness. The extreme case of two nutritionally essential resources implies that the resource intake vs per capita growth rate relationship differs qualitatively depending on which resource is limiting. Leon and Tumpson (1975) and Tilman (1980, 1982) discussed these sorts of essential resources, but did not have a proper analysis of how they affected a Lotka–Volterra approximation to the system. Tilman (1982) presented an erroneous derivation that purported to show that systems with two nutritionally essential resources could be reduced to a Lotka–Volterra form with fixed coefficients of the interaction terms. Abrams (1987b) later re-examined this case, and a related model in which adaptively flexible behaviour resulted in nonlinear competitive effects at all consumer densities. Chase and Leibold's (2003) widely cited book had a thorough discussion of the importance of relative resource conversion efficiencies in determining competitive effects. However, they assumed linear functional responses for each resource. Adaptive functional responses with nutritionally interacting resources can have a variety of forms (Abrams 1987c), and the models of competition between species with such adaptive functional responses can have a variety of unusual dynamics, some of which were touched upon in Chapter 6. Even without nutritional interactions, the presence of different ratios of energy or nutrient content to handling time in different resource types makes it advantageous to ignore lower-quality foods when the abundance of better foods is high (Pyke et al. 1977). This greatly changes the form of the functional response (Fryxell and Lundberg 1998; Ma et al. 2003). The resulting effects on interspecific competition have largely been ignored.

Most of the above discussion has assumed that competition was purely exploitative. Early empirical studies (reviewed in Schoener 1983) showed that this was far from universally true. Much evidence and purely logical arguments suggest that interference should frequently exist, both within species (DeAngelis et al. 1975; Abrams

1984b, 1992c, 2010b; Abrams and Ginzburg 2000), and between species (Schoener 1983; Abrams 1986a, 2010b). The already wide variety of potential competition–similarity relationships is increased by the possibility of an interference component to the interaction. Interference can affect any or all of the separate functions that make up the per capita growth rate of a consumer. In addition, interference is likely to have a wide range of functional forms for any single component of consumer growth. However, relatively little theory has been developed to describe this range of possibilities.

One of the more common approaches to modelling interference has been to make functional responses 'predator-dependent'. In the case of a type II response this usually involves adding a term proportional to predator abundance in the denominator of the response. This formulation arises when increased predator abundance interferes with the capture process (DeAngelis et al. 1975). Interference reduces capture rates generally in the model, which seems to imply that there is no benefit to the successful 'interfering' individual. However, 'stealing' a food item should increase the per capita food intake of the 'thief', even though it may decrease the mean capture rate for the consumer population. Any model that seeks to understand when interference will be observed or how it will evolve over time should consider these individual-level costs and benefits. However, they are seldom included in ecological models with predator interference. An even more important omission is the scarcity of theory dealing with the combined effects of both intra- and interspecific interference; the preponderance of experimental studies of functional responses deal with a single consumer and single resource (Jeschke et al. 2002).

Cases of two competing predators with interference have occasionally been modelled explicitly, but most examples have simply assumed an additive combination of independent effects of the two predators in the denominator of the functional response. Another possibility included in some models is that the interference increases per capita death rate (e.g., Abrams 1986a; McPeek 2019a), and here again, independence of effects is usually assumed. A broader approach would be to derive mathematical representations of the interference by assuming adaptively variable behaviours of the different species involved. In most cases, intraspecific interference enables coexistence of species having greater overlap in resource use, with interspecific interference having the opposite effect.

Another aspect omitted from most early theoretical works on competition–similarity relationships is the impact of different forms of spatial and temporal variation in resource uptake rates by the various consumers. These had been recognized as influencing the 'strength' of competition relatively early in the modern history of competition theory (Haigh and Maynard Smith 1972; Chesson and Warner 1981; Abrams 1984a), but are still relatively poorly understood (as documented in the next three chapters). In most cases, the measure of 'similarity' between species must take differences in temporal responses between consumers in all resource-related parameters into account. The effects of such differences will usually depend on other aspects of the resource dynamics that affect the rate of response of resource abundances to the temporal changes in interaction parameters. More details are provided in Chapters 8, 9, and 10.

7.1.4 Continued use of outdated similarity–competition relationships

Returning to the simpler case of a single homogeneous and temporally constant environment, the studies from the 70s and early 80s reviewed earlier in this section had shown that no single function based solely on the sets of per capita capture rates of all consumers (e.g., that of MacArthur 1972) would accurately represent competitive effects. The nonlinearity of effects also means that competition between two species is seldom measurable by a single number. Nevertheless, MacArthur's proposed function for an all-encompassing 'competition coefficient' is still being widely used, as noted by Leimar et al. (2013) and Falster et al. (2021).

This raises the question of why the early critiques were and still are being disregarded by so many. The answer to this question is not obvious. There is, of course, no universal answer. The widespread prior use of the Gaussian form has been given as the basis for its use by some of the few authors who have acknowledged that there are other possibilities (e.g. Day 2000). Once a certain amount of theory has accumulated, it gains a life of its own, and continues regardless of demonstrations of its inadequacies or limited range of applicability. And that is particularly true when (as is true here) there exists a broad range of alternatives, none of which has universal applicability. This situation is similar to the use of the logistic and chemostat equations to represent resource growth (as in many of the models discussed in this book).

In a recent article, Falster et al. (2021) cite D'Andrea and Ostling (2016) as showing that 'there are few models that relate traits to fitness via competition for a common resource without including a phenomenological competition function' (Falster et al. 2021, p. 264). This is definitely not the case, unless one adopts an abnormally narrow definition of 'trait'. Every consumer–resource model has a relationship between traits and fitness, at least if demographic and ecological parameter values are regarded as traits. Even if more easily measured traits such as body size are the focus, determining their effect on fitness requires knowledge of their relationship to consumer–resource parameters. The framework Falster et al. (2021) are attacking seems to be the one that I criticized above, consisting of independent (non-interacting) resources that can be arrayed along a one-dimensional axis, and that all have logistic growth and are consumed by predators with linear functional responses. However, there have been a large number of published models that do not assume this. Even by the time that the one-dimensional framework was first being applied by scientists other than MacArthur's students and collaborators, Schoener's (1974b) review of field studies had documented the fact that multi-dimensional niche separation was the norm in nature. While May (1973, 1974) had briefly discussed extending the Gaussian utilization-curve framework to two or more resource dimensions, this has yet to inspire a large body of theory based on multi-dimensional partitioning, in spite of its known predominance in natural systems (Schoener 1974b). At about the same time Roughgarden (1974) showed that different shapes of the consumers' utilization functions produced different shapes of the competition coefficient as a function of similarity. The shapes considered in that work often led to much greater permissible similarity of competitors than suggested by MacArthur (1972). Wilson (1975) showed that the MacArthur formula was inaccurate for several empirical systems.

The point of the Falster et al. (2021) article is that Lotka–Volterra models with MacArthur's competition function do not provide a reasonable description of similarity and competition in plants, which is certainly true in most if not all cases. However, this point (and the remedy of using consumer–resource models rather than the Lotka–Volterra) had been put forward 50 years ago in Grenny et al. (1973) and Williams (1972). Falster et al. (2021) state, 'The more promising approach will be to model depletion of resources by populations in relation to their traits, with its consequences for fitness landscapes and competitive exclusion'. However, nearly 50 years earlier, in the same journal W.E. Neill (1974, p. 407) had stated, 'it may be more profitable now to set the Lotka–Volterra models aside, and to begin to emphasize mechanisms and dynamic aspects of resource utilization as a reasonable alternative'. It is clear that history has been forgotten.

7.2 Laboratory studies of competition

The strength of competition should be easiest to measure in a comprehensive way using species that are small in size, have short generation times, and can be maintained under controlled conditions. These requirements led Gause (1936) to study competition between two protozoans in laboratory culture vessels, and Park (1948) to study competition between pairs of flour beetle species in small jars. Both authors observed large effects of interspecific competition on abundance, including exclusion. Several early laboratory studies of interspecific competition during the MacArthur era featured work on *Drosophila* species. The seminal work by Ayala (1969) has already been mentioned in Chapter 1. By quantifying relative short-term changes in responses from a wide range of starting densities, his results provided clear evidence of strong nonlinearity in the interaction. However, Ayala's initial interpretation of the results was that they invalidated the competitive exclusion principle. Gilpin (Gilpin and Ayala 1973) reinterpreted these experiments, and Gilpin led a group that initiated an expanded set of experiments involving almost thirty *Drosophila* species. These culminated in a study by Pomerantz et al. (1980) on intraspecific competition in close to 30 species and a related study by Gilpin et al. (1986) on a related interspecific system involving up to ten species in a single treatment. The intraspecific results demonstrated strong nonlinearity. Although Pomerantz (1981) argued otherwise, the intraspecific findings strongly suggested similar nonlinearity in interspecific competition (Abrams 1983a).

The interspecific competition experiments described in Gilpin et al. (1986) revealed that laboratory studies often do not escape many of the problems they attributed to studies in natural environments (see the quotation in Section 7.1). The experiments often could not be run for long enough to distinguish outcomes of highly unequal coexistence from exclusion. They also did not allow a large enough number of treatments to examine the full spectrum of competitive effects over a wide range of abundances for any given pair of species. Nevertheless, the results confirmed that alternative multi-species outcomes were possible in some cases, and that pairwise interactions were not independent of the presence of additional species. Coexistence of more than three species did not occur in the 30 treatments initialized with ten species present (Gilpin et al. 1986), and survival of only two species was the most common outcome. This probably reflected the relatively (but not completely) homogeneous resource conditions in the laboratory environment.

Other studies of laboratory microcosms from the 1970s confirmed that nonlinear effects were common. Neill (1974, 1975) constructed a series of replicated aquatic microcosms involving four microcrustacean consumers of algae. He studied the impacts of removal of different subsets of the species on the abundances of the remaining species. He also examined the effects of adding one or two predatory fish; each of these treatments had different per capita impacts on the four invertebrate species. The results of both types of manipulation were inconsistent with the Lotka–Volterra model in that effects of consumers on each other were non-additive, and the effects of different magnitudes of perturbation were not linearly related. An earlier study of competition between four protozoan species by Vandermeer (1969) had concluded that the Lotka–Volterra model provided a good description of their dynamics. This may have been true, but the experiments constituted a relatively weak test of this claim.

The laboratory systems described above all revealed high levels of competition between many pairs of species. However, these systems lacked the spatial variation that is found in almost all natural systems, and they were also characterized by a relatively small or unknown number of resource types. The earlier laboratory studies of competition (Gause 1936; Park 1948) also involved one or a very small number of resources.

More recent years have seen many laboratory studies of competition, and I will not attempt to review all of them here. A number of the studies have examined competition among groups of species where the small spatial scale in the lab is not so different from natural conditions. Bacterial communities have proven to be useful for such multi-species experiments (e.g., Friedman et al. 2017; Abreu et al. 2020). While not all results are consistent with Lotka–Volterra models, some studies have had a high degree of consistency with those simple models (Friedman et al. 2017). However, the resource utilization patterns of the bacteria are usually poorly known, and precisely controlled changes in the mortality of a single species/strain are usually not possible.

7.3 Field studies of competition

For most ecologists, the question of how much interspecific competition is experienced by species must be settled by controlled field manipulations. The review by

Adler et al. (2010), quoted in the first paragraph of this chapter, suggested that these studies had not been conclusive. This section will present a selective and brief review of these studies.

7.3.1 A historical review

There is little doubt that the first prominent experimental field study of competition, Connell's (1961) experimental work on two barnacle species, would not have been carried out had there not been highly suggestive evidence of a large effect of at least one of the species on the other. Connell's article was followed by many field studies on competition over the next two decades. Most of these involved addition or removal of a single species from a pair. The two comprehensive reviews of field studies of competition, which both appeared 22 years later (Schoener 1983; Connell 1983), found sizable competitive effects in a large majority of the studies. Schoener's review included 150 separate studies, and he remarked that the number seemed to be increasing faster than exponentially. Most of the studies measured the change in population size in one species following addition or removal of a second species. In most studies, the species were closely related taxonomically. However, the finding of many large effects was likely due to a selection bias for studying systems in which there was already some evidence suggesting large effects. In addition, experimental studies are almost always carried out on a small spatial scale, and these have uncertain applicability to effects on larger spatial scales (where effects are likely to be smaller).

Regardless of the large body of experimental studies published by 1983, many still argued against the importance of interspecific competition (many chapters in Strong et al. (1984)). Grover (1997) provides a more detailed review of field (and experimental) studies from the last millennium. Field experiments on competition have continued, and Denno et al. (1995) and Kaplan and Denno (2007) found a high frequency of large magnitude effects in studies of phytophagous insects, a group which had been thought to be relatively free from competition (Hairston et al. 1960).

MacArthur's own fieldwork primarily involved competition between bird species for food, but was purely observational, and did not provide estimates of competitive effects. Later research by his student, Martin Cody (1973), greatly expanded this observational approach to other bird communities. Unfortunately, long-term manipulative experiments in the field were impractical for the species he studied, so the applications largely involved the assumption that there was a single limiting category of resources, and that simple measures of overlap described competition. One of the more rigorous early observational studies involved different sets of species of seed-eating sparrows in different habitats (Pulliam 1975), and was published in *Science*. It appeared to support MacArthur's theory, but was later shown by the same author not to be consistent with it (Pulliam 1983).

Terrestrial plant communities have been favourite subjects for those studying competition (Keddy 2000; Kraft et al. 2015; and many others). Large interspecific effects have frequently been described in plant communities (Levine et al. 2017).

However, a recent meta-analysis (Adler et al. 2018) found that intraspecific competition exceeded interspecific on average by a factor of four to five. Unfortunately, most field studies have been very limited in their spatial and temporal scales, so have shed little light on the strength of interactions at more biologically relevant scales. Adler et al. (2018) was consistent with this generalization, in that the relative magnitude of intraspecific competition was greater in observational than experimental studies, and in studies that quantified growth over the full life cycle.

The literature on plant competition is too extensive to provide a short summary of it here. The immobility of the adult stage and the observable effects of one individual on other nearby individuals make competition especially evident in terrestrial plants. However, the involvement of spatial resource segregation at moderate to large spatial scales calls into question estimates of population-wide effects based on experiments involving much smaller spatial scales; i.e. the vast majority of studies. The spatial proximity of interacting individuals in terrestrial plant communities also allows a variety of positive effect pathways to co-occur with negative effects via other mechanisms (Bertness and Callaway 1992). As a result, estimating interactions at a proper spatial scale and studying all potential pathways of effects both are very difficult and seldom occur. Hart et al. (2017) have recently emphasized the importance of spatial scale in assessing the causes of coexistence, and Clark et al. (2010) describe several neglected aspects of niche partitioning in plant communities.

Space is not the only complicating feature of terrestrial plant competition. Pollination, particularly in insect-pollinated plants, provides another coexistence mechanism, with recent work suggesting that it often provides a rare-species advantage (Wei et al. 2021). Interactions of various sorts with soil-based fungi and microbes are increasingly being recognized as having important and varied roles in determining the outcome(s) of competition (Klironomos 2002; Cardinaux et al. 2018; Ke and Wan 2020). Lekberg et al. (2018) have recently carried out a meta-analysis suggesting that plant–soil feedback could promote coexistence, but its extent is usually not known. Plants are also subject to herbivory by a variety of organisms, which may often have a decisive effect on coexistence (Levi et al. 2019).

In the aquatic realm, phytoplankton species are more easily studied, and have a much shorter generation time than most terrestrial plants. Here there exists an unusual mix of theory and experimental work that has provided good insight into competitive interactions in some systems (Litchman and Klausmeier 2008; Litchman et al. 2010), although coexistence at large spatial scales has again received too little attention.

7.3.2 An illustrative example: competition between hermit crabs

Among animals, most studies of competition have failed to explore resource dynamics in spatially extended systems, and the effects that this has on competition. Such an approach is particularly necessary in many marine species, which share with plants the characteristic of dispersal during a life history stage during which competition is unlikely to occur. Hermit crabs present a case where resource use is easy to observe and manipulate, and the resources are not self-reproducing. Early in my graduate studies, I decided that I would study competition for shells between hermit crab species for my Ph.D. Recent work by Richard Vance (1972) had documented that empty shells were a major limiting resource for a set of three species, and because crabs ceased growing when their shell was too small, there seemed to be little possibility of other major limiting resources. The number of species present in hermit crab communities varied on a large spatial scale, but the same communities occurred over large geographic zones, and all had sufficiently few species that community-wide studies seemed possible. The outer coast of the Pacific Northwest had similar species from Central California to Alaska.

The fact that hermit crabs do not kill snails to obtain shells means that the resources (shells of different shapes and sizes) had purely abiotic dynamics. The ability to follow marked shells made it possible to experimentally study the process of exploitative competition, and exchanges of shells between species via 'shell fighting'. The system seemed much more amenable to estimating magnitudes of competitive effects than systems where competition was based on resources whose 'consumption' could not be so easily observed or whose dynamics were much more complicated. The abiotic nature of the resource dynamics made the likely range of consumer dynamics simpler, and the ability to determine shell preference via selection experiments in the laboratory simplified analysis of the relative values of different shell types, and the degree of shell limitation for different size classes. All of this seemed to suggest that competition would be far easier to quantify for hermit crabs than for birds and other common subjects of studies of competition.

While my Ph.D. thesis ended up being purely theoretical, most of my early academic career was largely devoted to working on competition between hermit crab species for empty shells. The majority of the work was done with little grant support, and this contributed to several limitations to the results. The large spatial ranges of all the species made it impossible to have random range-wide sampling, and the extent of shell exchanges (due to 'shell fighting') in the field was difficult to determine over broad spatial scales. Nevertheless, it seemed clear that intraspecific competition was by far the predominant form in all sets of species that were studied in detail. Shell use could be followed by introducing marked shells, both in the field and in laboratory tanks. Intertidal communities studied ranged in species number from two (the southern end of the Great Barrier Reef (Abrams 1981a) to nine in Micronesia—studied primarily in Guam and Enewetak (Abrams 1981b). It was also possible to compare the subset of three species found in relatively sheltered intertidal zones in the Pacific Northwest to the set of six species found in outer coast locations (Abrams 1987e, h), which included the three protected-water species. Communities having more species tended to have somewhat larger estimates of average pairwise effects, particularly comparing the Southern Great Barrier Reef to Micronesia. In the Pacific Northwest intertidal species assemblages, the two species with relatively high total interspecific effects in the more speciose outer coast community were low intertidal species that also occurred in the subtidal. For these species, the subtidal part of the range was not sampled as extensively, which likely increased the estimates of relative size of interspecific to intraspecific effects.

Estimates of relative strengths of inter- and intraspecific competition produced from analysis of the occupancy of released marked shells (performed for some, but not all of the communities studied) closely matched that based on overlap in habitat and shell type. The most thoroughly studied system consisted of three intertidal species found in relatively protected shorelines in the Pacific Northwest region of North America (Abrams 1987e). This was one of several communities in which competition was studied by marked shell releases as well as size-specific resource overlap measures. The others were the intertidal species on the Pacific coast of Panama (Abrams 1980c, 1981c) and a pair of species found at the southern end of the Great Barrier Reef (Abrams, 1981a). All species experienced close to or more than an order of magnitude more intraspecific than interspecific competition in those size classes constituting the greatest part of the reproductive output of the population. For most species in other communities that I studied the estimated intraspecific fraction of all competitive effects ranged from several-fold to more than an order of magnitude greater than the summed effects of interspecific competition from all other species in the community. The largest estimates of interspecific competition were for a subtidal community (Abrams et al. 1986), where most of the sampling was carried out by trawling (Nyblade 1974). This blurs small-scale spatial segregation, and probably led to overestimates of interspecific competition.

There are many unknowns about the population dynamics of every species of hermit crab involved in the studies discussed above. The long pelagic larval stage of most hermit crab species makes it difficult to determine the movement rates between spatial locations. It is possible that parasites play a significant role in population dynamics for some species of hermit crabs. Shell exchange between crabs (usually called 'shell fighting') is another complicating factor. Bertness (1981) argued that this had a significant effect on the overall interaction of the Panamanian intertidal species. Shell fighting does have the potential to make interactions asymmetric, but is unlikely to have greatly changed estimates of interaction strength in this community (Abrams 1981c). Hazlett (1978, 2013) has repeatedly found that shell exchanges due to shell fighting between conspecifics are usually mutually beneficial. Even if exchanges were strictly competitive, the mean level of competition would still be limited by the high degree of spatial segregation documented in the studies reviewed above. None of the unknowns in these studies seems likely to significantly change the general conclusion that interspecific competition is much less than intraspecific.

7.3.3 Current status of field studies of competition

Little attention has been paid to quantifying interspecific competition in hermit crabs since the 1980s. It may have been regarded as too small and obscure a group of organisms to produce generalities. However, the 1980s was also the time when attempts to measure competition in any manner other than long-term manipulations became unpopular, as documented in Strong et al. (1984). Interest in hermit crabs may have been a victim of this trend.

Notwithstanding many decades of research on competition, only a small fraction of the extant species in any higher-level taxonomic group have been studied with the goal of quantifying the interspecific competition they experience. While we can hope to have a more unbiased sample of species in future field studies, the ecological research community is too small to provide a comprehensive answer in the next couple of decades. This means that theory is likely to provide our best indication of when and where competition is expected to be strong for some time to come. One aspect of such theory must be a better understanding of how overlap in resource use is related to the population-level effects of competition. It is unlikely that theory focused on the Lotka–Volterra or simplified MacArthur models will be sufficient for these tasks.

7.4 Does competitive neutrality occur?

One perspective on the structure of communities that rose to prominence about 20 years ago is the idea that many communities consist of ecologically equivalent competitors that persist for long time periods only undergoing random changes in population size. This idea was not accepted in the last millennium; Armstrong and McGehee's influential 1980 article on coexistence actually defined such competitive neutrality as exclusion under the assumption that the ultimate outcome of a single-species system would be reached relatively quickly. However, Hubbell (2001) promoted the proposition that most rainforest trees (and probably other sets of species) are competitively equivalent. He argued that species-abundance distributions provided strong evidence of this, and that speciation could balance the relatively high rate of extinction that such a system would likely exhibit. However, the number of species on a given trophic level is influenced by all of their interspecific interactions, as well as the available species pool, and temporal fluctuations in abundance. It is therefore difficult to assign a single mechanism underlying a distribution of abundances. As with most generalized statistics in large biological communities, we know very little about the factors influencing the spectrum of abundances expected in a complete community or in segments thereof. The statistical distribution of species abundances is likely influenced by many factors other than competition (Abrams 2001b).

Leibold and McPeek (2006) and Velland (2010, 2016; Velland et al. 2014), among others, have supported aspects of Hubbell's argument in a more qualified manner. However, there is still no conclusive evidence for neutral coexistence in any pair of species, let alone large groups of species. On the other hand, there are many reasons for rejecting the idea. First, the accumulation of deleterious mutations occurs more rapidly in species with smaller population sizes (Lynch et al. 1995a, b). This provides an automatic advantage to the most common in a set of otherwise equivalent competing species, and would convert neutrality to a case of exclusion with a priority effect.

Secondly, species-specific diseases, particularly sexually transmitted ones, provide a stabilizing force that by itself would convert neutrality to coexistence. Third, sexual behaviours themselves can cause density-dependent mortality or density-dependent reduction in fertility (Gómez-Llano et al. 2021), again converting neutrality to stable coexistence. Fourth, adaptive evolution occurs at different rates depending on population size, and, even when it produces convergence, it in general is not expected to equalize competitive abilities (Pásztor et al. 2020). While some of these forces act in opposition, there is no reason for them to exactly cancel out at exactly the same population size in two or more species. A variety of differences in spatial distribution and differences in associations with mutualists provide other possible mechanisms producing coexistence among species having no differences in 'resource use', when the latter is defined narrowly. Although evolution may favour convergence of resource use in some systems, it is not expected to give rise to exact equality of competitive abilities.

McPeek and Gomulkiewicz (2005) and Leibold and McPeek (2006) cite a relatively small number of sets of species having similar resource use in their sympathetic reviews of the idea of neutral coexistence. However, for none of these is the full life cycle studied, nor are spatial differences in species distribution considered thoroughly. Even Pásztor et al. (2020), who criticize the idea of neutrality, seem to agree with McPeek and Gomulkiewicz (2005) in their assertion that, 'The vast number of co-occurring, seemingly identical, ecologically equivalent sister and cryptic species provide empirical grounds for challenging conventional ecological wisdom'. (They meant the wisdom that neutrality was unlikely.) If the number of such sets of species were really 'vast', and if other mechanisms affecting competition (e.g., sexually transmitted diseases, sexual selection and/or conflict, and hidden spatial resources) had been studied and ruled out, the cases cited might indicate high levels of competition. Neither of these circumstances applies to our current knowledge.

It is important to note that McPeek and Siepielski (2019) and Ousterhout et al. (2019) have significantly revised the interpretation of coexistence in *Enallagma* damselflies, one of the primary examples of possible neutral coexistence in McPeek's earlier work. The more recent studies argue that interspecific effects tended to be weak and/or asymmetrical, and that at least some species exhibit much stronger intraspecific than interspecific competition. Pásztor et al. (2020) also end up concluding that neutrality is unlikely. Leibold et al. (2019) have recently suggested that the appearance of neutrality at large spatial scales actually represents the outcome of adaptive evolution of niche-partitioning at local scales.

In other recent work, Gómez-Llano et al. (2021), and McPeek and Siepielski (2019) argued that reproductive interactions could result in intraspecific density dependence, and that this might explain cases of coexistence with high overlap in consumed resources. Reproductive competition is one of a large number of potential 'limiting factors' whose importance in coexistence was stressed by Levin (1970), and was something I regarded as common knowledge when I noted their potential role in character displacement 35 years ago (Abrams 1986a, p.111). Gómez-Llano et al. (2021) are correct in arguing that the details of several of the various mechanisms of sex-related density dependence have received relatively little study and deserve more attention. However, they do not support an argument for neutrality.

Neutrality would greatly simplify competitive relationships if it occurred. This may be in part responsible for the popularity of the idea. Another simplifying assumption in most competition theory (including many examples in this book) is symmetry (roughly that the effect of species 1 on species 2 is similar to the reverse effect of 2 on 1.) In many cases, symmetry of competitive interactions in simple models is required to obtain an analytical solution for the equilibrium. In other cases, it is required so that the analytical solution is sufficiently simple to understand. In any case, the reality is that there exists no theoretical reason to expect symmetry, and even the earliest reviews of empirical work revealed that a large proportion of relationships are asymmetrical. More than 80% of the studies reviewed by Schoener (1983) that both found competition and examined the issue of symmetry found that interactions were asymmetric.

Asymmetry has many potential effects beyond this simple inequality of the effects between a pair of species. Abrams and Cortez (2015a) examined asymmetrical cases of the widely used 2-consumer–2-resource MacArthur model with the addition of asymmetric Lotka–Volterra competition between the two resources. Asymmetrical resource competition, as expected, caused asymmetrical interactions between the consumers. However, it also frequently changed the sign of some or all of the commonly used measures of inter-consumer effects, and frequently produced cyclic dynamics in the two-consumer system in cases where the single-consumer subsystems are both stable and the analogous symmetric case is stable.

7.5 Interspecific competition in a food web context

The focus on competitive coexistence in the ecological literature would lead one to believe that competing species have the greatest effect on whether a species can persist. The important part of the word 'coexistence' is 'existence'. In fact, higher and lower trophic levels are also important in determining whether a given species is able to persist. There appears to be little if any empirical work examining the relative importance of competition in food webs with abundant species on the trophic levels above the competitors. Thus, this section will be entirely theoretical. It will use a simple 3-trophic level model to examine the relative effects of various other species on the abundance of a focal species on the middle level.

A question related to existence is how those species at other trophic levels affect the competitive interactions between species on a particular trophic level. Should predators on a higher level be regarded as a force that reduces or eliminates competitive effects on abundances? These questions are discussed briefly below. Not surprisingly, the answer depends on the measure of competition that is chosen. The model explored below is arguably the simplest possible one in which the effects of competition in a food web can be explored. Nevertheless, it is not simple.

The 6-species model used here has two species on each of the three trophic levels. A similar model was used by Abrams (1993) to assess the generality of top-down and bottom-up effects in food webs. This model was also used by Matsuda and Abrams (2006) to evaluate optimal harvesting strategies for exploiting fish communities. In this food web, the two species on the middle level ('consumers') interact indirectly both via the predators at the top trophic level and the resources at the lowest level. The resources are assumed to be biotic. The equations given below in Box 7.1 have type II functional responses for consumer species and include intraspecific density dependence for the predator species. Setting handling times to zero and eliminating the top predators' direct density dependence results in the simple case of all-linear per capita growth rates, used by MacArthur in his 2-level model.

The simplest version of this model is one in which all handling times are zero and there is no direct intraspecific interference on the top trophic level (i.e., all Y values are zero). This produces a 'MacArthur' model. The interactions at each trophic level can be characterized by the response of each species to a small change in the per capita mortality rate of the other species at that trophic level. (Of course, this is not a complete description of the variety of possible responses to a larger range of perturbation parameters and magnitudes.) The starting conditions in the

---

Box 7.1 Competition in a simple 6-species, 3-trophic level model

The full model used here has the following form:

$$\frac{dR_1}{dt} = R_1\left(r_1 - k_{11}R_1 - k_{12}R_2\right) - \frac{c_{11}R_1N_1}{1+c_{11}h_{11}R_1+c_{12}h_{12}R_2} - \frac{c_{21}R_1N_2}{1+c_{21}h_{21}R_1+c_{22}h_{22}R_2}$$

$$\frac{dR_2}{dt} = R_2\left(r_2 - k_{21}R_1 - k_{22}R_2\right) - \frac{c_{12}R_2N_1}{1+c_{11}h_{11}R_1+c_{12}h_{12}R_2} - \frac{c_{22}R_2N_2}{1+c_{21}h_{21}R_1+c_{22}h_{22}R_2}$$

$$\frac{dN_1}{dt} = N_1\left(\frac{b_{11}c_{11}R_1+b_{12}c_{12}R_2}{1+c_{11}h_{11}R_1+c_{12}h_{12}R_2} - d_1\right) - \left(\frac{C_{11}N_1P_1}{1+C_{11}H_{11}N_1+C_{12}H_{12}N_2}\right) - \left(\frac{C_{21}N_1P_2}{1+C_{21}H_{21}N_1+C_{22}H_{22}N_2}\right)$$

$$\frac{dN_2}{dt} = N_2\left(\frac{b_{21}c_{21}R_1+b_{22}c_{22}R_2}{1+c_{21}h_{21}R_1+c_{22}h_{22}R_2} - d_2\right) - \left(\frac{C_{12}N_2P_1}{1+C_{11}H_{11}N_1+C_{12}H_{12}N_2}\right) - \left(\frac{C_{22}N_2P_2}{1+C_{21}H_{21}N_1+C_{22}H_{22}N_2}\right)$$

$$\frac{dP_1}{dt} = P_1\left(\frac{B_{11}C_{11}N_1+B_{12}C_{12}N_2}{1+C_{11}H_{11}N_1+C_{12}H_{12}N_2} - D_1 - Y_{11}P_1 - Y_{12}P_2\right)$$

$$\frac{dP_2}{dt} = P_2\left(\frac{B_{21}C_{21}N_1+B_{22}C_{22}N_2}{1+C_{21}H_{21}N_1+C_{22}H_{22}N_2} - D_2 - Y_{21}P_1 - Y_{22}P_2\right)$$

The parameters in the above equations have the meanings given in similar equations in previous chapters. The resource density dependence parameter, k, has been given dual subscripts to distinguish intraspecific (k_{ii}) from interspecific (k_{ij}) effects on each species i. The double subscripts involving N or P parameters give the identifier for the consumer/predator first and resource/prey second. The parameters Y_{ij} are the 'interference' coefficients of the predator, and the first subscript gives the affected species and the second gives the effector. The presence of interference has dynamic consequences that are in some ways similar to those produced by an additional (higher) trophic level.

present simple example assume that the two species at each of the top two lev-els are equivalent to each other except for their consumption rates of resources. The two species on the mid and top level are also characterized by having mirror image consumption rates of their two prey species. In the numerical results given below, these parameters are also assumed to be scaled so the sum of the top con-sumers' two consumption rate constants is unity; i.e., $C_{ii} = 1 - C_{ij}$, and $c_{ii} = c_r(1 - c_{ij})$, where c_r is the ratio of the mean capture rate of the mid-level predators to that of the top-level predators. The competition coefficients at the lowest level (k_{ij}) are also assumed to be equal to each other. The k values for each resource are adjusted to maintain the summed equilibrium resource densities constant as the competition between them increases (in the absence of predators). This is done by assuming that $k_{ii} = 1/(1 + \alpha)k_i^*$ and $k_{ij} = \alpha/(1 + \alpha)k_i^*$, where k_i^* is $k_{ii} + k_{ij}$.

The interactions at each level in the simple MacArthur style version of the model (no handling time or predator interference) are as follows:

1. The response of each of the top-level species to mortality in the other top-level species can be positive or negative. If there is no interspecific competition on the lowest level ($k_{ij} = 0$ for $i \neq j$, meaning $\alpha = 0$), the results of Levine (1976) apply to the top and middle levels. This means that the top consumers have mutually positive effects when resource overlap of the mid-level species is greater than resource over-lap of the top-level species (i.e., c is closer to 1/2 than is C). If competition between the two bottom-level species were allowed, high enough competition could reverse the sign of the interaction between top-level species.

2. If the top two levels have equivalent prey partitioning ($C = c$), then there is no com-petition between the species on the top level, unless the resources compete with each other. With some between-resource competition, given $C = c$, the between-predator effect is again negative (P_i increases with D_j), and the magnitude of the positive effect of D_j on P_i increases as the level of between-resource competition increases (again provided that all species persist).

3. The second level (N_1, N_2) has no interaction based on response to a small change in the per capita mortality of the other species at that level, and this does not depend on any parameter, so long as all six species persist. (Recall the assumption of no interference competition ($Y_{ij} = 0$) in either top-level species.) The equilib-rium abundances of the two mid-level species are completely determined by the characteristics of the two species on the top level (provided that all six species persist).

4. The bottom-level species (R_1, R_2) respond to changes in each other's neutral parameter r as though the higher-level species were absent. They have zero effect on each other unless they compete directly with each other ($\alpha > 0$; i.e., $k_{ij} > 0$). This is because the equilibrium abundances of their consumers (the mid-level species) are controlled by the top-level species, eliminating the apparent competition that would otherwise occur. This is again true only so long as all species persist.

In summary, if all three levels share resources in this simple case of a MacArthur style model, the bottom-level species are the only ones for which mutually negative effects on equilibrium abundance revealed by a neutral perturbation can be assumed when there is 'direct' competition between them ($\alpha > 0$). The middle level has no interaction between its two species based on their lack of change in response to a small shift in a neutral parameter of the other species. This result for the middle level remains true for nonlinear functional responses of the top predators, provided that the equilibrium point is stable, and the predators are solely limited by their food consumption rate (i.e., the Y parameters are zero). When the bottom resources do not interact, the interaction between the top-level species is determined by whether the top-level predators are more generalized than the mid-level predators; i.e., C is closer to 1/2 than is c (negative effects) or C is further from 1/2 than c (positive effects). When the bottom-level resources interact, the symmetrical case with $C = c$ is always characterized by competition between the two top predators.

The effect of a large magnitude change in the mortality rate of one intermediate (consumer) species can be extinction. This result is inevitable if the mortality rate of one consumer is increased sufficiently. The disappearance of one of the mid-level species will generally also entail extinction of one of the top predators in the absence of strong interference at the top level. Extinction of a top predator may also precede extinction of the second consumer. A sufficient decrease in the mortality rate of one competitor will always produce extinction of the other competitor in a 2-species Lotka–Volterra system. However, that is not the case here. Relatively low productivity systems (low r values) are required for extinction to be the consequence of a low enough mortality rate of the superior consumer on the intermediate trophic level. Once this occurs, coexistence of the top two consumers becomes impossible unless one or both have interference coefficients ($Y_{ii} > 0$). When the Y coefficient of the remaining consumer is zero, then the abundance of the mid-level consumer again becomes independent of its own per capita death rate. Increases in its death rate simply reduce the equilibrium of the remaining top consumer.

The full model in Box 7.1 allows type II responses and direct density dependence of top-consumer death rates. These often alter the results for the simple case described above. Type II responses on the top level do not change the independence of mid-level equilibrium abundances on the parameters of lower-level species, and on the mortality rates of the mid-level species themselves. However, type II responses, even when they do not produce population cycles, alter the interaction of the top-level species. It is no longer true that these species do not interact in a symmetrical system with equal partitioning on the top two levels ($C = c$) and no competition between resources.

If the handling time produces population cycles, the mean population sizes usually respond differently to parameter changes than do the equilibrium population sizes. Independent of the handling time, self-limitation on the top trophic level also can cause differences in the signs of responses to parameter change. If the self-limitation terms (Y_{ij}) are non-zero, this means that equilibrium abundances of the middle level will be influenced by their consumption rates of the bottom-level species, and by other lower level parameters. Each mid-level species' abundance is also affected by mortality in the other mid-level species. However, these effects are likely to be relatively small if the top-level species' interference parameters (Y) are themselves small.

A comprehensive analysis of the cases with both handling time and self-limitation has not been carried out. Two related examples with handling time and the potential for cycles are provided here, simply as an illustration of the large range of effects that may occur. Both are cases in which the top two trophic levels consist of two species with 'mirror image' consumption rates of their two prey. As a result, the equilibrium has equal abundances of the two species at each level. The two examples both assume identical partitioning for the two top levels. Equal partitioning in the corresponding model with no handling time, no top predator interference, and no interspecific resource competition, would imply no competitive effect of a small change in mortality for either of the species on each of the top two levels.

Figure 7.2 shows the population dynamics of the four species on the top two trophic levels in an example that can exhibit anti-synchronized dynamics. The mean abundances of the two species on a given level are the same for all three levels. There is an alternative attractor for this system (not illustrated) with much smaller amplitude and perfectly synchronized cycles. Starting at the anti-synchronized attractor shown in Figure 7.2, a 10% increase ($d_2 = 0.275$) in the mortality rate of one the mid-level consumers (which has no effect on the equilibrium abundances of either of the mid-level species) produces small decreases in the mean abundances of both mid-level consumers (with a larger decrease for the species suffering increased mortality). This mid-level mortality perturbation produces much larger changes in the abundances of the top-level species, with the top predator more specialized on the perturbed prey decreasing by a relatively large amount, while the other predator increases by a somewhat smaller amount. Both changes are much greater in magnitude than those of the second-level species. If mid-level mortality d_2 is increased to 0.3, the system becomes stable, and the abundances of the two intermediate predators equalize. Further increases in d_2 do not alter these abundances until d_2 is sufficient to cause extinction of top predator 2 at approximately $d_2 = 0.46$. This causes the system to undergo limit cycles, and it also causes a large increase in the mean abundance of top predator 1 (to 0.6289), a large decrease in the mean abundance of intermediate predator 2 (to 0.1084), and a modest increase in the mean abundance of intermediate predator 1 (to 0.2178). A slight further increase in d_2 to 0.47 results in the extinction of intermediate predator 2 from all initial conditions, producing a system with cyclic dynamics of top predator 1, intermediate predator 1, and the two resources.

Figure 7.3 shows the dynamics of the top two trophic levels for a second example with symmetrical attack rate values and the same relative specialization for both the top two levels (as in the preceding figure). Here, the handling times for the top-level predators are increased to $H_{ij} = 1.5$ and the initial death rates of species at that level are decreased $D_1 = D_2 = 0.05$ (and the other parameters are identical to those of Figure 7.2; in particular $d_1 = d_2 = 0.25$ in the initial symmetrical condition). This set of parameters produces quasi-periodic cycles, which makes it difficult to obtain precise measures of the mean values. However, the effect of a small increase in the mortality rate of one of the intermediate predators again decreases the mean abundance of both predators by small amounts, while having larger effects (one positive, one negative) on the abundances of the top predators. In this case increasing d_2 never stabilizes the equilibrium, although the cycles become simpler in form when d_2 is sufficiently large.

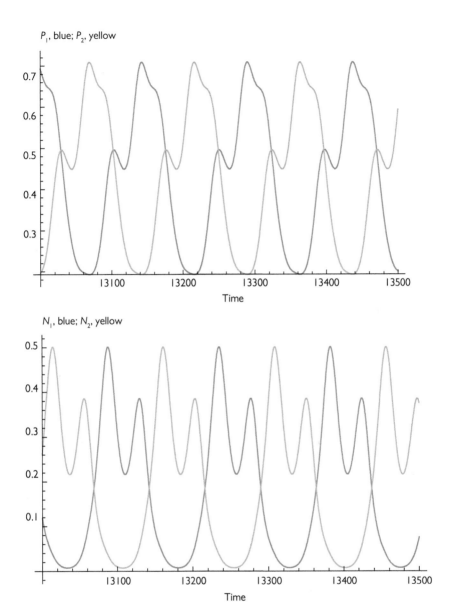

Fig. 7.2 Dynamics of the top two trophic levels for the model given in Box 7.1. The consumers have 'mirror image' consumption rates of the two resources they consume. The parameter values are: $r_1 = r_2 = 1$; $k_{11} = k_{22} = 1$; $k_{12} = k_{21} = 0$, $b_{ij} = 1$; $c_{11} = c_{22} = 0.7$; $c_{12} = c_{21} = 0.3$; $d_1 = d_2 = 0.25$; $B_{ij} = 0.5$; $C_{11} = C_{22} = 0.7$; $C_{12} = C_{21} = 0.3$; $H_{ij} = 1.25$; $D_1 = D_2 = 0.075$; $Y_{ij} = 0$.

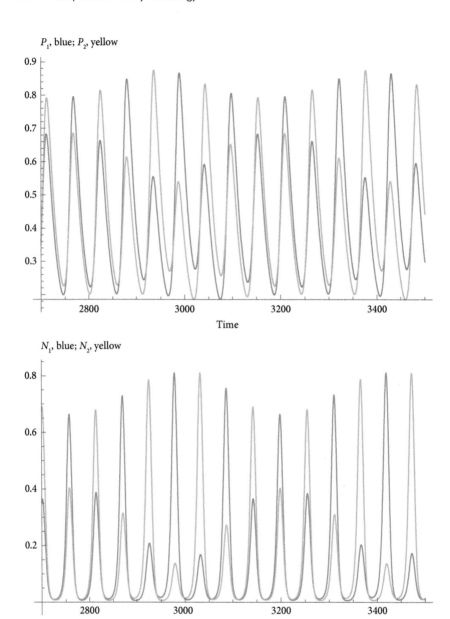

Fig. 7.3 Dynamics of the top two trophic levels for the model given in Box 7.1. Most of the parameters are identical to those given in Figure 7.2. The parameters that differ are; $H_{ij} = 1.5$, and $D_1 = D_2 = 0.05$. These changes alter the nature of the population cycles and increase their amplitude.

The mortality on the intermediate predator 2 (d_2) primarily reduces the mean density of top predator, P_2, and only causes small changes in the mean abundances of N_1 and N_2. However, an increase in d_2 increases the mean abundance of N_1, indicating a negative effect of mid-level species 2 on species 1. When d_2 is raised to 0.46, the ratio of the mean P_1 to the mean P_2 becomes 114.26, while the corresponding ratio of mean N_1 to mean N_2 is 1.2205. This again reflects the much larger effect on predators than on the one competitor of this mid-level species. When $d_2 = 0.47$, top predator species 2 goes extinct and the ratio of mean densities of N_1 relative to N_2 grows to 5.893. At $d_2 = 0.48$, N_2 also goes extinct. In the six-species system with all species present, the impact of a mid-level predator's mortality on the abundances of the higher-level species is much greater than its impact on its competitor. The effect on the competitor is only non-zero in cycling systems, or in the presence of interference terms on the top level.

Another question that can be addressed with the model in Box 7.1 is how the persistence of a given species is impacted by another species on the same trophic level vs those on other trophic levels. This is central to the issue of whether existence of a given species is more sensitive to changes in a competitor, a predator, or a food species. It also bears on the question of whether direct effects measured by changes in equilibrium abundance are needed to produce indirect effects. All these questions are raised by the phenomenon described above—i.e., the fact that the two middle-level species do not affect each other's abundance when 'effect' is measured by the sensitivity to a small change in a neutral parameter of the other species. In spite of the lack of a measured effect on their own population sizes, the two mid-level species transmit an effect of a mortality change in either of the top predator species to the bottom-level resources. Thus, increasing top predator 1's mortality D_1 will usually affect both of the lower-level resources, and at least one of those resources will increase.

Of course, the above is one of the simplest possible food web models that includes competition and more than two trophic levels. Even here, I have only discussed some highly simplified or highly symmetrical cases.

7.6 Competition between species in theory and reality

This chapter has documented why there is no universally applicable measure of interspecific competition that can be calculated based on measurements of relative consumption rates of different resources. This means that observations of niche overlap do not provide good guesses at competitive effects. The chapter has also argued that there is a vanishingly small probability that any two competing species in a natural community are ever competitively equivalent. Even near-equivalence is likely to be rare. Although ecologists have made measurements of competition over small spatial and temporal scales, these do not usually allow an estimate of population-wide effects of one species on another. The bottom line is that we are ignorant of the extent of competition in the vast majority of communities. Some of the natural systems in which competitive effects are most likely to be calculable have suggested that levels of interspecific competition are much lower than intraspecific. Food web theory suggests that a given level of similarity in resource use may be consistent with many different types and levels of interaction depending on the food web context of the focal competitor species. This again means that simple measures of similarity in sets of resource capture/intake rates cannot be translated into a risk of competitive exclusion.

8

Competition in seasonal environments:
temporal overlap

8.1 Introduction

This chapter deals with a topic that has not received enough attention from theoretical ecologists, i.e., the implications of seasonality for competitive interactions. It explores seasonality in a very simple model of competition and finds surprisingly complex results. Many of those results are new, and are given in greater detail than for most of the other chapters. Readers who are less interested in these details could skip Sections 8.2.7 through 8.3.3, or simply go directly to the discussion (Section 8.4). The topic is important, both because of the seasonal nature of most environments, and because adding seasonality to the simplest and most often used consumer–resource model (the MacArthur system) drastically changes its dynamics and the more general principles about the relationship between similarity and competition that follow from those dynamics.

Periodic variation (seasonality) has long been recognized as an important determinant of interspecific interactions in most natural environments (e.g., Fretwell 1972; Sommer 1984; Holt 2008; McMeans et al. 2020; Wollrab et al. 2021). In spite of this, relatively little theory deals with competing species in seasonal environments, with notable exceptions, reviewed in Section 8.1.1 below. Seasonal predator–prey interactions have received more attention (including Rinaldi et al. 1993; Abrams 1997; King and Schaffer 1999; Barraquand et al. 2017; Sauve et al. 2020). These articles all point out the possibilities for complex population fluctuations, abrupt changes in dynamics with continuous parameter change, and for two or more possible dynamical attractors in a given system. When resources are living, competition involves coupled predator-prey systems, so it should have a similarly wide range of outcomes. The present chapter will address the impact of seasonality in simple models of resource competition between consumer species. Numerical analysis of simple models will show that an equally wide range of dynamical outcomes is possible, and that many of the outcomes are inconsistent with currently popular ideas about interspecific competition and coexistence.

The present analysis adopts a consumer–resource framework for studying seasonal competition, and it begins by arguing why this framework is required. The analysis

Competition Theory. Peter A. Abrams, Oxford University Press. © Peter A. Abrams (2022).
DOI: 10.1093/oso/9780192895523.003.0008

concentrates on competition between two consumers for a single resource, although some cases with more species of either type are also considered. It also focuses on a seasonal version of the simplest possible consumer–resource system; the one introduced by MacArthur (1970, 1972). The analyses in this chapter differ from most previous works on seasonal consumer–resource competition in two respects; not assuming rapid resource dynamics, and not concentrating exclusively on scenarios where seasonality promotes coexistence.

The analysis is important in assessing a number of generalizations that appear to be widely accepted in the recent literature: (1) mutual invasibility is needed for, and always implies, coexistence; (2) temporal differences in resource use make coexistence more likely; (3) coexistence can be understood as being a result of separate stabilizing and equalizing processes; (4) the four qualitative outcomes that characterize the 2-species LV model apply to all 2-species competition models; (5) the effects of input or removal of individuals of one competitor on the abundance of another are relatively insensitive to their initial abundances.

Two likely reasons why competition in seasonal environments has received limited attention are the need to address most questions numerically and the complexity of the outcomes. This chapter will present numerical results for a relatively broad range of systems and questions, but will not produce a comprehensive analysis of any single model. The goal of the analysis is to alert ecologists to a range of possibilities that seem to have escaped attention, and that are inconsistent with some widely used generalizations in competition theory. The models considered here deserve a more complete exploration in the future.

8.1.1 A brief history of work on seasonal competition

In the 1970s, it was generally thought that coexistence was possible if consumer species 'partitioned' resources. Partitioning involved differences in the inherent characteristics of the resource items used by a set of two or more consumer species, differences in where those resources were consumed, and/or differences in when they were consumed. Schoener (1974b) reviewed these three types of partitioning, and concluded that the third type—temporal partitioning—had been observed less commonly than partitioning involving space or the inherent characteristics of the resource. Nevertheless, he found that this type of partitioning was reasonably common among predatory species. The term 'partitioning' is usually applied to predictable differences in use, rather than the outcome of independent stochastic variation in different species. This chapter will only treat regular periodic temporal variation. Recent work on autocorrelated stochastic variation (Schreiber 2021) suggests that such variation can have qualitatively similar effects. Li and Chesson (2016) used the main model considered here to show that, when seasonal variability promotes coexistence, at least one form of stochastic variation can also do so.

At the time of Schoener's (1974b) study there was very little data to connect partitioning to quantitative descriptions of population-level interactions, and this was particularly true of temporal partitioning. More than a decade earlier, Hutchinson (1961) had introduced the idea that temporal partitioning could promote coexistence.

However, he did not support his verbal argument with any explicit mathematical analysis (as discussed by Li and Chesson 2016). It is still frequently stated that temporal differences in resource use are 'stabilizing' or that they promote coexistence (e.g. Manlick and Pauli, 2020), and this possibility has been supported by recent theory (e.g., Li and Chesson 2016; Miller and Klausmeier 2017; McMeans et al. 2020).

Most of the early analyses of the effects of seasonal variation on competition adopted one or the other of two major simplifications. The first was to leave resources out of the model entirely, usually basing the analysis on periodic temporal variation in parameters of Lotka–Volterra (LV) models. The second was to assume that resources change rapidly enough with respect to consumer abundance that one can substitute 'quasi-equilibrium' formulas for the resource in the consumer equations. Early examples of seasonal LV models include: Cushing (1980); Namba (1984); and Namba and Takahashi (1993). Namba and Takahashi (1993) used an LV model to examine cases in which different parameters of a given species had asynchronous responses to the periodic environmental variation. They showed that this could cause exclusion in cases where it would not otherwise occur. They also showed that some systems could have three alternative outcomes (exclusion of species 1, exclusion of species 2, or coexistence) depending on the initial densities and timing of introduction of the two species. Two alternative states of coexistence also occurred frequently in their model. These outcomes required asynchronous seasonal changes in different parameters of each of the competitors' per capita growth rates. However, it is not immediately clear how or why such within-species asynchrony should arise from the process of competition. More recent analyses of coexistence based on modified versions of the LV model include Scranton and Vasseur (2016) and Picoche and Barraquand (2019), both of which concentrated on the potential for coexistence to be enabled by seasonal variation.

Unfortunately, the LV model is so simplified that it is unclear how seasonal variation should be incorporated into different parameters of the model. As early as 1970, MacArthur (1970, 1972) had suggested that models of competition in which only the competitor abundances were represented as dynamic variables should be consistent with models in which the resources under competition were represented explicitly. Cushing (1986) addressed the implications of consumer–resource dynamics for the correlations between variable parameters in the LV model and MacArthur's consumer–resource model. However, Cushing's approach was based on the assumption of no resource extinction, which has many special properties (Abrams 1998). More importantly, Cushing used long-term temporal averages to obtain the parameter values in the LV system from the consumer–resource model. This does not replicate the type of time lags produced in actual consumer–resource models. Most of the subsequent published analyses of competition in periodically varying environments that incorporated resource dynamics assumed that the resource populations change much more rapidly than the consumer populations (e.g. Miller and Klausmeier 2017; Kremer and Klausmeier 2017), or simply that they adopt the equilibrium value for current consumer densities at each point in time (Smith and Amarasekare 2018; Amarasekare and Simon 2020). Others have chosen to examine cases in which seasonal variation promotes coexistence (Tredennick et al. 2017; Hening and Nguyen 2020; McMeans et al. 2020). Relatively rapid resource dynamics have also been assumed in most studies of the impacts of stochastic variation on competitive outcomes in consumer–resource models (e.g., Abrams 1984a; Loreau 1992; Barabás et al. 2012; Li and Chesson 2016), all of which found that variation promoted coexistence.

It is certainly true that seasonally variable uptake rates of resources can allow coexistence in situations where it would not occur without seasonality (Li and Chesson 2016). Descamps-Julien and Gonzalez (2005) provide an empirical example from a laboratory system with two competing diatoms differing in their optimal temperatures for nutrient uptake. They found that coexistence did not occur at a fixed temperature, but did occur with imposed variation in temperature. However, there has been sufficiently little empirical work on this topic to determine relative prevalence of such coexistence-promoting effects of seasonal variation. A recent article (Abreu et al. 2020) examined competition between two bacteria in a chemostat under fluctuating dilution rates. They found that the outcome was similar to that in an environment characterized by the average dilution rate. However, dilution rate (or any other parameter with effects on per capita growth that are independent of a competitor species' abundance) should have little effect on competitive interactions (Chesson and Huntly 1997).

8.1.2 Aspects of seasonal variation in competition treated here

The use of consumer–resource models is essential in addressing temporal niche segregation because the interaction of these two dynamic entities usually results in some delay between when a consumer species experiences a changed value of a demographic parameter, and when that change has its maximum effect on resource abundance. Consumer–resource systems in a constant environment frequently have an oscillatory approach to equilibrium (May 1973), in which the damped cycles of each species/component are temporally displaced (see Figure 2.1). This difference in timing influences other aspects of the dynamics of seasonally variable systems; it enables temporal partitioning to make coexistence less or more likely, depending on details of the dynamics. The assumption of quasi-instantaneous equilibrium by MacArthur (1972) and many later works eliminates this lag.

In the following sections, I examine the consequences of sinusoidal variation in different consumer growth parameters for a simple and widely used family of consumer–resource models of competition. The present work will not treat the case of seasonality that only affects resource growth, which was examined in Abrams (2004b), and will be treated in Chapter 9. This chapter concentrates on systems with a single resource population, but also has a more limited consideration of systems with two resources. Most of the analysis is based on MacArthur's consumer–resource model (1970, 1972), with a logistically growing resource and one or more consumer species with linear functional and numerical responses. Alternative models explored later in the chapter either assume abiotic (non-living) resources or consumers with type II functional responses.

Temporal partitioning of resources requires some environmental variation that affects the resource uptake or conversion rates of different consumer species in different ways. There can also be indirect temporal partitioning due to seasonal differences in mortality rates of different consumers. However, variation in mortality does not affect the ability of species to coexist in most simple consumer resource models with a single resource (Chesson and Huntly 1997), because variation in a density- and resource- independent mortality rate does not alter the mean resource abundance that produces zero population growth. Early claims that stochastic variation in mortality or loss rates always favoured exclusion (May and MacArthur 1972; May 1973, 1974) proved to be incorrect (Abrams 1976; Turelli 1977, 1978, 1981), and stochastic variation in various difference equation models promoted coexistence, as first shown in Chesson and Warner (1981). The models considered here are differential equation models with seasonal variation in consumer uptake or conversion parameters. However, the results are also relevant to parameters influencing mortality when the per capita mortality rate is a function of resource intake rate.

8.1.3 Why are the dynamics of seasonal systems important?

A brief answer to this question is simply that most environments are seasonal (Fretwell 1972). However, much of the structure of current competition theory is built upon constant environment models. Seasonal systems also have implications for the currently popular tri-partite classification of mechanisms of coexistence suggested by Chesson (1994, 2020a, c). The first of Chesson's categories is 'classical' methods of coexistence ('resource partitioning'), which involve differences in the inherent characteristics of the resource units/items that affect their relative rates of capture or conversion. The second and third categories require sustained temporal variation. The second is the 'temporal storage effect', under which a population is able to continue to benefit from favourable periods during subsequent unfavourable ones. Chesson later expanded storage to be coexistence under temporally variable competition via an appropriate difference between species in their covariance of 'environment' with 'competition'. Li and Chesson (2016) used this theory to explain coexistence in seasonal environments in the MacArthur model. The third category is 'relative non-linearity', in which population fluctuations allow a competitor with a higher resource requirement to coexist by having more rapid growth during periods of high resource availability because of a functional or numerical response that increases more rapidly with increasing resource abundance (e.g. Armstrong and McGehee 1976a,b, 1980). Barabás et al. (2018) and Amarasekare (2020) review and support this classification of mechanisms of coexistence, and Ellner et al. (2019) propose a numerical approach to quantifying the temporal storage mechanism.

One problem with this body of work is that temporal resource partitioning can actually make coexistence less likely. While this can be interpreted as negative 'storage', common usage of this word does not provide a clear meaning for such a negative. Another problem is that the initial densities of consumers and the timing of their arrival in the system may determine whether coexistence is promoted or inhibited. Moreover, a wide range of different competitive effects occurs within each of the coexistence-promoting and coexistence-inhibiting types of periodic variation. The focus on coexistence in a 2-competitor non-seasonal context seems to have inhibited more general work on the wide variety of impacts of seasonal variation on the nature of competition.

The present analysis has implications for the widely used practice (reviewed in and promoted by Grainger et al. (2019)) of inferring coexistence by looking at the ability of each consumer to increase when it is very rare and the other consumer(s) are at their limiting dynamics ('invasion analysis'). For the models considered here, invasion analysis often fails to predict the ability of species to persist together indefinitely in seasonal environments. It is neither necessary nor sufficient for coexistence. These models also raise problems for the popular assertion that coexistence arises from stabilizing factors ('niche differences') and equalizing factors (following Chesson (2000a)). Competitors that are equal in all demographic parameters and have maximal temporal niche differences often fail to coexist. In some cases, unequal mean demographic parameters are required for coexistence.

The presence of seasonal variation changes almost any measure of interspecific effects between competing consumers, and has a major effect on the range of other parameters allowing coexistence in those cases where seasonality promotes coexistence. These more quantitative aspects of competition, and their dependence on resource dynamics, will also be examined in this chapter.

8.2 A modelling framework and a seasonal MacArthur system

This section presents numerical analyses of examples of commonly used consumer–resource models. The selection of specific examples is designed to illustrate the operation of general mechanisms using models that are familiar in a non-seasonal context.

8.2.1 General features of the models

The basic model has a single resource with abundance R, and a number of consumer species with abundance N_i for species i. It is:

$$\frac{dR}{dt} = f(R) - \sum_i c_i(t) R N_i g\left(c_i(t) R\right)$$

$$\frac{dN_i}{dt} = N_i \left[b_i(t) c_i(t) R N g\left(c_i(t) R\right) - d_i(t) - z_i N_i\right] \tag{8.1a, b}$$

The function g represents effects of resource abundance on the baseline attack rate c_i; the most common of these is the effect of handling time or satiation. If the consumers have linear functional responses $g = 1$; $g = 1/(1 + c(t)hR)$ when the response is described by Holling's formula for a type II response, where h is the handling time required for each resource item. The function g is assumed to be identical across species to rule out differences in the linearity of the consumers' responses as a factor generating coexistence. The three consumer parameters (b, c, d) are respectively resource conversion efficiency, resource attack rate (maximal per capita consumption rate), and per capita mortality rate. The analysis will examine seasonal variation in each of these three parameters. The parameter z_i represents a per capita negative effect of species i on its own per capita growth rate. Such intraspecific interference terms can allow coexistence even in the absence of seasonality. In most of the analysis these z_i parameters are assumed to equal zero. However, interference terms can be used to measure the strength of exclusion in cases where it occurs in the absence of interference. The function f is the resource growth rate. In the MacArthur model, f is the logistic growth function, $g = 1$, and $z = 0$.

The variation in parameters is modelled by multiplying the mean parameter (here denoted by capitals: B_i, C_i, or D_i) by the following function:

$$\left(1 + \gamma_i \sin \left(\frac{2\pi(t - L_i)}{q} \right) \right) \tag{8.2}$$

The parameter q is the period of the cycle, and L_i is the lag in the variation experienced by species i relative to that experienced by species 1 ($L_1 = 0$ and $L_i < q$, by definition). The parameter γ (with $0 \leq \gamma \leq 1$) adjusts the magnitude of the sinusoidal variation, from non-existent ($\gamma = 0$) to a value ($\gamma = 1$) large enough reduce the parameter to zero at the low point of the cycle. This framework implies a maximum parameter value of twice the mean. An extension of this model considered briefly below has two resources with equivalent growth parameters but different vulnerabilities to the consumer; this reveals the interaction of (structural) resource partitioning with seasonal partitioning. In the 2-resource analysis below, expression (8.2) multiplies each of the two mean capture rate parameters for species i, C_{i1} and C_{i2}.

For many, if not most, species the annual cycle of change in climatic variables causes changes in population growth parameters. Thus, some previous work (Li and Chesson 2016) has measured time in years and has been limited to the case of $q = 1$. However, there are multi-year cycles and quasi-cycles in climatic variable (e.g. the Pacific decadal oscillation). Even for annual cycles, q obviously takes on values greater than one if the timescale used to measure the other rate constants in the model is less than annual. The fundamental issue is the relative speeds of the population dynamics (of all species) and seasonal change. Greater change in population abundances within a season can be produced by either longer seasons or more rapid population dynamics (of both species). An increase in q is equivalent to maintaining $q = 1$ and increasing the set of rate constants, r, k, C, and d, by the same factor; q is thus simply a scaling factor for the four temporal rate constants. In Hutchinson's (1961) famous example of plankton, an annual cycle often involves several reversals in the relative abundances of all of the species concerned due to their rapid dynamics (see Figure 1 of Sommer et al. 2012). This implies $q \gg 1$. I concentrate on anti-synchronized cycles; $L = q/2$. This minimizes the temporal parameter overlap of two species with sinusoidal variation in resource-related parameters.

The most commonly used functions for resource growth are the logistic and the linear abiotic ('chemostat') models. The versions of these models used here are respectively given by $f(R) = m + R(r - kR)$, and $f(R) = I - ER$. In the logistic model r is the maximum per capita rate of increase, and k is the per capita density-dependent reduction in that rate; m is an input due to external immigration, and a (small) positive value of m is appropriate when the system is part of a larger metacommunity. In the abiotic model, I is the rate of input, and E is the per capita loss rate. (Any external input from other patches can be included in I.) Because of its prominence in previous theory, I will concentrate on the 'MacArthur' system (MacArthur 1970, 1972; Chesson 1990, 2020a, b), which assumes a logistic resource and a linear consumer functional response. Some models with two resources will be considered to compare the impacts of temporal partitioning due to seasonality to the impacts of strict resource-type partitioning. These simple models are inadequate representations of almost any natural system (Abrams 1988a, 2010b), but there is no reason to believe they produce an unusually wide range of sensitivities to seasonal forcing.

Understanding the impact of environmental variation is complicated because of the interaction of the periodicity of the environment with the inherent periodicity of the consumer–resource interaction. These interacting periodicities produce the complicated range of dynamics found in seasonal predator–prey models (Rinaldi et al. 1993; King and Schaffer 1999; Sauve et al. 2020). As shown in the following section (8.2.2), it is possible for the environmental fluctuations to produce a positive effect on a single resident consumer when the recovery of the resource (prey) from heavy consumption coincides with the consumer's (predator's) period of greatest consumption or conversion rate. Such situations are reflected in a lower mean resource abundance than in a constant environment. The analysis begins with an examination of lags and resource dynamics in a system having only a single consumer (and a single logistic resource).

8.2.2 Resource lags and mutual invasibility of MacArthur systems

The response of both resource and consumer populations to an instantaneous change in a consumer rate parameter takes a finite amount of time to be fully realized. The duration of these lags and the pattern of seasonal variation jointly determine the resource abundance at any given time in the future. A rough indication of the effect of lagged resource changes on consumer dynamics can be seen in Figure 8.1, which plots the resource abundance and the consumer's scaled per capita capture rate $(c(t))$ in eqs (8.1) as a function of time in a periodic environment for the MacArthur consumer–resource model, with both species exhibiting their limiting dynamics. Resource abundance is shown in blue, resident consumer $c(t)$ is dashed-black.

The plot also shows the $c(t)$ curve of a rare competitor (dashed-red) whose response to the environment is 180 degrees behind that of the first species. (The c-curves have been multiplied by 10 for visual comparison with the much larger resource density.)

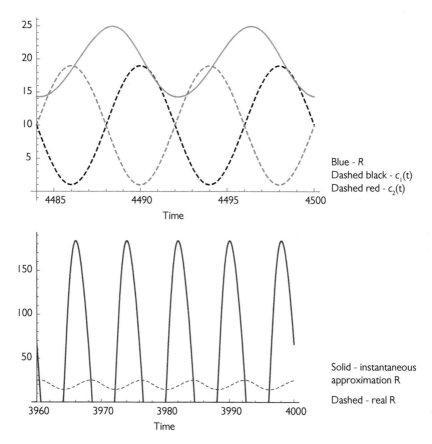

Blue - R
Dashed black - $c_1(t)$
Dashed red - $c_2(t)$

Solid - instantaneous approximation R

Dashed - real R

Fig. 8.1 Panel A. Resource dynamics (solid blue) and the dynamics of the intake rate parameters $c(t)$ for two anti-synchronized consumer species in the MacArthur model; the resident consumer's $c(t)$ is given by the black dashed line, and the invader by the red dashed line; both have been multiplied by a factor of 10 to make them easily visible on the scale of the graph. The parameters are: $r = 0.25$; $k = 0.00125$; $\gamma_1 = \gamma_2 = 0.9$; $C_1 = C_2 = 1$; $b_1 = b_2 = 0.01$; $d_1 = d_2 = 0.2$; $m = 0.001$; $q = 8$, $L = 4$.

Panel B. This gives the resource curve from panel A (dashed) together with the hypothetical resource curve that would apply if the resource achieved equilibrium instantaneously with respect to the actual consumer abundance at all times in the system from panel A (negative values are omitted). The (hypothetical) negative resource abundances occur because the consumer abundance reaches higher levels than it would if the resource actually did achieve equilibrium instantaneously. This shows the inadequacy of the assumption of immediate resource equilibration in this system.

The resident's $c(t)$ curve is more closely correlated with resource abundance than is the invader's $c(t)$; this can be seen by comparing the resource abundance at the peaks of the two $c(t)$ curves. The resulting lower resource intake of the invader implies a negative mean per capita growth rate, since the resident's mean is zero at its limiting dynamics. Thus the invader is excluded. The roles of resident and invader are arbitrary here, meaning that either species can exclude the other. Figure 8.1B shows the resource abundances that would occur if the resource abundance were to immediately reach its equilibrium for the actual consumer abundances in this system; this is presented to show how misleading such an assumption would be here.

The per capita rate of increase of an invader with a different lag (but otherwise equivalent parameters) can easily be calculated from the resource abundance curve shown in Figure 8.1A. That invasion curve is shown in Figure 8.2A. This figure shows that the invader (say species 2) must have a lag of slightly greater than 4.5 to increase when rare; this is greater than 1/2 the period length ($q = 8$). Because the invader and resident roles are interchangeable with a $q/2$ phase shift under this scenario, if invasion of one competitor is impossible for all lags $\leq q/2$, symmetry implies that mutual invasion of two otherwise equivalent species is impossible for any lag. If invasion density is low enough, the starting point of the environmental cycle will not significantly affect invasion success. In Figure 8.2, an invader that has a lag somewhat greater than 1/4 period ($L = 2.2767$ in Figure 8.1) has the lowest fitness of any invading type. The invasion curve implies that invasion of a 'later' consumer becomes more difficult as lags increase from 0 to 2.2767; a still larger L increases invasion fitness, which becomes positive at a lag of 4.552. Figure 8.2A shows that, if species 2 had a mortality rate that was lower than that of species 1 by 0.02, it could invade if its lag were less than approximately 1.20, but not for any lags between 1.20 and 3.42. In other words, it is possible for greater differences in seasonal timing to make invasion impossible when shorter lags make it possible. The reverse is possible as well. In general, the failure of mutual invasion need not rule out coexistence, although it is impossible in the system illustrated in Figure 8.2. No coexistence attractor was observed for any integer $q < 20$ in this system.

Coexistence with mutual invasibility is possible for a range of lags when resource dynamics are more rapid relative to consumer dynamics. If the resource parameters r and k are each multiplied by 10 in the Figure 8.2A example, the invasion curve becomes that shown in Figure 8.2B. This figure implies that lags between approximately 1.5 and 6.5 allow mutual invasibility. Plots like those in Figures 8.2A, B differ in form depending on the length of the environmental period, and the demographic parameters of both species. If the dynamics of both species are sufficiently rapid (equivalently, if the period q is sufficiently long) then species with anti-synchronized seasonal cycles in their attack rates will always coexist in this model. However, for a broad range of short to moderate period lengths (with other parameters as in Figures 8.2A, B), coexistence of any pair of seasonally offset but otherwise equivalent competitors is not possible.

Temporal correlation between relatively high resource abundance and relatively high consumer capture rate in a seasonally varying consumer species produces an

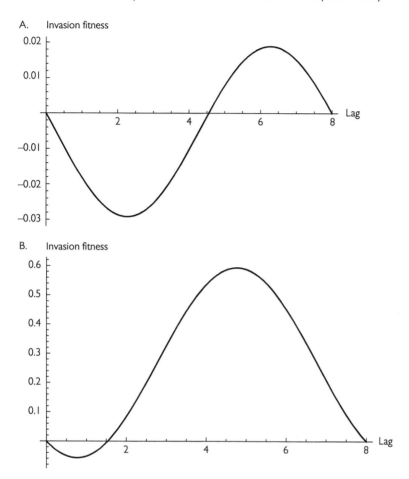

Fig. 8.2 Panel A. The per capita growth rate of a very rare (lagged) invader in the MacArthur system (eqs (8.1)) with the parameters (other than L) given in Figure 8.1, having a resident species undergoing its limiting population cycle. The invasion rate is given for invaders characterized by different lags, L.
Panel B. As in panel A, except that r and k have been multiplied by 10. The invasion rate is given for invaders characterized by different lags, L.

average resource abundance that is lower than the equilibrium value for an equivalent aseasonal consumer, preventing its invasion. It also prevents invasion for a range of seasonal consumer types that differ from the resident only in the lag of their seasonal $c(t)$ curves, including an anti-synchronized species. However, it is always possible for an otherwise identical consumer with a slightly earlier $c(t)$ curve (L slightly less than q) to invade and displace the resident, unless it suffers some lowered mean parameter as a consequence of that earlier timing. If the timing of the original resident is adaptive this implies that early individuals will have lower mean values of one or more fitness parameters, and this may be large enough to outweigh the advantage of earlier resource exploitation. Lags that are small enough relative to q always result in exclusion of the lagged invader unless it has some advantage in mean fitness parameters.

The ability of an anti-synchronized invader to increase when rare in the MacArthur model can be determined from the mean resource abundance in the resident-only system, once it has reached its limiting dynamics. Mean resource abundance less than the equilibrium for an otherwise equivalent aseasonal species implies that the aseasonal invader would be unable to increase when rare. As noted above, the mean resource abundance is also an accurate indicator of the outcome of competition with an anti-synchronized competitor, which will be discussed first. This equivalence is due to the fact that the per capita growth rate function is linear, and the seasonal term is equal and opposite at all points in time for the resident and anti-synchronized invader. For the resident undergoing its limiting dynamics, the mean per capita growth rate must be zero. Thus, mean$[R] = d/BC$−mean$[\gamma\sin[2\pi t /q])R]$. If the mean R in the resident-only system is less than its aseasonal equilibrium of d/BC, this implies that mean$[\gamma\sin[2\pi t /q])R] > 0$. The anti-synchronized invader's per capita growth is $BC(1 + \gamma\sin[2\pi(t-(q/2))/q])R]-d$. Because the value of the anti-synchronized sine function is the negative of its original value, the invasion condition becomes $BC(1-\gamma\sin[2\pi t/q])R-d > 0$. This cannot be satisfied if mean$[R]$ is less than its aseasonal equilibrium of $d/(BC)$. Note that the amplitude of the change in $c(t)$, given by γ, does not alter the condition for invasion from low density for an otherwise equivalent anti-synchronized competitor, although it does modify the magnitude of the invasion growth rate.

Whether the mean resource density is greater or less than the aseasonal equilibrium value is determined by the seasonal period, q. (Recall that a larger q is equivalent to proportional increases in all rate parameters of the two species; $r, k, c,$ and d.) The mean R, in turn, determines whether there are mutual exclusion outcomes for two anti-synchronized competitors. Figure 8.3 shows the mean R for a range of period lengths (q) in a single-consumer system that is otherwise identical to the one assumed in Figures 8.1 and 8.2A. The range of periods shown encompasses the range of qualitatively different dynamics produced. The period lengths associated with two or more dots at a given q have alternative attractors, characterized by different mean R. The alternative attractors for each period were found by continuing all known attractors occurring at higher or lower q and by starting each parameter set from five randomly chosen starting abundances from the interval between 0 and twice the aseasonal equilibrium population size for each species. This numerical search method does not ensure that all possible attractors have been located, but it has a good chance of locating most of those with significant basins of attraction. The population dynamics observed for the lowest range of period lengths $(q \leq 10)$ consists of a simple cycle with period q. The majority of the longer periods illustrated, starting with $q = 11$, have at least two alternative population cycles.

The environmental period lengths, q, associated with alternative attractors often have an attractor with period q and another which is a small integer multiple of q.

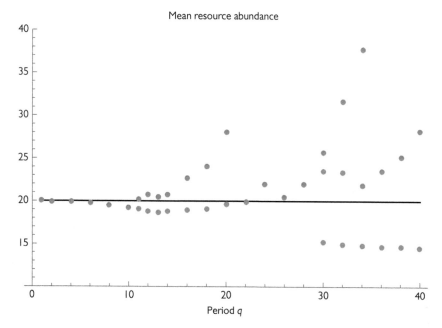

Fig. 8.3 The mean resource abundance (blue dots) in a consumer–resource (predator–prey) system for a range of seasonal periods, q. This system is otherwise identical to that in Figures 8.1 and 8.2. (Parameters are: $r = 0.25$; $k = 0.00125$; $\gamma_1 = 0.9$; $C_1 = 1$; $b_1 = 0.01$; $d_1 = 0.2$; $m = 0.001$) The black line gives the equilibrium resource abundance for the same system in the absence of any variation. Note that only integer periods were studied, and only even integers were simulated for most of the range shown.

Aperiodic population variation also occurs for some values of q. The seasonal period lengths with a mean resource density much lower than the aseasonal equilibrium of $R = 20$ that characterizes q between 30 and 40 are all simple cycles with period q. The attractors having longer period population cycles within this range of q have mean resource densities > 20; this implies that they can be invaded by an anti-synchronized competitor. Because of the irregular forms of the dynamics that occur for $q = 18$ through 23, determining an accurate mean abundance may require averaging over an extremely long temporal interval, and for some significant part of this interval, the mean resource abundance may be significantly greater or less than the long-term mean. This has the potential to make invasion success dependent upon initial abundance of the invader.

Periods moderately longer than the largest q ($= 40$) shown in Figure 8.3 are mostly characterized by population cycles with the environmental period, q. For sufficiently long periods, there are two or more local population maxima per cycle in each consumer species. An environmental period of 60 is the smallest integer period > 40 that lacks any attractor with a resource density less than the equilibrium of 20. All $q > 60$

have mean resource densities > 20, implying coexistence with an anti-synchronized competitor. Integer values of q of 69 through 76 have two or three alternative attractors with one, two, or four local maxima in abundance per period. Periods of 77 and 78 appear to be the largest environmental periods having population cycles that are longer than q ($2q$ in this case). Environmental periods of 79 and greater imply at least two local maxima in consumer population size per period.

8.2.3 Coexistence in a 2-consumer MacArthur system

The dynamics of single-consumer systems described in Section 8.2.2 suggest that alternative exclusion occurs at sufficiently short periods (equivalently, slow dynamics of both species), and coexistence for sufficiently long periods. Outcomes for intermediate period lengths are more complicated, and require numerical exploration. The outcomes of simulations for a range of q in the anti-synchronized Figure 8.3 example are shown in Table 8.1. Figure 8.4 illustrates a small number of these cases. The most common qualitative outcomes are: (1) alternative exclusion only, for small q; (2) coexistence with irregular dynamics for low intermediate q; (3) alternative exclusion or coexistence with equal mean densities for high intermediate q; and (4) coexistence with equal mean densities for sufficiently large q. The dynamics within each category can exhibit a wide range of forms and cycle lengths and can be aperiodic. There are a number of values of q that do not fit this rough categorization; a range of period lengths (30 through 34) between categories (2) and (3) only exhibit alternative exclusion. As noted above, when the 1-consumer system has a single attractor for a given environmental period, the ability of an anti-synchronized competitor to invade from low density is determined by whether the resident produces a mean resource above or below the aseasonal equilibrium (20 in this case). Note that the initial population size and time of arrival for the invading species can be important in the outcome (given two or more attractors), since an initially much less abundant species may become more abundant during the first part of the post-invasion period if conditions at $t = 0$ give it a sufficient advantage over its competitor.

There is often a single outcome of these 2-consumer competition models for cases in which the corresponding single-species model has alternative outcomes. For periods of $q = 11$ through 18, there are two attractors for the initial resident consumer species, having mean resource abundances on either side of the equilibrium value of 20. If the resident occupies an attractor with mean resource abundance > 20, the invader initially increases. However, numerical integration shows that the growth of the invader shifts the dynamics of the resident to its other attractor, which then causes extinction of the invader. This makes it impossible for the initial invader to re-invade. For example, when $q = 12$ in a single-consumer system, there is an attractor with a period 12 cycle in abundances characterized by mean $R < 20$, and a second attractor with a higher-amplitude period 36 population cycle characterized by a mean R of 20.73. The initial increase of an anti-synchronized rare invader into this second resident attractor eventually shifts the resident dynamics as shown in Figure 8.5

(roughly around $t = 200$). This results in the exclusion of the invader. The outcome is similar for all periods $11 \le q \le 18$, which have two attractors with mean resource abundances above and below $R = 20$. Thus, all systems having integer periods $q < 20$ in 2-consumer systems exhibit eventual exclusion of a rare invader species.

Table 8.1 Outcomes of 2-consumer competition in a MacArthur model with periodic variation in c, and anti-synchronized but otherwise identical competitors The parameters in eqs (8.1) are: $\{r = 0.25, k = 0.00125, \gamma_1 = 0.9, \gamma_2 = 0.9, C_1 = 1, C_2 = 1, B_1 = 0.01, B_2 = 0.01, d_1 = 0.2, d_2 = 0.2, m = 0.001\}$ Periods (q) explored were: (1) all integer periods ≤ 12; (2) even integer periods from 14 through 60; (3) all integer multiples of 5 from 60 through 100; (4) integer multiples of 10 from 100 through 200. Some additional cases were simulated, but not included in the table.

Period, q	Outcome	Population period	Population dynamics*
12 or less	alt. exclusion	q	simple cycle
14	alt. exclusion	$2q$	2-cycle
16	alt. exclusion	$2q$	simple cycle
18	alt. exclusion	$16q$	8-cycle
20	alt. unequal coex.	$4q$	2-cycle**
22–28	coexistence	aperiodic	chaos-intervals of near exclusion
30–34	alt. exclusion	q	simple cycle**
36–58	alt. exclusion	q	simple cycle
	or coexistence	q	simple cycle
60–100	coexistence	q	simple cycle
110	altern. unequal coexistence	q	2-cycles
	or equal coexistence	q	simple cycle
120	coexistence	q or aperiodic	simple cycle, or chaos
130, 140	coexistence	aperiodic	chaos
150	coexistence	$2q$	8-cycle
160–190	coexistence	q	4-cycle
200	coexistence	$3q$	complex cycle

* Periodic dynamics described by a number of cycles are classified by the number of distinct population peaks of the resource species during the course of a single cycle. A 'simple cycle' involves one local maximum in the abundance of the resource per cycle with approximately anti-phase dynamics of the two consumers. The winner in the 'alternative exclusion' outcomes is determined by numerical dominance during the first environmental cycle; this is influenced both by abundance at time zero and by which species is favoured by the initial environmental conditions. 'Coexistence' implies equal mean densities of the consumers unless it is labelled as 'unequal'. Unequal cases of coexistence involve two equivalent attractors with the consumer identities reversed.

** $q = 20$ and $q = 30$ can exhibit extremely long transient periods of chaotic dynamics. In the former case, the chaos is characterized by one species being almost extinct, while in the latter there is alternating extreme dominance.

Consumer species 1 is the solid line; species 2 is the dashed line

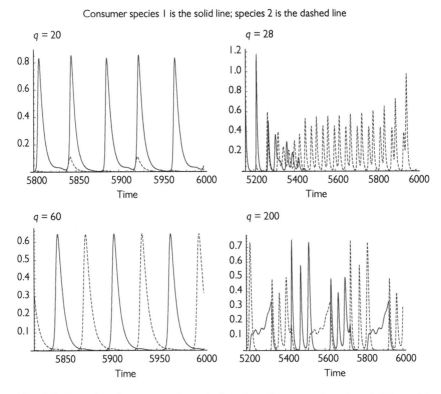

Fig. 8.4 Examples of consumer dynamics based on the parameters given in Figure 8.3 and the second consumer having a lag, L of $q/2$. The panels show examples characterized by period lengths, q.

A variant of this 2-consumer model adds equal low rates of immigration of each consumer. This is a rough approximation to modelling a single patch within a meta-community when that patch receives low levels of immigration from many other (potential unsynchronized) patches. This immigration means that there will always be non-zero abundance of all species in the focal patch, but coexistence can then be defined operationally as having mean abundances greater than some very small fraction of the equilibrium abundance. In the above example, an immigration rate 4–5 orders of magnitude less than the mean total consumer abundance did not change the qualitative dynamics described in Table 8.1 for periods listed, except for $q = 20$. In this case, the alternative cyclic attractors are replaced by a single chaotic attractor (with consumer immigration rates of 0.00001). For periods $q = 22$ through 28, which have chaotic dynamics, the stretches of time characterized by pronounced numerical dominance by one species were shortened and the minimum densities (which were many orders of magnitude less than the equilibrium) were increased by low-level immigration. This immigration also has a significant effect on the dynamics of

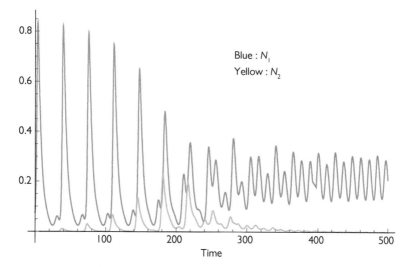

Fig. 8.5 The time course of consumer densities in a system where consumer species 1 (N_1, blue) has the parameters given under Figure 8.3. The seasonal period is $q = 12$. Consumer 2 has growth parameters identical to species 1 ($\gamma_2 = 0.9$; $C_2 = 1$; $b_2 = 0.01$; $d_2 = 0.2$), except it is anti-synchronous; i.e., $L = 6$ for species 2. The plot shows the outcome when a small number of consumer 2 is introduced into a system where consumer 1 is undergoing the period-36 cycles represented by the higher dot above $q = 12$ in Figure 8.3.

systems that are close to, but outside of, the range of periods allowing coexistence in the absence of consumer immigration. For example, if $q = 19$, immigration at a rate of 0.00001 for both consumers allows the initially less common species to persist at an average abundance approximately 1/16 that of the dominant species, rather than being excluded. Holt et al. (2003) discussed similar cases, in which a low level of immigration of a species into a 'sink' habitat can maintain significant abundances.

Larger rates of consumer immigration can eliminate the possibility of (quasi-) exclusion outcomes for some periods in the interval from 36 to 58, which has alternative exclusion and coexistence outcomes in the absence of consumer immigration. In this range of periods, greater immigration and longer periods both act to eliminate the exclusion outcomes. For example, $q = 40$ produces coexistence from all initial conditions when the consumer immigration rate is 0.001. However, when the immigration rate is 0.0001, coexistence and exclusion are both possibilities. Even when immigration is 0.0001, exclusion does not occur for $q \geq 52$.

This complicated catalogue of dynamic outcomes has many features that are incompatible with techniques of 'modern competition theory', such as invasion analysis. They also show that temporal resource segregation has many possible relationships to coexistence having no analogue in traditional partitioning in temporally constant environments.

8.2.4 *How robust and representative is the example?*

The above results are a 'best case scenario' for coexistence, in that consumers have equal parameters and anti-synchronized seasonality. The robustness of the exclusion outcomes may be explored by adding direct consumer interference ($z > 0$ in eq. (8.1)). When each of two or more consumers has purely intraspecific interference, this favours coexistence, and the magnitude of the interference, z, required for coexistence provides an index of the robustness of the exclusion outcomes. In the system considered in Figure 8.3, if $q = 30$, the maximum z that permits exclusion is 0.2251; at this point interference represents an increase of approximately 16% to the per capita death rate. Short environmental periods (or rapid dynamics of both species) produce small effects on the mean resource abundance, and these can be overcome by relatively small intraspecific interference effects. For example, a period of $q = 6$ always produces coexistence if $z > 0.0135$, which represents only a 3% increase in the mortality rate.

All of the above is based on a single set of parameters for the underlying aseasonal model. Li and Chesson (2016) have used the same MacArthur system with sinusoidal seasonal variation in c to illustrate how such variation promotes coexistence. An obvious question is whether the negative effects on coexistence shown here are restricted to a narrow range of the other parameters. Because a large enough q always produces coexistence, I examined this question for $q = 10$, as this is in the middle of the range of relatively low environmental periods that only exhibited alternative exclusion. I will describe how coexistence is affected by changing the consumer parameters (B, C, d) from the values they have in the example above. Each of these parameters may be increased or decreased by a large amount without changing the qualitative form of the outcome. Values of d from 0.02 to 1.8 were examined (the original system assumed $d = 0.2$). All d values within this range only produced alternative exclusion outcomes. Values of B between 0.00125 and 0.1 were simulated (baseline $B = 0.01$); only large B values between roughly 0.08 and 0.1 produced coexistence. The original C ($= 1$) was changed to values between 0.1 and 10. Only $C \leq 0.125$ and $C \geq 8$ produced coexistence. The highest value ($C = 10$) produces an equilibrium resource abundance less than 2% of its carrying capacity, which seems unlikely in most natural systems; persistence of the consumer would be threatened in such systems. Slower consumer dynamics may be modelled by reducing both B and d by equal proportions. In the example considered in Figure 8.3 with $q = 10$, the outcome remains alternative exclusion until B and d are decreased to 1/8 or less of their original value ($B \leq 0.00125$; $d \leq 0.025$). At these values, consumer abundance only varies from approximately 4.9% below its mean value to approximately 3.5% above it. Such a small level of seasonal variation would be difficult to detect in most natural systems. These results suggest that the phenomenon of exclusion due to seasonal partitioning is not restricted to a very narrow range of parameters.

As an example that is unrelated to the case in Figure 8.3, consider the Li and Chesson (2016) parameter set modified by assuming a lag of 1/2 rather than their 1/4 of a seasonal period (and $q = 1$). This change makes coexistence easier than in their analysis. Otherwise, the parameters are equivalent to those used by Li and Chesson (2016) except that the capture rate constant, C, has been doubled. The parameters are: $r = 2.5$, $k = 0.0125$, $\gamma_1 = \gamma_2 = 1$, $C_1 = C_2 = 0.1$, $B_1 = B_2 = 0.25$, $d_1 = d_2 = 1$, $m = 0$, $q = 1$, and $L = 0.5$. The common mortality rate falls within a range where Li and Chesson (2016) obtained coexistence over a broad range of relative mortalities. The dynamics of this system, like theirs, is characterized by small amplitude oscillations of consumers around the resource equilibrium. However, the mean resource density here is 39.5056, compared to the aseasonal equilibrium of 40. This implies alternative exclusion outcomes. Exclusion of a rare invading type occurs until the invader has a d less than its competitor by 0.024718. Larger values of the consumption rates make coexistence less likely; $C = 0.2$ requires a d-advantage of 0.042055 for invasion, while $C = 3$ requires a d-advantage of 0.28801. Proportional increases in both C and d, or in both B and d, favour exclusion; both of these changes produce lower mean resource abundance relative to the aseasonal equilibrium.

This system has much more rapid consumer dynamics than does the system explored in Figure 8.3. As a result, the qualitative dynamics change more rapidly with increasing q. In the single-species system, the simple cycles of period q change to a 2-cycle with period $2q$ for $q \geq 2.2$. Coexistence becomes possible at periods just above $q = 2.3$ (two alternative 2-cycles having very unequal densities), and this persists until $q \approx 3.3$, where the pattern reverts to a simple period q cycle and alternative exclusion. At $q \approx 4$ this changes to coexistence with simple anti-synchronized period q cycles. Anti-synchronized coexistence continues for larger q, although the dynamics change to a 2-cycle at approximately $q = 12$. Although the system does not have as wide a range of different outcomes across different seasonal periods as in Figure 8.3, it does show that periodic variation may produce exclusion or alternative asymmetrical coexistence outcomes, neither of which is consistent with the approximately anti-synchronized coexistence produced by very rapid resource dynamics.

8.2.5 Coexistence of a seasonal and an aseasonal consumer

In many real-world seasonal systems, the variation is an annual cycle in which part of the year is worse for all consumers. This suggests that there may be many more cases in which species differ in the amplitude rather than the timing of their responses. The range of potential outcomes in this scenario may be explored by considering the extreme case in which one species lacks any seasonal response.

The single-consumer system from Figure 8.3 above may be modified to understand this case by assuming that γ is close to its maximum of 1 for one consumer and $\gamma = 0$ for the other, but all other parameters are equal in the two consumers. The example examined here assumes $\gamma_1 = 0.9$ and $\gamma_2 = 0$. Given equal values for all other consumer parameters, the seasonal species 1 then has a zero rate of increase when it is an extremely rare invader and species 2 is at its stable equilibrium, and this is true regardless of the period or amplitude of fluctuations in $c_1(t)$. The seasonal species 1 may be able to increase if the small amount of seasonal variation in the resource produced by its small initial numbers favours its own growth; however, this must be determined numerically. When the aseasonal species 2 is the invader, it can increase when rare if and only if species 1 produces an average resource abundance greater than the aseasonal equilibrium resource requirement. Invasion must result in persistence of an invading species 2 in cases where the seasonal species 1 has only a single attractor when it is alone, but numerical analysis is again necessary to determine what occurs after invasion when there are two or more attractors for the seasonal species when it occurs alone. Table 8.2 presents the outcomes for alternative scenarios in which the aseasonal species either has a slight advantage or a slight disadvantage. This inequality prevents the neutrality that characterizes the case of invasion by a very small number of seasonal species. Here I consider two cases: $d_2 = 0.1999$ or $d_2 = 0.2001$. In the first case (aseasonal advantage; $d_2 = 0.1999$), exclusion of the seasonal species is always a possible outcome when the aseasonal species is at its single-consumer equilibrium. However, exclusion of the aseasonal species 2 is possible if the seasonal species 1 reaches a dynamic attractor with a mean resource density less than 0.1999. Figure 8.3 shows that this is possible for $q < 22$, as well as a range of intermediate $q \geq 30$. In this example coexistence is only possible for periods around $q = 24$ or 26. In these cases, the seasonal species, when alone, can exhibit a longer period high amplitude fluctuation with a mean $R > 20$; this can be invaded

Table 8.2 Outcomes of 2-consumer competition with periodic variation in c for consumer species 1 and no variation in the otherwise nearly identical consumer species 2 The parameters in eqs (8.1) are: {$r = 0.25$, k = 0.00125, $\gamma_1 = 0.9$, $\gamma_2 = 0$, $C_1 = 1$, $C_2 = 1$, $B_1 = 0.01$, $B_2 = 0.01$, $d_1 = 0.2$, $m = 0.001$. In 2A $d_2 = 0.1999$; in 2B $d_2 = 0.2001$} Periods explored were: (1) all integer periods ≤ 12; (2) even integer periods from 14 through 60; (3) all integer multiples of 5 from 60 through 100; (4) integer multiples of 10 from 100 through 200. Some additional cases were simulated, but not included in the table.

2A. Aseasonal species has a slight advantage ($d_2 = 0.1999$)

Period, q	Outcome
22 or less	alt. exclusion (stable if sp. 2 wins; various dynamics for different q if sp. 1 wins)
24, 26	coexistence, aperiodic, alternating near-exclusion
28	species 2 wins (after lengthy transient if it is the invader)
30–56	alt. exclusion (simple cycle if sp. 1 wins; stable if sp. 2 wins)
≥ 58	species 2 wins (stable equilibrium)

2B. Aseasonal species has a slight disadvantage ($d_2 = 0.2001$)

Period, q	Outcome
18 or less	species 2 wins
20	coexistence; $5q$ 2-cycle; species 1 dominant
22	species 1 wins; chaotic
24, 26	coexistence; chaotic; alternating dominance
28	coexistence; (30:1 dominance by species 2)
30–54	species 1 wins; or coex. with species 2 dominant
≥ 56	coexistence; species 2 numerically dominant

by the aseasonal species. However, invasion shifts the seasonal species to a shorter period fluctuation (period q), which results in temporary near-exclusion of the aseasonal type. This near exclusion shifts the dynamics of the seasonal species back to the long period fluctuations, which permits another period of its own near-exclusion. The resulting chaotic long-term dynamics are illustrated in Figure 8.6.

The second case considered in Table 8.2 considers a slight mortality disadvantage for the aseasonal species ($d_2 = 0.2001$). This means that the seasonal species can always increase when it is a rare invader. The result is exclusion of the aseasonal species regardless of initial densities when the period, q is relatively small. The chaotic coexistence that occurs for $q = 24$ has a pattern that is almost identical to that for the marginally lower d_2 illustrated in Figure 8.6. Periods of 30–54 have alternative outcomes, but species 1 is never excluded completely. For large q (56 and larger) the aseasonal species 2 is always the numerical dominant.

The single-seasonal-consumer results in Figure 8.3 might suggest more complicated outcomes for periods between $q = 12$ and $q = 18$, where there are alternative attractors. However, as in the case of two seasonal species, invasion of an aseasonal species with the resident at its higher-mean-R attractor shifts the resident to its low-R attractor, and the invader is then excluded.

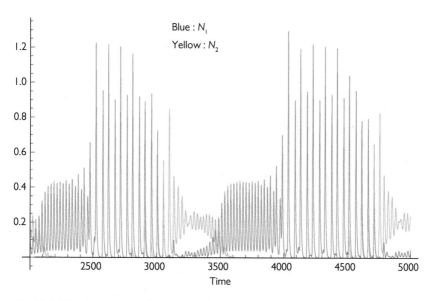

Fig. 8.6 The time course of consumer dynamics in a 2-consumer system with consumer 1 having seasonal variation in c and consumer 2 having no seasonal variation ($\gamma_1 = 0.9; \gamma_2 = 0$). The environmental period is $q = 24$. Both species have identical values of the other parameters, which are given in the legend of Figure 8.3. The dynamics are aperiodic.

8.2.6 A more complete description of seasonal interactions

The analysis thus far has not examined the quantitative nature of the interaction; how much does one consumer change in abundance with some change that only affects the other consumer? How much does abundance change (and in what direction) if the other consumer is removed? A related set of questions deals with the robustness of coexistence generated by seasonality when it occurs. In other words, how far can a fitness-related parameter in one species be changed without causing extinction of the other species (or of itself)? In the case of these unequal systems, how does the similarity in seasonal response curves affect the ability of species to coexist? Can species that are unable to coexist in completely symmetrical cases be made to coexist by giving one or the other an advantage or disadvantage? These questions all require numerical approaches, so there is only room for a few illustrative examples below.

The robustness of coexistence, when it occurs in the symmetrical system, can be measured by the maximum disadvantage in one species' fitness parameter(s) (or advantage in its competitor's parameter(s)) that allows both to persist. A convenient parameter in the models considered here (and in many natural systems) is the per capita death rate of one of the consumer species. Changing d, or any parameter that only affects the per capita growth rate of the focal species, is also the way that indirect interactions are traditionally measured (Yodzis 1988). In the traditional (aseasonal) MacArthur model, the change in population sizes of both species produced by a defined change in the mortality of one of them is independent of the current population sizes or mortality rates (provided the comparison involves no change in the number of consumers or resources). The next two figures show that this is not the case in the seasonal model considered here.

Figure 8.7 shows the impact of changing the mortality rate of consumer species 1 in the example of anti-synchronized species with variation in $c(t)$. The parameters are as in Figure 8.3, with $q = 60$ and $L = 30$. Both consumers have an initial mortality of 0.2. Species 1 (blue dots) goes extinct when its mortality reaches 0.29, and species 2 (red dots) goes extinct when species 1's mortality drops to 0.07. Notice that both extinctions represent discontinuous changes in population size; in both cases, the drop is larger than the difference between the equilibrium size (when $d_1 = 0.2$) and the population size for the mortality rate immediately lower than the extinction value ($d_1 = 0.08$ for extinction of species 2 and $d_1 = 0.28$ for extinction of species 1). Near the original equilibrium, the effects of mortality are very small, and consumer 2's abundance actually decreases as consumer 1's mortality increases for initial mortalities between $d_1 = 0.18$ through and including $d_1 = 0.24$. Thus, for close to 1/3 of the possible range of mortalities, species 1 has a positive effect on species 2, according to the standard (Yodzis 1988) method of measuring indirect effects. In fact, $d_1 = 0.27$ is the first mortality rate greater than the equilibrium of 0.2 that produces a consumer 2 population greater than its original equilibrium value (at $d_1 = d_2 = 0.2$). At $d_1 = 0.25$ the two consumers have nearly identical mean population sizes. Increasing the mortality of one species does not have a large effect on the abundance of the other until the mortality of the manipulated species is close to that which causes its

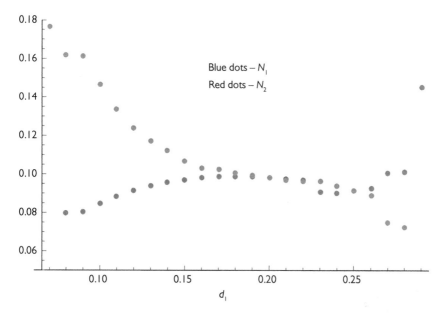

Fig. 8.7 The response of mean consumer abundances to increased or decreased per capita mortality of consumer 1. The system is based on the parameters given in Figure 8.3 with $q = 60$, and both consumers having the same parameters except that consumer 2's c value is lagged by 30 time units.

own extinction. Mortalities of 0.25 through 0.28 are associated with very long period or irregular population fluctuations.

The range of mortality rates allowing coexistence for the system in Figure 8.7 is relatively large. A good comparison case is an analogous non-seasonal system with standard resource partitioning. Here the resource population consists of two species, each with half the equilibrium abundance of the preceding case (due to a k value twice as large). The two consumers are assumed to have opposite specializations on the two resources, but have the same mean C of 1; i.e. a lower value of C_L and a higher value of C_H where $C_L + C_H = 2$. Moderate specialization with $C_L = 0.8$ and $C_H = 1.2$ yields coexistence for $d_1 = 0.14$ through 0.29, a range approximately 3/4 that in Figure 8.7. Greater partitioning with $C_L = 0.7$ and $C_H = 1.3$ yields coexistence for $d_1 = 0.12$ through 0.36, which is 6/5 times the range in Figure 8.7. The extreme values of mortality that limit coexistence in this non-seasonal model correspond to cases in which the lower-d species causes extinction of one of the resources, and this produces a discontinuous change in the abundance of the surviving consumer species.

Another example of responses to mortality (i.e., 'interaction shape') is provided by a seasonal MacArthur model with a single resource, anti-synchronized consumers, and a period length of $q = 60$. The other initial parameters are $r = 1$; $k = 0.01$,

$m = 0.001$, $\gamma_1 = \gamma_2 = 1$; $C_1 = C_2 = 0.25$; $B_1 = B_2 = 0.2$; $d_1 = d_2 = 0.25$. These parameters produce anti-synchronized dynamics of the two consumers with equal mean densities of 1.6928. Figure 8.8 shows how increasing the per capita mortality rate of species 1 alters the temporal mean abundances of both species. In Figure 8.8A, the mean abundance of consumer 1 is given by blue dots, and that of consumer 2 by black dots. Figure 8.8B shows the corresponding estimated competition coefficient at each mortality rate calculated from the ratio of the population change in the unmanipulated species to that in the manipulated one, whose death rate is increased to the next larger value. Most of the larger jumps in population sizes or the competition coefficient occur as a result of qualitative changes in the dynamics as the death rate of consumer 1 is increased. Exclusion of species 1 occurs when d_1 reaches approximately 2.1. Positive values of the competition coefficient imply that the mean densities of both species change in the same direction in response to increased mortality of species 1; there are cases in which this involves increases in both, and others in which both decrease following higher mortality of consumer 1. The frequent competition coefficients that are much larger than 1 in absolute magnitude indicate a much greater effect of the parameter change of species 1 on the population size of its competitor than on its own population size. This occurs in spite of maximal temporal partitioning and equal mean values of all of the parameters.

If the system described in Figures 8.8A and 8.8B had a shorter period, q, the relationships would change considerably. The system only exhibits alternative exclusion for integer periods of 6 and smaller. The period-60 case illustrated in Figure 8.8 is thus well above the $q = 7$ required to produce coexistence. In general, the extent of competition decreases with longer periods. In spite of the long periods, the density of each species alone (4.489) is much greater than the sum of the two densities when they occur together with equal mortalities (3.384). This increase is greater than the largest effect possible under an LV model of two exploitative competitors.

Another potential consequence of seasonal variation in systems with equal mean parameter values is the presence of alternative coexistence attractors having unequal mean abundances of the two consumers. In these cases, the effect of one species on another (as measured by the response to altered mortality of the first) differs greatly depending on which of the alternative states is originally occupied. An example of such a scenario based on the MacArthur system is exhibited by the following parameter set: {$r = 1$, $k = 0.01$, $\gamma_1 = \gamma_2 = 1$, $C_1 = C_2 = 0.25$, $B_1 = B_2 = 0.1$, $d_1 = d_2 = 0.25$, $m = 0.001$, $q = 7$, $L = 3.5$}. This has two alternative long-period cycles in which the mean abundance of one consumer is approximately three times greater than that of the other. Initial conditions determine which is more abundant. Not surprisingly, any quantitative measure of the interaction will depend on what attractor characterizes the initial state, and whether the perturbation used to measure the interaction shifts the system to the other attractor.

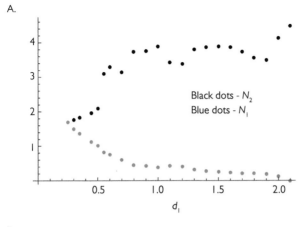

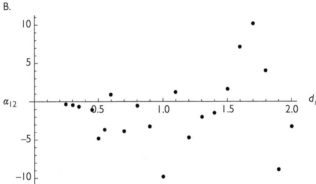

Fig. 8.8 Panel A. The mean population sizes of two competing consumers as a function of the mortality rate of consumer 1 for a MacArthur system in which the common parameter values are: $q = 60$; $L = 30$; $r = 1$; $k = 0.01$; $m = 0.001$; $\gamma_1 = \gamma_2 = 1$; $C_1 = C_2 = 0.25$; $B_1 = B_2 = 0.2$; $d_2 = 0.25$. The initial mortality of consumer 1 is identical to that of species 2 ($d_1 = 0.25$), and the graph shows the changes in both species as the mortality rate of consumer 1 is increased.
Panel B. The competition coefficient based on the changes shown in panel A; the method of calculation is described in the text.

8.2.7 Seasonality resource conversion efficiency, b

The attack rate, c, has been assumed to be the seasonally changing characteristic in most previous analyses of seasonally varying competition (Abrams 1984a, Loreau 1992, Li and Chesson 2016). An alternative is that the consumers' resource conversion efficiencies, b, vary seasonally. Variation in b differs from variation in c qualitatively,

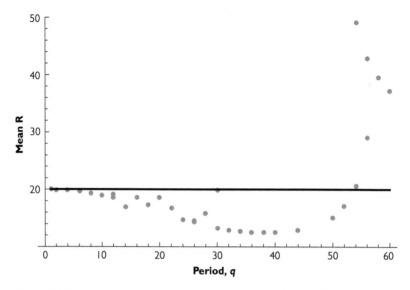

Fig. 8.9 The mean resource density as a function of period length for a system with a single consumer species. The parameters are as in Figure 8.3, but the variation is in the conversion efficiency, b.

in that the variable factor in this case only has a direct effect on the consumer's dynamics. Assume b varies periodically in a single-consumer system that has the same mean parameter values as in Figure 8.3. This produces mean resource densities for a range of period lengths, q, given in Figure 8.9. These results show that the single-consumer system undergoes a sequence of changes in dynamics as the environmental period is lengthened.

In the single-consumer system, the lowest period that has an attractor with a mean resource density greater than the constant equilibrium value of 20 is $q = 54$. Thus periods < 54 produce alternative exclusion. Periods above 60 (not shown) also had a single attractor characterized by a mean resource density significantly above 20 ($q = 120$ was the longest period that was examined). The results for competition between anti-synchronized consumers over a wide range of periods are summarized in Table 8.3. Alternative exclusion happens over a broad range of environmental periods; coexistence is not a possible outcome until $q \geq 54$. All periods above 54 exhibited coexistence; $q = 54$ was the only case examined where coexistence, exclusion of consumer 1, and exclusion of consumer 2 were all possible outcomes.

The main differences between competition with b-variation in this example and the corresponding system having variation in c are: (1) alternative coexistence and exclusion outcomes for a particular system are less common; and (2) there is only a single transition between the two types of outcome as the period length is changed in a directional manner; alternative exclusion occurs for $q < 54$, and coexistence for $q > 54$. These two qualitative differences have been observed with periodic b for several other parameter sets, but their generality remains unknown. The system described

Table 8.3 Outcomes of 2-consumer competition with periodic variation in b. The parameters in eqs (8.1) are $\{r = 0.25, k = 0.00125, \gamma_1 = 0.9, \gamma_2 = 0.9, C_1 = 1, C_2 = 1, B_1 = 0.01, B_2 = 0.01, d_1 = 0.2, d_2 = 0.2, m = 0.001\}$

Period, q	Outcome	Population period	Population dynamics[*]
12 or less	altern. exclusion	q	simple cycle
14	altern. exclusion	$2q$	simple cycle
16	altern. exclusion	aperiodic	chaos
18	altern. exclusion	$12q$	4-cycle
20	altern. exclusion	$2q$	simple-cycle
22	altern. exclusion	aperiodic	chaos
24	altern. exclusion	$4q$	2-cycle
26–54	altern. exclusion	q	simple cycle
54 alt.	coexistence	aperiodic	chaos
56, 58	coexistence	aperiodic	chaos
60	coexistence	$6q$	alternative 6-cycles
70	coexistence	$3q$	alternative 4-cycles
80	coexistence	aperiodic	irregular
90	coexistence	aperiodic	irregular
100	coexistence	q	simple cycle
120	coexistence	q	simple cycle

[*] Dynamics are classified by the number of distinct population peaks of the resource during the course of a single cycle

in Table 8.3, and most other cases of b-variation examined also had mutual exclusion outcomes at low q for parameter sets where exclusion was a possibility.

8.2.8 A 3-consumer system with variation in c

The analysis has thus far only considered two consumers. A wide range of outcomes also occurs in the case of three temporally displaced, but otherwise equal competitors. Table 8.4 is based on the same mean parameter values as Figure 8.3, except that there are three competitors, with the second and third having lags of 1/3 and 2/3 of a period relative to consumer 1.

The population trajectories for one of the three attractors for $q = 300$ are illustrated in Figure 8.10. The other two attractors are identical in form and in their temporal order of species dominance, but differ in which of the consumers is identified with a particular dynamic role. Each line in the figure may characterize any one of the three species. The species identities associated with the three trajectories are determined by initial conditions. Note that the attractor is either aperiodic or has a cycle with a very long period.

Table 8.4 Competitive outcomes with three consumers and variation in *c*. The species have *c*-curves that are offset by 120 degrees (i.e., $q/3$). Other parameters are given in the text

Period, *q*	Outcome	Population period	Population dynamics
12 or less	altern. exclusion*	*q*	simple cycle
13	altern. exclusion	2*q*	two cycle
14,15	altern. exclusion	2*q*	asymm. cycle
16	altern. exclusion	irregular, then 2*q*	asymm. cycle
17	altern. exclusion	irregular, then 4*q*	2-cycle
(All of above outcomes are preceded by ever-lengthening periods of dominance by different species with ever-decreasing population minima for the other two species; each species is largely replaced by the species having the next earlier peak in $c(t)$)			
18	coexistence	irregular	chaos or very long cycle
20	coexistence	irregular	chaotic; varying dominance times
22	coexistence	irregular	chaotic; lengthening dom. times
24	coexistence	irregular	chaotic; order of dominance shifts
28	coexistence	irregular	irregular dominance intervals
30	coexistence	irregular	chaos-lengthening dominance
35, 40, 50	altern. exclusion	*q*	asymm. cycle
60	coexistence	*q*(short term)	long periods; altern. dominance
70	coexistence	5*q*	4-cycle; earlier replaces later
80	coexistence	11*q*	16-cycle; earlier replaces later
90,	coexistence	2*q*	simple cycles; earlier replaces later
120, 150	coexistence	*q*	simple cycles; later replaces earlier
300	coexistence	4*q*	three complex cycles that differ only in species identities

* 'altern. exclusion' means any one of the three species can be the one that persists.

In summary of the Table 8.4 example, coexistence of two or three species is impossible for short seasonal period lengths ($q < 18$; equivalent to slow population dynamics), and for a range of intermediate period lengths that includes the range of *q* between 35 and 50. Coexistence of all three is possible for all periods of 60 and above, as well as some more limited range of period lengths from approximately 18 to approximately 30. For very long periods (e.g., $q = 300$; Figure 8.10) each consumer species undergoes several local peaks in abundance during its period of dominance, but each species has a different pattern of variation. The order of dominance of the coexisting species at periods of 120 and above corresponds to the temporal order of their lags (1 replaced by 2, 2 by 3, and 3 by 1). There are three distinct trajectories over time, with each species capable of exhibiting any trajectory. Seasonal periods significantly longer than $q = 300$ were not explored systematically.

Consumer species 1 (lag 0) is the solid line, species 2 (lag 100) is short dashed, and species 3 (lag 200) is long dashed.

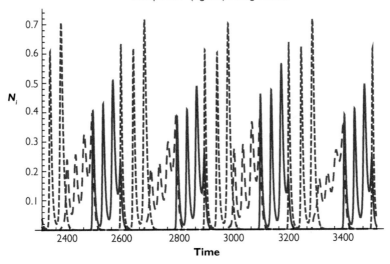

Fig. 8.10 The dynamics of a 3-competitor system with seasonality in attack rates. Here, $q = 300$, and consumers 2 and 3 have lags of 100 and 200 time units in their attack rates relative to consumer 1. The parameter values are given in the text.

8.2.9 A 2-resource system with temporal and non-temporal partitioning

The invasion curves shown early in this analysis (Figures 8.2A, B) suggest that the disadvantage to a rare invader with a lag is maximal for intermediate lags between 0 and 180 degrees. Given the symmetry of the system and the variation, the same applies to the effects (often an advantage for the invader) of lags between 180 and 360 degrees ($q/2 - q$). This should imply that, if the aseasonal system is characterized by coexistence due to resource partitioning or direct negative self-effects, coexistence in the seasonal system will occur more often when the two consumers have either very similar or very different patterns of temporal variation.

Introducing a second resource makes it possible to examine the impact of seasonal variation in systems that have partitioning by resource type. The example from Figure 8.3 may be converted into an equivalent 2-resource model by introducing a second resource, with the capture rate constants of consumer i on resource j denoted C_{ij}. The mean of the two C_{ij} values of each consumer is equal to the C_i value used in Figure 8.3. The two r values are the same as those in Figure 8.3, and the k value of each resource is multiplied by a factor of 2 to maintain the same equilibrium total resource population in the absence of consumption. When the B and d parameters are each equal for the two consumers, the dynamics of each consumer and the summed resource populations are identical to those in the single-resource system of Figure 8.3.

Table 8.5 Lags producing one or more outcomes of coexistence for a 2-resource system based on the Figure 8.3 example with period $q = 36$

Overlap[*]		
0.9231	Low L coexistence	Low L coexistence or exclusion
	$0 \leq L \leq 3.65$	$3.56 < L \leq 3.65$
	Medium L coexistence	Medium L coexistence or exclusion
	$16.6 < L < 19.4$	$15.9 \leq L \leq 16.6$ and $19.4 \leq L \leq 21.1$
	High L coexistence	High L coexistence or exclusion
	$32.45 \leq L \leq 36$	$32.33 < L \leq 32.44$
0.8	Low L coexistence	Low L coexistence or exclusion
	$0 \leq L \leq 6.46$	$6.47 \leq L \leq 6.53$
	Medium L coexistence	
	$12.41 \leq L \leq 23.59$	
	High L coexistence	
	$29.54 \leq L \leq 36$	
0.72414	Coexistence for all lags	

[*] Calculated based on the MacArthur formula; see text

Resource partitioning may be introduced by allowing C_{i1} and C_{i2} to differ from each other. Table 8.5 gives the ranges of lags L that permit coexistence of two consumers in systems having three different levels of resource partitioning, determined by the C_{ij}. The table considers low resource partitioning ($C_{ii} = 1.2$; $C_{ij} = 0.8$; MacArthur $\alpha = 0.9231$) and two higher levels of partitioning ($C_{ii} = 1.333$; $C_{ij} = 0.6667$; MacArthur $\alpha = 0.8$; and $C_{ii} = 1.4$; $C_{ij} = 0.6$; MacArthur $\alpha = 0.72414$). The table lists the lags that allow coexistence of the two consumers. Values of L close to the limits of coexistence are in some cases characterized by alternative outcomes of exclusion of the disfavoured species or coexistence. The disfavoured species is the lagged one for $L < 18$, and the non-lagged species when $L > 18$. For all lags other than $L = 18$, the abundances are unequal. Recall that in the corresponding system with one resource and a lag of exactly $q/2$, coexistence is impossible for q between 30 and 35, while both coexistence and exclusion of either species were possible for $36 \leq q < 60$.

8.3 Competition in other 2-consumer–1-resource models

While the MacArthur consumer–resource model with all-linear component functions is very widely used, at least two other types of components have frequently been employed in competition theory; (1) models with abiotic (chemostat) resource dynamics; and (2) models in which the consumer functional response is type II. The former has fewer possibilities for large delays in the resource dynamics, while the latter has more such possibilities, as well as the possibility of self-sustaining cycles in the absence of seasonality. When either of these alternative scenarios is adopted in seasonal models, the dynamics change significantly. The combination of biotic resources and type II responses creates an interaction of the exogenous and endogenous periodicities, which is known to produce complex dynamics in 1-consumer–1-resource systems (Rinaldi et al. 1993, Sauve et al. 2020). In the 2-consumer–2-resource case, both of these alternative assumptions also produce a wide range of dynamical behaviours, including cases having alternative exclusion as the only outcome, and other cases with both exclusion and coexistence outcomes. Exclusion due to temporal partitioning is again associated with consumer dynamics that are not much slower than the resource dynamics and a relatively short seasonal period (equivalently, slow dynamics of all species). Here I will again just provide a few examples that illustrate some of the similarities and differences of these alternative models from the ones examined earlier in the chapter. All of these models deserve a much more comprehensive analysis.

8.3.1 Systems with abiotic resources

The dynamics of competitive systems having abiotic resources are usually simpler than those with biotic resources, and differences in seasonal responses are more likely to promote coexistence. However, there is still a lag in the response of the resource abundance to temporal cycles in consumption, and this can still lead to a failure of seasonal partitioning to produce coexistence, as well as producing various dynamic complexities. Relatively high resource exit rate parameters imply strong stabilization, so the following examples have low values of E in the abiotic growth function, $I - ER$. Table 8.6 describes the dynamical outcomes for an example that displays a wide range of outcomes as the period length q is increased. The qualitative outcomes here have many similarities with the logistic resource results presented in Section 8.2. A broad range of relatively low period lengths produce only alternative exclusion, and sufficiently long periods produce coexistence with anti-synchronized cycles. Intermediate periods can produce two alternative coexistence equilibria in which one species has a greater average abundance than the other, or a single coexistence attractor with aperiodic variation.

If the Table 8.6 system is changed to have a smaller death rate (equal for both consumers), the upper boundary of the range of q producing alternative exclusion is reduced; i.e., coexistence is favoured. The opposite is true for death rates greater than that in Table 8.6. As in the model with logistic resources, there are abrupt changes in the limiting dynamics with changes in the mortality rate of one consumer species. Even without abrupt changes, measures of interspecific effects change greatly

Table 8.6 Outcomes in the abiotic resource-linear functional response model {parameters: $I = 0.5$, $E = 0.005$, $\gamma_1 = \gamma_2 = 1$; $C_1 = C_2 = 0.2$, $B_1 = B_2 = 0.25$; $D_1 = D_2 = 0.5$}

Period, q	Outcome	Cycle period	Population dynamics[*]
≤ 51	alternative exclusion	q	simple cycle
52–55	alternative unequal coexist.	q	simple-cycle
56,57	alternative unequal coexist.	$2q$	2-cycle
58	coexistence; two mirror-image cycles with unequal abundance or unequal chaos	q or aperiodic	2-cycle or chaos
59, 60	coexistence; single chaotic (equal mean abundance)	aperiodic	chaos
61–64	coexistence; single complex or chaotic cycle (equal mean abundance)	$\gg q$ or aperiodic	complex cycle
64, 65	coexistence; mirror image unequal cycles	$2q$	2-cycles
66-70	coexistence; one equal and two unequal attractors	equal-period $3q$ unequals $2q$	one 3-cycle; two simple cycles
71–74	coexistence; two equal cycles	q or $3q$	simple or 2-cycle
≥ 75	coexistence; antisynchronous equivalent cycles	q	simple until q exceeds approximately 120

[*] Cycles are classified by number of maxima in abundance for the more abundant (or both) consumer species during the course of the cycle (where 'simple' denotes 1 peak per species per period q). Equal and unequal refer to the mean consumer abundances.

as the mortality rate of one species is increased from the minimum to the maximum allowing coexistence.

Cases similar to that considered in Table 8.6 are also characterized by similar possibilities for exclusion or coexistence when an aseasonal species competes with a seasonal one that is otherwise equivalent. The seasonal species can prevent invasion by the aseasonal species or vice versa in the abiotic resource model provided that $q < 52$; for lower q, a seasonal consumer produces a lower mean resource density than the equilibrium resource density for the aseasonal one. However, for $q \geq 52$, the aseasonal species always excludes the seasonal. This outcome for long environmental periods is sensitive to the exact equality of all of the mean fixed parameter values; if the aseasonal species has a slight disadvantage in one of the mean parameters (resulting in a higher mean resource abundance in a single-consumer system), it can coexist with the seasonal species in environments with long environmental periods (or equivalently, with rapid consumer and resource dynamical rates).

8.3.2 Biotic resources with type II functional responses

Another simple model assumes logistic resource growth and consumers that have type II (disc equation) functional responses. Here the resource capture rate of a consumer individual is given by $CR/(1 + ChR)$, where h is handling time. This adds the possibility of sustained cycles in the absence of seasonality. Seasonal systems with a single type II consumer species and a living resource are known to have a wide range of complicated dynamical behaviours due to interactions of the seasonal forcing and the forcing due to the inherent predator–prey cycle (King and Schaffer 1999; Sauve et al. 2020). This suggests that the range of dynamical patterns that can be observed in the 2-consumer case is likely greater than for either of the linear functional response models considered thus far in this chapter. However, the possibility of sustained or very slowly damped cycles in non-seasonal logistic-type II systems raises the possibility that these systems could display different responses to seasonal variation. Sustained cycles in an aseasonal single-consumer system occur when

$$d < \frac{B\,(Chr - k)}{h(Chr + k)} \tag{8.3}$$

Thus, sustained cycles do not occur for any mortality rate when $Chr/k < 1$. For example if $C = 1$, this implies that an aseasonal system is stable for all d when $h < k/r$; i.e., $h \le 0.005$ when the other parameters are as in the Figure 8.3 example.

A type II response alters the nature of cycles in seasonal 1-consumer systems, even if sustained cycles do not occur in the aseasonal case having the same parameters. However, the broad pattern of coexistence and exclusion in a system of two anti-synchronized competitors described above is often similar, at least for systems that have mortality rates slightly greater than the stability threshold given by eq. (8.3). If the Figure 8.3 example of a MacArthur model is changed to have $h = 0.005$ for each consumer, relatively short periods, q, usually produce alternative exclusion outcomes, and relatively long periods produce coexistence. The dividing line between these two outcomes occurs between periods of 67 and 68. The dynamics of this system are more often aperiodic than for the linear functional response system, and there are more frequent changes in the qualitative pattern of the dynamics with increasing q. The presence of a significant but low-level consumer immigration allows coexistence for a range of intermediate period lengths (roughly q between 22 and 28), and allows a range of alternative coexistence or exclusion outcomes for periods between 50 and 60. However, periods > 60 only produce coexistence. Doubling h to 0.01 in this example (with no consumer immigration) produces cyclic dynamics even in the absence of seasonal variation. However, the general impact of seasonal cycles on coexistence is similar to the previous case of $h = 0,005$; alternative exclusion for shorter periods, and coexistence for longer periods, with the switch occurring between $q = 77$ and $q = 78$.

8.3.3 Abiotic resources with type II functional responses

Here, the type II functional response does not destabilize the equilibrium in the absence of seasonal variation. The general pattern of alternative exclusion for low environmental periods and coexistence for greater periods continues to hold in this case for a system broadly similar to the previously considered linear functional response–abiotic resource model. It has parameters: $I = 0.5$, $E = 0.005$, $\gamma_1 = \gamma_2 = 0.9$; $C_1 = C_2 = 0.2$, $h_1 = h_2 = 0.05$; $B_1 = B_2 = 0.05$; $D_1 = D_2 = 0.2$, with $L = q/2$. Alternative exclusion outcomes occur from period lengths of $q = 2$ through $q = 180$; periods of 190 through 248 yield two 'mirror image' attractors having unequal mean abundances for the two consumers. Periods of 250 and above yield coexistence with anti-synchronized cycles and equal mean consumer abundances.

8.4 Discussion

When two consumers share a single limiting resource, different responses to seasonal variation by otherwise equivalent consumer species can allow or prohibit coexistence. The set of outcomes possible in the models considered here is surprisingly complicated. The present numerical analysis of a range of very simple models needs to be extended to a more complete and rigorous analysis of a wider range of consumer–resource systems. Nevertheless seasonality clearly has the potential to make coexistence more difficult and/or to increase competitive effects between species. Seasonality can also make coexistence possible, and/or produce multiple alternative outcomes. Coexistence may become more difficult with greater difference between the consumer species in their seasonal response curves. Seasonal competition may also produce highly nonlinear, and even nonmonotonic, responses of both species to monotonic changes in the density-independent mortality rate of one of them. There are many oversimplified features of the models considered here, but their component functions are used in most of the current theoretical literature about competition. It is notable that many of the same complexities in outcome described here were observed by Namba and Takahashi (1993) in an LV competition model with asynchronous variation both between species and between different parameters within a species.

This analysis challenges the standard answers to a number of common questions about competition:

(1) Does mutual invasion imply coexistence, and is it needed for coexistence? Most research seems to assume 'yes' answers to both of these, but there are many situations in the models considered here where each answer is 'no'.

(2) Do temporal differences in resource use always promote coexistence? While previous work has shown that not all differences promote coexistence (e.g., Li and Chesson 2016), there seems to have been a lack of appreciation of the fact that some temporal differences in resource use may actually prevent coexistence.

Greater temporal differences may prohibit coexistence in cases where it is possible for little or no difference. Seasonality in trait values related to resource use can also prevent coexistence in cases where resource partitioning or negative self-effects would have allowed it in a constant environment.

(3) Does 'modern coexistence theory' help to understand this case? The two aspects of this theory trace back to the 2000 *Annual Review* article by Peter Chesson (Chesson 2000a), although, as noted before, he did not propose the name. These aspects are the division of competition into equalizing and stabilizing factors, and the division of the stabilizing factors into resource partitioning, relative nonlinearity, and the temporal storage effect. The seasonal MacArthur model with anti-synchronized variation is a case in which all mean parameters can be equal, but coexistence may be impossible for wide ranges of parameter values because of the temporal partitioning. This is also possible for several other commonly used consumer–resource models. If coexistence occurs because of resource partitioning in a 2-resource model or because of intraspecific density dependence, then there are many circumstances under which adding or increasing seasonal partitioning makes coexistence less likely. Differences in mean fitness parameters can allow coexistence where it would otherwise be impossible (e.g., seasonal vs aseasonal species). While Chesson's technical definition of storage can be applied to cases of coexistence revealed here, there is no actual 'storage' of anything in the consumers; the delayed positive effects of competition in a single-consumer system are due to lags in the resource response. 'Storage' would have to be expanded to something that might be termed 'negative' or 'inverse' storage to apply to the scenarios where seasonal partitioning makes coexistence less likely.

(4) What range of alternative outcomes is possible in 2-species competition? This question is usually answered by the four outcomes that are possible in the LV model. In the LV model, the only case with more than one outcome is alternative exclusion, which is thought to require interference competition (or relatively unlikely relationships between relative capture rates and conversion efficiencies) in a consumer–resource context. The seasonal model considered here exhibits at least three alternative outcomes in many cases. When there are two alternatives, they may both be coexistence with different relative abundances of the two competitors. Unlike constant-environment models, purely exploitative competition produces alternative outcomes in a wide range of cases in this seasonal 2-consumer–1-resource model.

(5) Are per capita competitive effects relatively insensitive to the abundances of the competitors? The per capita effect of one species on a second in the LV model is independent of the abundance of either species. This remains true for the MacArthur consumer–resource system, provided the perturbation used to measure the effect does not cause extinction of any resources (Abrams 1998). However, it is definitely not the case for the seasonal version of MacArthur's model considered here.

There is some disagreement on the need for mutual invasibility to be present in order to label a situation as 'coexistence'. This shortcut has been widely used (Turelli 1981; Abrams 2006a; Chesson 2020a; and many others), but it frequently does not correspond to the actual possibility of coexistence. Importantly, the existence of a strong Allee Effect in a single-species system has not been taken as implying that the species cannot exist. Thus, it would seem inappropriate to adopt a more stringent requirement for (co)existence in systems with two or more species. Barabás et al. (2018) have recently reminded ecologists that successful invasion is not needed to produce coexistence in 2-competitor systems with Allee Effects (see also Meszéna et al. 2006). It is unclear how often success of invasion from very low density is needed for either existence or coexistence. However, given the multiple mechanisms implying that such invasibility is not needed, assuming otherwise in most cases does not seem appropriate.

The results presented here provide an additional reason against using 'ability to invade from near-zero density' as a universal criterion for coexistence. These are cases where invasion from low density succeeds, but subsequent population growth produces changes in system dynamics that reverse the initial increase, and the invader goes extinct. Ability to increase from low densities without being able to persist was discussed for a resource-based 2-consumer competition model by Abrams and Shen (1989) and for the multi-species LV model by Case (1995). Mylius and Diekmann (2001) discussed the mechanism causing invasion analysis to fail in these sorts of systems as cases of 'the resident strikes back'. All of these examples, as well as those presented here, involve qualitative shifts in the residents' dynamics caused by the initial increase of the invader.

The preceding two paragraphs suggest that a rigorous definition of coexistence should not be based on mutual invasion. The systems considered here include cases in which an initially successful invader is excluded by shifting the dynamics of the resident as well as cases in which a range of (non-equilibrium) initial densities would allow coexistence of species that are each incapable of invasion from low density when the resident exhibits its limiting dynamics. The current results also provide examples where the timing of introduction, as well as the number of individuals introduced, affects the ability of a species to increase and become established in a community (see Yamamichi et al. 2014 for examples from other systems). These issues mean that there are likely to be many more exceptions to 'invasion analysis' than have already been recognized.

Temporal segregation does not require that the two consumers have phase shifts in their seasonal parameter curves. Simply having different amplitudes of fluctuation in a resource-related parameter is sufficient to affect coexistence. The extreme case of this scenario is the model of a seasonal vs an aseasonal consumer, which again exhibits cases where the difference in seasonal response promotes coexistence, and others in which it inhibits coexistence. Variation was once assumed to have a negative impact on populations, due the fact that growth upward from a lower-than-equilibrium abundance is slower than negative growth from a comparably higher abundance (May 1973). However, as shown here, variation can also permit a greater mean level of resource depletion, so traits that magnify the response to seasonal environmental variation may be favoured.

While the exclusion outcomes documented here might be interpreted as 'negative' or 'inverse' storage, this classification does not aid in determining when such outcomes are likely to occur. Small differences in resource dynamics or the parameters of either species' demographic functions can produce large differences in the resulting system dynamics, which would be very hard to predict without a quantitative analysis involving the linked dynamics of consumers and resources. The frequent presence of alternative attractors makes it possible for the 'sign' of storage to differ depending on the timing and initial densities of the competing species. There are also problems in categorizing the correlation of competition and the environment, as the 'environment' is influenced by both the seasonal changes in $c(t)$ and the seasonal changes in R, but the resource variation is influenced by both consumers and by initial conditions.

The presence of alternative exclusion and coexistence outcomes in a single system is not restricted to (and may not even be most common in) seasonal environment models. The case of consumers with Allee Effects in each competitor was mentioned above. The possibility of two alternative exclusion outcomes as well as a coexistence outcome has been noted in at least one empirical system (without seasonal variation). This is Edmunds et al.'s (2003) analysis of a stage-structured model of competition between two *Tribolium* flour beetle species, representing the system studied experimentally by Park (1948). The outcome involves population cycles and the mechanism requires stage-structured consumer populations. It is likely that there are many more mechanisms producing, and more empirical examples of, multiple alternative equilibria in a given competitive system.

The lag in the responses of resources and consumers to changes in each other's abundance is an essential feature determining whether fluctuations help or hinder coexistence of competing consumers. Unfortunately, there are few studies of the timing and dynamics of these mutual responses. Similarly, as recently noted by McMeans et al. (2020), there are few well-studied examples of seasonal changes in resource-related population dynamical parameters. Thus, it will likely be some time before we have an empirical basis for judging the role of seasonality in promoting or inhibiting coexistence.

Real environmental variation is a mix of various seasonal and stochastic components. Li and Chesson (2016) showed, using the MacArthur model, that at least some stochastic variation can have coexistence-promoting effects similar to those of seasonality. More recently, Schreiber (2021) has shown some results similar to those described here using models with temporally autocorrelated stochastic variation. It thus seems likely that stochastic seasonality can also have coexistence-inhibiting effects similar to those described here. Sauve et al. (2020) stressed the need to examine non-sinusoidal variation in the parameters of consumer–resource models. A wide variety of more realistic models and combinations of several seasonal parameters in simple models remain to be studied.

9

Relative nonlinearity and seasonality

9.1 Introduction

Chapter 8 showed there are many pathways by which temporal resource partitioning can make coexistence more difficult. Regardless of its effects on coexistence, temporal variation alters the quantitative effects of one competitor on another. This chapter will examine several other types of temporal variation, and their effects on the nature and outcomes of competition, again largely in the rather special context of two consumers and one resource. It will concentrate on systems in which there are differences in the shape of the two consumers' functional or numerical responses. Such differences were not present in the models in Chapter 8, but were a key component of the earliest models that established the possibility of coexistence of two consumers on a single resource. The implications of these simple models for more realistic systems are also discussed.

The conclusion of this chapter will again be that different types of temporal variation can have a wide range of different effects on the outcome(s) of competition between species that differ in their functional or numerical response shapes. Four more specific results will be stressed. First, temporal variation can make exclusion either more likely or less likely depending on the details of the mechanism(s) driving that variation. Secondly, mutual invasibility of single consumer subsystems is often not required for coexistence of two consumers. Third, alternative attractors are a frequent occurrence. Finally, different period lengths for the temporal environmental variation can result in qualitatively different effects on coexistence. These findings also arose from the analyses of interspecific differences in the responses to seasonal environmental variation considered in Chapter 8. The combined findings of both chapters argue for greater understanding of the mechanistic details of consumer–resource interactions as a prerequisite for predicting the consequences of variable environments for interspecific competition.

Much recent work on competition in temporally varying environments has dealt with classifying mechanisms of coexistence into the two categories in Chesson's (1994) three-way categorization of all coexistence mechanisms that deal specifically with temporal variation (e.g. Letten et al. 2018; Ellner et al. 2019). The first mechanism is the storage effect, which is usually described in terms of a positive covariance of competition and the environment, measured when each of the species is a very rare invader in a system in which the other species is/are characterized by their

Competition Theory in Ecology. Peter A. Abrams, Oxford University Press. © Peter A. Abrams (2022).
DOI: 10.1093/oso/9780192895523.003.0009

limiting dynamics. All of the examples of coexistence in Chapter 8 would be classified as examples of the storage effect. The second category of coexistence associated with temporal variation is 'relative nonlinearity', in which variation affects two species in different ways because of differences in the shape of their functional or numerical responses. The analysis of simple models in this chapter suggests that these two mechanisms interact with each other in a wide variety of ways and are difficult to separate. The combination of differences in linearity with sustained variation in population sizes can hinder or promote coexistence. Thus, models with differently shaped consumer demographic and ecological functions provide more arguments against a research focus on a single simple classification of coexistence mechanisms. Such a focus is likely to restrict rather than expand our understanding of the full range of possible competitive outcomes, and the pathways underlying these outcomes.

The environmental variation considered in Chapter 8 had differing effects on consumer species that were identical in their mean abilities to capture or convert resources. This chapter will take up a similar exploration of how variation affects competition, but it will consider consumer species that differ in the shape of their relationship between resource abundance and immediate per capita growth rate. Such species cannot coexist on a single resource in the absence of temporal variation. However, because of the differences in the shapes of their responses to resource abundance, temporal variation changes the nature of the interaction. This chapter looks at how variation affects both the nature of the interaction and the conditions for coexistence. The three forms of temporal variation treated here are: (1) variation arising as a consequence of an unstable consumer–resource interaction; (2) variation in the environment that only directly affects resource growth parameters; and (3) variation in the environment that affects neutral consumer parameters (for example, per capita mortality rates in most of the simple models in this chapter). These three types of variation are all scenarios under which differences in the nonlinearity of consumer per capita growth rate functions alter competitive effects.

9.2 Inherently unstable consumer–resource interactions

Earlier chapters have already shown that the shapes of consumer species' functional responses have a major effect on the relationship between their overlap in use of different resources and several measures of competition between them, even when both species have the same response shapes. However, most species differ in response shapes. Armstrong and McGehee (1976a) showed that such differences could lead to coexistence. They examined competition for a single biotic resource between a consumer with a linear functional response and a similar one with a type II response. This did not affect coexistence unless there were sustained population cycles, driven by the consumer–resource interaction involving the type II response. Armstrong and McGehee (1976b, 1980) later extended this analysis to systems with more resources and included the possibility of abiotic resource growth. At the time of their original work, the main issue of interest to ecologists was whether two or more consumers could coexist while limited by a single resource. The presence of a saturating functional response (one having a negative second derivative for a significant range of resource abundances) had been shown to produce population cycles in a simple predator–prey model by Rosenzweig and MacArthur (1963). The mechanism driving coexistence in the Armstrong–McGehee model is that a species with a less saturating response (including linear or accelerating responses) has a higher mean resource intake requirement for positive population growth. The species with the more strongly saturating response is characterized by a lower intake requirement for zero population growth. However, the population cycles that the saturating species generates when it is numerically dominant increase the mean abundance of the resource. This larger mean abundance can be sufficient that the less efficient but less saturating consumer is able to persist in the system. The two consumer species then coexist in a system involving resource cycles that are smaller in amplitude than those generated when the saturating species is the only consumer present.

The above results were widely known when Armstrong and McGehee (1980) published their then-comprehensive review of coexistence. However, the likelihood that this mechanism explained many cases of coexistence was not assessed, empirically or theoretically for many years. Abrams and Holt (2002) used extensive numerical results to determine the range of potential parameter space over which coexistence occurred in several models. They concentrated on the simplest scenario exhibiting this mechanism (logistic resource growth; two consumers, one with a linear, and the other with a Holling type II functional response). The range of a neutral parameter (the per capita mortality rate) allowing coexistence in the 2-species system was compared to the range of that parameter allowing existence in a single-species system. Armstrong (1976) was the first to use this approach; he called the measure a 'coexistence bandwidth'. The bandwidth in the single-resource system could then be compared to similar systems in which species used two resources and exhibited various amounts of resource partitioning.

The main results of the Abrams and Holt (2002) analysis were:

(1) Coexistence bandwidth for two consumers can be substantial when one consumer has a linear functional response and the other has a type II response that reaches one half of its maximum value when the prey abundance is a relatively small fraction (on the order of 1/5 or less) of its carrying capacity (see Fig. 2, p. 285 in Abrams and Holt (2002)). The coexistence bandwidth could be increased substantially if the species with a linear functional response also had an accelerating numerical response (per capita growth rate increases at faster than linearly as the rate of consumption increases).

(2) It is possible, but difficult (very narrow bandwidth) to get coexistence of more than two consumer species on a single resource in this model.

(3) If each consumer utilizes two or more resources, the mechanism driven by different functional response shapes is less likely to operate. This is because the

high efficiency that leads to cycles in the single-resource model frequently causes extinction or quasi-extinction of the most vulnerable resource species in a 2-or-more resource system with a single consumer (see Fig. 5 in Abrams and Holt 2002). This reduces that system to one having a single resource, and one that is likely to be stable because of its lower vulnerability to the consumer. Thus, the mechanism is less likely in part because the required interaction-generated population cycles are less likely. The lower probability of limit cycle dynamics in multi-prey systems was also made in connection with purely intraspecific competition (Abrams 2009b). Even if both resources persist, the amplitude of cycles is usually reduced by the difference in their vulnerabilities, which reduces the impact of the cycles on coexistence (Abrams and Holt 2002). In addition, substantial partitioning of resources alone produces a large coexistence band-width, and thereby reduces the maximum extent to which the bandwidth can be increased by any additional mechanism.

If we move beyond the issue of coexistence, the interaction-driven cycles of the Armstrong–McGehee system provide an example of how the quantitative form of competition between two coexisting consumers is changed when their dynamics are characterized by interaction-driven instability. The standard description of interspecific effects used in previous chapters has been to determine how a neutral parameter in one species affects the mean abundance of a competitor, relative to the effect on its own abundance. Abrams et al. (2003) examined this issue for the Armstrong–McGehee model, as well as examining the nature of population dynamics over the complete parameter space over which two consumers could coexist. Figures 6 and 7 in that work show a variety of complicated responses of both species to the per capita death rate of one of them. One phenomenon shown was non-monotonic responses of each consumer's abundance to the per capita mortality rate of the other consumer, meaning that interactions could not always be classified as (–, –). Another possibility in this system was that the mean density of a species could increase with an increase in its own mortality (later termed a 'hydra effect'). Abrams et al. (2003) also showed that it was possible for an increased death rate of the nonlinear consumer to change coexistence to exclusion of the linear species by stabilizing the system. Alternative dynamical attractors characterized by the opposite direction of response to a given species' mortality also exist over some ranges of mortality rates. Cases with two attractors are associated with abrupt changes in abundances with a very small parameter change when one attractor disappears. While there is still little evidence that that the simple model considered in Abrams et al. (2003) is a good description of many pairs of species in natural communities, it would be surprising if differences in the linearity of per capita growth rate functions did not occur in, and contribute to coexistence of, many sets of competing species. The non-traditional responses to neutral parameters predicted by the simple model are also likely to occur in some of these.

The Armstrong–McGehee scenario, based on consumption-driven cycles with no variability in the physical environment, is narrower than Chesson's (1994, 2018) concept of relative nonlinearity. However, it has been the subject of the majority of theoretical work on the relative nonlinearity mechanism of coexistence. The rest of this chapter shows how differences in the linearity of the consumers' functional or numerical responses affect coexistence and other aspects of competition when cycles are driven by qualitatively different mechanisms. These two other sources of cycles are: (1) periodic variation in the resource per capita growth rate; and (2) periodic variation in consumer mortality rates or other parameters that are not directly involved in their uptake or conversion of resources. These two additional drivers of consumer–resource cycles may operate alone, or in systems with interaction-caused instability. How combinations of different drivers of instability affect the likelihood of coexistence and the nature of interactions is only treated briefly, as there has been little previous work on this topic, and there are many different scenarios for combining different sources of instability. The chapter concentrates on the case of two consumers utilizing a single resource.

9.3 Differences in nonlinearity with seasonal resource growth

Variation is a pervasive feature of the natural world (Pimm 1991). A major source of such variation in consumer species is variation in one or more growth parameters of one or more of their resources. In a system with a single resource, variation in resource growth will only affect the consumer species differently if they have different shapes of their combined functional-numerical response or if the consumers also have temporal variation in their resource uptake or conversion rate parameters. To allow coexistence of two consumer species on a single resource, the consumer that is harmed to a greater extent by the fluctuations—i.e. the one with a more strongly saturating response—must also increase those fluctuations (or increase them to a greater extent).

Abrams (2004b) examined competition using a number of simple 2-consumer-1-resource models with different types of variability in the resource growth rate. Abiotic as well as biotic resource growth models were explored in a setting where one consumer had a linear functional response and the second had a type II response. The models considered here are very similar, but they allow both species to have type II responses, and add the possibility that one or more of the consumer death rates vary sinusoidally. Consumer mortality rate variation is not considered until Section 9.4, and the analysis in both sections will be limited to logistic resource growth with an added immigration rate, m. The dynamics are given by:

$$\frac{dN_1}{dt} = N_1 \left(\frac{B_1 C_1 R}{1 + C_1 h_1 R} - d_1 \left(1 + \gamma_{N1} \sin \left(\frac{2\pi t}{q} \right) \right) \right)$$

$$\frac{dN_2}{dt} = N_2 \left(\frac{B_2 C_2 R}{1 + C_2 h_2 R} - d_2 \left(1 + \gamma_{N2} \sin \left(\frac{2\pi t}{q} \right) \right) \right)$$

$$\frac{dR}{dt} = m + R \left[r \left(1 + \gamma_R \sin \left(\frac{2\pi t}{q} \right) \right) - kR \right] - \frac{C_1 N_1 R}{1 + C_1 h_1 R} - \frac{C_2 N_2 R}{1 + C_2 h_2 R}$$

$$(9.1a, b, c)$$

As in Chapter 8, the parameters γ_i ($0 \leq \gamma_i \leq 1$) measure the amplitude of the sinusoidal variation, and the parameter q is the period length. Models with sufficiently low handling times do not allow sustained population cycles (or coexistence) in the absence of a temporally variable parameter. If one consumer is characterized by $Ch(r/k) > 1$, then a sufficiently low mortality rate of that consumer will produce cycles in the absence of environmental variation (assuming immigration, m, is also very small). Abrams (2004b) showed that sinusoidal variation in the resource's maximum per capita growth rate, r, could allow coexistence on a single resource type in a variety of 2-consumer models. That work showed that abiotic resource growth implied a significantly lower ability of resource-driven cycles to allow coexistence on a single resource than was true of biotic resources. Variable biotic growth of resources frequently produced a wide coexistence bandwidth, but a sufficiently short period, q, produced bandwidths close to zero.

Although Abrams (2004b) stressed coexistence outcomes, the difference in the linearity of two consumers' per capita growth rates is not sufficient to produce coexistence in all 2-consumer–1-resource systems having environmentally driven variation in resource growth rate, and may even prevent coexistence (as is shown below). In addition, models with variable resource growth can produce alternative outcomes that depend on starting conditions. As in the systems explored in Chapter 8 (where species did not differ in functional or numerical response shape), there may be an inherent periodicity in the dynamics of the underlying deterministic system, which interacts with the variation in resource growth rate to produce complicated dynamics. This leads to the possibility that the initial abundances determine whether two consumers coexist or not. Initial abundances in other cases determine which of two distinct coexistence outcomes prevails.

A diversity of outcomes for particular parameter sets characterizes a system that was the subject of Figure 4 in the Abrams (2004b) analysis. That figure examined the coexistence bandwidth for a system with a single logistic resource. The nonlinear consumer had a handling time that implied an approximate 90% reduction in the consumer's per capita capture rate when the resource was at carrying capacity (relative to when the resource was rare). The results in Abrams (2004b) suggested that all periods of variation produced a significant coexistence bandwidth. Those results also suggested that for the consumer 1 parameters assumed, only three qualitatively different outcomes were possible for different per capita mortality rates of consumer 2. These were: exclusion of 1 by 2 from all starting conditions at low d_2; coexistence for a range of intermediate d_2; and exclusion of consumer 2 by consumer 1 at sufficiently high d_2. I was unable to replicate some parts of that figure in a recent reanalysis. In fact, this example has cases with coexistence and exclusion attractors for the same parameter set, and cases in which each consumer could exclude the other (alternative exclusion). Although the model in Abrams (2004b) lacked the low-level immigration into the resource population that are present in eqs (9.1), most of the differing results appear to be due to simulations that were not run long enough and a failure to look for alternative attractors. The results of a more accurate and more complete set of simulations for a small number of period lengths q are provided in Table 9.1.

Table 9.1 Resource growth period length and consumer competition outcome; nonlinear functional response (eqs (9.1) with $h_2 = 0$)

Period q	Outcomes	Parameter range(s)
No variation	Exclusion of sp. 1	$d_2 < 0.150$
	Coexistence	$0.15 < d_2 < 0.507$
	Exclusion of sp. 2	$d_2 > 0.507$
5	Exclusion of sp. 1	$d_2 < 0.185$
	Coexistence	$0.185 < d_2 < 0.508$
	Exclusion of sp. 2	$d_2 > 0.508$
10	Exclusion of sp. 1	$d_2 < 0.354$
	Excl. of 1 or quasi-excl.* of 2	$0.354 < d_2 < 0.391$
	Exclusion of sp. 1 or sp. 2	$0.391 < d_2 < 0.399$
	Coex. or exclusion of sp. 2	$0.399 < d_2 < 0.421$
	Exclusion of sp. 2	$d_2 > 0.421$
15	Exclusion of sp. 1	$d_2 < 0.467$
	Exclusion of sp. 1 or sp. 2	$0.467 < d_2 < 0.502$
	Coex. or exclusion of sp. 2	$0.502 < d_2 < 0.539$
	Exclusion of sp. 2	$d_2 > 0.539$
20	Exclusion of sp. 1	$d_2 < 0.457$
	Coexistence	$0.457 < d_2 < 0.579$
	Coex. or exclusion of sp. 2	$0.579 < d_2 < 0.606$
	Exclusion of sp. 2	$d_2 > 0.606$
30	Exclusion of sp. 1	$d_2 < 0.340$
	Coexistence	$0.340 < d_2 < 0.695$
	Exclusion of sp. 2	$d_2 > 0.695$
60	Exclusion of sp. 1	$d_2 < 0.270$
	Coexistence	$0.270 < d_2 < 0.811$
	Exclusion of sp. 2	$d_2 > 0.811$
100	Exclusion of sp. 1	$d_2 < 0.232$
	Coexistence	$0.232 < d_2 < 0.860$
	Exclusion of sp. 2	$d_2 > 0.860$

All of the systems considered above are characterized by these parameters in eqs. (9.1): $b_1 = b_2 = C_1 = C_2 = 1$; $h_1 = 10$; $h_2 = 0$; $r = k = 1$; $d_1 = 0.06$. The amplitude of the variation in r is 1.
· quasi-exclusion denotes persistence with a mean density at least two orders of magnitude less than that of the dominant species

The system with no intrinsic resource variation allows coexistence via the Armstrong–McGehee mechanism; the nonlinear consumer generates cycles over a range of d_2 values that spans more than a third (35.7%) of the total range of d_2 values that allow consumer 2 to persist in the absence of consumer 1 (see Table 9.1).

The minimum of $d_2 = 0.150$ for coexistence is because lower mortalities imply that consumer species 2 can persist at a lower constant resource density than can consumer 1, and will therefore exclude it. Intrinsic variation of the resource growth rate with the relatively short period of $q = 5$ contracts, rather than expands, the range of mortalities of species 2 that allow coexistence. However that value of q is sufficient to produce enough resource variation that it reduces the per capita rate of increase of the nonlinear consumer 1 even when both consumers are very rare. Because of its rarity, the linear consumer 2 has little influence on the resource dynamics at the upper end of the coexistence range for d_2. Here, the long-period consumer resource cycles, similar to what would occur with no variation in resource growth, dominate the dynamics. Thus the upper limit of d_2 is not changed greatly by the presence of variable resource growth.

The dynamics become more complicated when $q = 10$. This period makes it possible for the resource growth cycles to suppress the interaction-driven cycles caused by consumer 1 under some circumstances. This results in the possibility of two or three different outcomes for a given system, depending on initial conditions (i.e., initial abundances and the initial phase in the resource growth cycle). A period of $q = 10$ results in a significant range of mortalities of the linear species ($0.354 < d_2 < 0.391$) that are characterized by coexistence, but at very low relative abundance of consumer 2. At slightly higher d_2 there are only alternative exclusion outcomes. Figure 9.1 illustrates these alternatives for $d_2 = 0.395$, and shows their associated different resource dynamics. Dominance by either consumer produces resource dynamics that favour it over its competitor. For a wide range of slightly higher mortalities of consumer 2, the alternatives are either coexistence or exclusion of 2. These alternatives are shown in Figure 9.2. If the species occupy the coexistence attractor at $d_2 = 0.420$, the mean density of species 2 is approximately 61% that of species 1, but its equilibrium abundance drops to zero if d_2 is raised marginally to 0.421.

The range and nature of outcomes with a resource growth period of $q = 15$ is also complicated. For a significant range of intermediate d_2 values (0.467–0.502), coexistence is impossible, and either consumer excludes the other, depending on initial densities. The next higher range of d_2 (0.502–0.539) exhibits either coexistence of both species or exclusion of species 2. Still higher periods, $q > 0.539$, only produce exclusion of consumer 2. A resource growth cycle period of $q = 20$ allows coexistence over roughly 15% of the full range of mortalities that permit existence of species 2 in the absence of its competitor. However, a significant fraction of this range also has the alternative outcome of exclusion of species 2. The remainder of the periods explored in Table 9.1 ($q = 30$ and above) never resulted in alternative outcomes for any mortality rate. All of these longer periods produced the three outcomes of exclusion of 1 (at low d_2), coexistence (at intermediate d_2), and exclusion of species 2 (at high d_2). The total coexistence bandwidth increases over this range of period lengths from $q = 30$ to $q = 100$. Longer periods increase the strength of intraspecific relative to interspecific competition by lengthening the stretches of time characterized by an advantage of one consumer over the other.

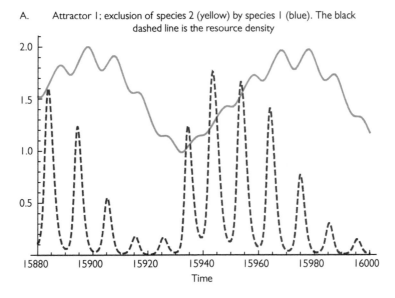

A. Attractor 1; exclusion of species 2 (yellow) by species 1 (blue). The black dashed line is the resource density

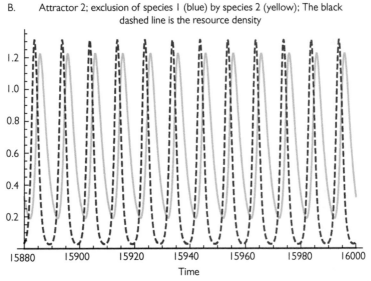

B. Attractor 2; exclusion of species 1 (blue) by species 2 (yellow); The black dashed line is the resource density

Fig. 9.1 Dynamics of consumer 1 and the resource, or consumer 2 and the resource for $q = 10$ in the Table 9.1 nonlinear functional response model with $d_2 = 0.395$. These are alternative exclusion outcomes. In panel A, the nonlinear consumer 1 dominates the dynamics and produces irregular oscillations in population sizes with the larger amplitude fluctuations having a period greater than 80. The result is a low mean resource abundance that excludes consumer 2. If consumer 2 is initially sufficiently abundant the final dynamics are as shown in panel B. Here there is a simple period-10 cycle, with a high amplitude rapid fluctuation in resource abundance that excludes the nonlinear consumer 1.

A.

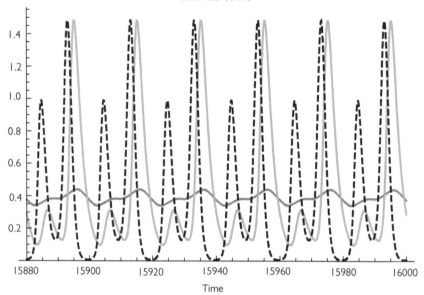

Attractor 1 – Coexistence; The abundance of consumer species 1 is the blue line; the abundance of species 2 is the yellow line; the resource abundance is the black dashed line

B.

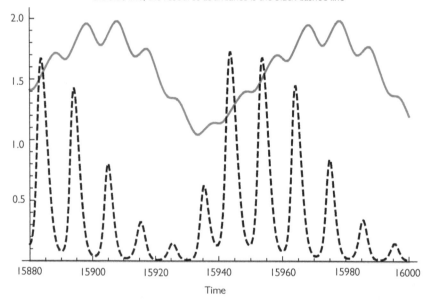

Attractor 2 – Exclusion of consumer species 2; The abundance of species 1 is the blue line; the resource abundance is the black dashed line

Fig. 9.2 The potential dynamics for the same Table 9.1 system with $q = 10$ illustrated in Figure 9.1, but with mortality increased by about 3.8% to $d_2 = 0.41$. Rather than alternative exclusion outcomes, this system exhibits either coexistence or exclusion of species 2. The coexistence outcome shown in panel A has a period of 20, or twice the period of the resource growth rate variation.

In terms of classifying coexistence mechanisms, the Armstrong–McGehee system (with or without variation in resource growth) exhibits differences in the linearity of the functional responses, so it is a case of 'relative nonlinearity'. However, it also exhibits periodic variation in the per capita intake rate of the nonlinear consumer on the resource, due to the cycles in resource abundance, which make the functional response more or less saturated at different points in time. This makes the 'advantage' in per capita growth rate switch periodically between the two consumer species. A similar variation in the relative attack rates (and thus relative per capita growth rates) of the two consumers in the systems considered in the previous chapter is what allows coexistence for some parameter values in that system. However, the systems in Chapter 8 lack any differences involving the nonlinearity of consumer functions. Both the systems considered immediately above and those in Chapter 8 involve periodic shifts in competitive advantage. Thus, it is unclear whether the 'mechanisms producing competition' are truly distinct in these two cases.

There are several values of q with parameter ranges for which each competitor is able to prevent the other from increasing from low abundance. In all of these cases coexistence would also be impossible if the system (eqs (9.1)) were modified so that each consumer had a very small direct negative effect on its per capita growth rate. This means that environmental variation in resource growth is capable of preventing coexistence in cases where it would otherwise be possible, for at least some ranges of both species' mortality rates.

The example illustrated in Figures 9.1 and 9.2 is based on the scaled parameter values used in Abrams (2004b). There has not yet been a systematic exploration of the model behaviours across all potentially plausible parameter combinations in the system with variation in resource growth. However, a wide range of other parameter values exhibit alternative attractors involving coexistence and exclusion. Figures 9.3 and 9.4 illustrate two such cases for two systems with parameter values more similar to those employed in the main examples used in Chapter 8 (which, in turn, are similar to those adopted in Li and Chesson (2016)).

It is true that a difference in the linearity of at least one fitness component (the functional responses of the consumer species) in the Armstrong–McGehee model is an essential component of the 'coexistence mechanism'. However, it is not sufficient that there be a difference in nonlinearity for coexistence to be possible for some combinations of values of the consumer mortality rates. Nor do all differences in non-linearity between two consumers allow coexistence for some values of the mortality rates. Moreover, the nature of interspecific effects and the coexistence bandwidth can differ greatly, depending on the mechanistic basis of the interaction. This may be illustrated with a few examples employing often used, but seldom combined, model components. It should be noted that the definition of 'relative nonlinearity' has evolved since Chesson (1994) first defined it (e.g. Kang and Chesson 2010), and the latter paper acknowledges that it may not always favour coexistence. There is still not a clear verbal definition of the mechanism labelled as 'relative nonlinearity'.

Coexistence in the system described above cannot be attributed solely to differences in the shapes of the consumer growth functions. This is because the difference in nonlinearity is due to differences in the functional response shapes.

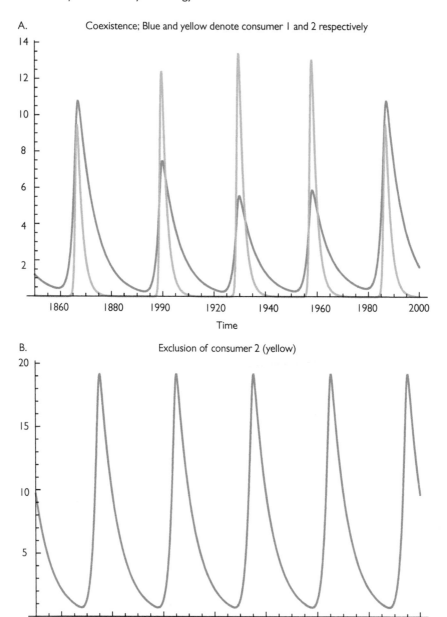

Fig. 9.3 Alternative attractors for eqs (9.1) with a logistic resource, variable resource growth, and parameter values: $r = 1$, $k = 0.01$, $C_1 = 0.25$, $C_2 = 0.25$, $h_1 = 0.1$, $B_1 = B_2 = 0.1$, $d_1 = 0.15$, $d_2 = 0.5$, $m = 0.0001$, $q = 30$, and $\gamma_R = 0.75$. Panel A has consumer coexistence with cycles of period 120 ($= 4q$). The mean abundances of N_1, N_2, and R are: 2.2654, 1.2807, and 20. Panel B has population cycles matching the environmental period ($q = 30$) and mean abundances N_1, N_2, and R are respectively 5.8133, 0, and 19.191.

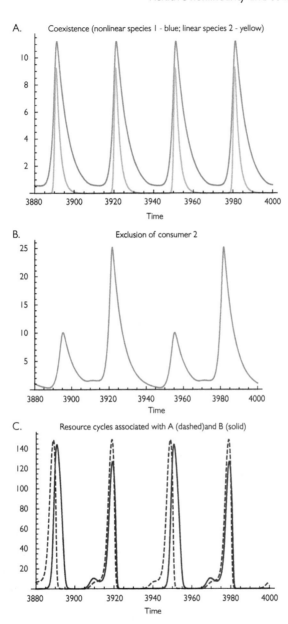

Fig. 9.4 Alternative attractors for eqs (9.1) with a logistic resource, and parameter values: $r = 1$, $k = 0.01$, $C_1 = 0.25$, $C_2 = 0.25$, $h_1 = 0.1$, $h_2 = 0$, $B_1 = B_2 = 0.1$, $d_1 = 0.175$, $d_2 = 0.51$, $m = 0.0001$, $q = 15$, and $\gamma_R = 0.75$. Panel A shows consumer densities for the coexistence attractor, and panel B shows the abundance of consumer 1 for the exclusion attractor. Panel C shows the resource abundances corresponding to panels A and B, with the solid line corresponding to panel B.

The potential instability of a system that lacks variable parameters is due to the impact of the consumer's nonlinear functional response on the resource dynamics (e.g., May 1973). A sufficiently saturating response results in a net positive effect of increased resource abundance on the resource per capita growth rate near the equilibrium point in unstable systems. Thus, it is of interest to determine whether differences in the nonlinearity of consumer growth per se are able to generate coexistence of two consumers in the absence of direct effects of that nonlinearity on the stability of resource dynamics.

One can explore the role of differences in consumer nonlinearity per se by analysing a system in which one consumer has a nonlinear numerical response, the other has a linear numerical response, and both have linear functional responses. The shapes of numerical responses have largely escaped empirical study, but it is well known that physical constraints place upper limits on birth rates, so saturation of numerical responses is inevitable. The nonlinearity in the numerical response can bring about coexistence if there are cycles driven by temporal variation in some other parameter. The nonlinear (saturating) species is more adversely affected by variation in resource density. Thus, if the nonlinear species is able to increase when rare, and it also causes greater amplitude cycles in resource abundance when it is common, both species should be able to persist. The next section considers variation in the resource per capita growth rate, to make it as similar as possible to the systems with a nonlinear functional response discussed above and in Abrams (2004b).

9.4 Differences in the nonlinearity of numerical responses

Most saturating functional responses are capable of generating cycles in most models with purely biotic resource growth. This is not true for a saturating numerical response in which consumer growth rate is solely a function of resource abundance. The form of the numerical response has received remarkably little empirical attention (as noted in Getz (1993) and Abrams (1997)). However, saturation should be a general property due to the restrictions body size and other factors place on maximum reproductive rate. A simple form for such a model is given below, for a system where each species has a linear functional response but species one has a nonlinear numerical response with the same (Michaelis–Menten) shape as Holling's type II disc equation. The resource may have temporal variation in its maximum per capita growth parameter, r. The resulting model has the form:

$$\frac{dR}{dt} = R\left(r(t) - kR\right) - C_1 R N_1 - C_2 R N_2$$

$$\frac{dN_1}{dt} = N_1\left(\frac{B_1 C_1 R}{1 + zC_1 R} - d_1\left(t\right)\right) \qquad\qquad (9.2a, b, c)$$

$$\frac{dN_2}{dt} = N_2\left(B_2 C_2 R - d_2\left(t\right)\right)$$

The parameter z plays the same role in determining the per capita growth rate function for consumer 1 as does the 'handling time' in a type II functional response (h in eqs (9.1)). However, here z reflects limitations on the growth and/or birth rate of the species rather than limitation on time available for searching for resources. A positive value of z cannot bring about coexistence of the two consumers in the absence of some other factor that generates sustained cycles. In eqs (9.2), there are three potentially time-dependent parameters, which will be examined below. The variation in each time-dependent parameter is assumed not to affect the mean value of the parameter. The simulations described below impose variation using the same multiplicative sinusoidal function with a mean value of 1 given explicitly in eqs (9.1). If all of the functions of time in eqs (9.2) were constants, the system could only have a stable equilibrium with a single consumer present (ignoring the theoretical possibility of a neutral line of equilibria). The consumer characterized by a lower equilibrium resource density would exclude the other.

The first case considered here is variation in the resource per capita growth rate. Here the parameter r changes sinusoidally and this variation in resource growth leads to sustained cycles in the abundances of all three species. The convex shape of consumer 1's growth function makes it more likely that any type of variation will increase both the mean resource abundance required for the consumer to increase when rare, as well as the long-term mean resource abundance when the nonlinear consumer is present alone. Variation in either or both consumer death rates also cause sustained density fluctuations; their effects on the interaction between consumers will be considered later, in Section 9.5.

The parameters used here are equivalent to those used in the main example of eqs (9.1) discussed in Section 9.3. These are: $B_1 = B_2 = C_1 = C_2 = 1$; $z = 10$; $r = k = 1$; $d_1 = 0.06$. The amplitude of the cycle in r is assumed to have its maximum value, implying sinusoidal variation with a minimum $r(t) = 0$ and a maximum $r(t) = 2$. In the system where consumer 1 has a nonlinear functional response (eqs (9.1)) and no variation in r, these parameter values imply large amplitude consumer–resource cycles and a coexistence bandwidth (range of d_2) of approximately 0.357. This compares with the maximum range of 1.0 for d_2 that allows existence of the linear species in the absence of competition. On the other hand, a difference in the nonlinearity of the two numerical responses never allows coexistence; the only possibilities in this example are exclusion of consumer 2 for $d_2 > 0.15$ and exclusion of consumer 1 for $d_2 < 0.15$.

Table 9.2 describes the outcomes of competition in a system that is identical to that of Table 9.1 except that it is the numerical rather than the functional response of consumer 1 that is nonlinear. Table 9.2 shows that coexistence is impossible for the three lowest environmental periods studied; $q = 5$, 10, or 15. Periods of 10 and 15 have a significant range of d_2 values that produce exclusion of either species, roughly corresponding to initial abundances. (The 'roughly' arises from the influence of the initial direction of change in resource growth, which can reverse the relative population size during the initial time span.) If small magnitude direct negative effects of each consumer's density on its own growth rate were added to eqs (9.2) this would

Table 9.2 Resource growth period length and consumer competition outcome; nonlinear numerical response model (eqs. 9.2)

Period q	Outcomes	Parameter range(s)
No variation	Exclusion of sp. 1	$d_2 < 0.150$
	Exclusion of sp. 2	$d_2 > 0.150$
5	Exclusion of sp. 1	$d_2 < 0.182$
	Exclusion of sp. 2	$d_2 > 0.182$
10	Exclusion of sp. 1	$d_2 < 0.281$
	Exclusion of sp. 1 or sp. 2	$0.281 < d_2 < 0.400$
	Exclusion of sp. 2	$d_2 > 0.400$
15	Exclusion of sp. 1	$d_2 < 0.400$
	Exclusion of sp. 1 or sp. 2	$0.400 < d_2 < 0.500$
	Exclusion of sp. 2	$d_2 > 0.500$
20	Exclusion of sp. 1	$d_2 < 0.459$
	Coexistence	$0.459 < d_2 < 0.484$
	Coex. or exclusion of sp 2	$0.484 < d_2 < 0.529$
	Exclusion of sp. 2	$d_2 > 0.529$
30	Exclusion of sp. 1	$d_2 < 0.341$
	Coexistence	$0.341 < d_2 < 0.584$
	Coex. or exclusion of sp. 2	$0.584 < d_2 < 0.588$
	Exclusion of sp. 2	$d_2 > 0.588$
60	Exclusion of sp. 1	$d_2 < 0.270$
	Coexistence	$0.270 < d_2 < 0.704$
	Exclusion of sp. 2	$d_2 > 0.704$
100	Exclusion of sp. 1	$d_2 < 0.232$
	Coexistence	$0.232 < d_2 < 0.750$
	Exclusion of sp. 2	$d_2 > 0.750$

The parameters shared with the Table 9.1 model have values identical to those in Table 9.1; the one new parameter, z, has a value of 10.

allow coexistence for some range of relative mortality rates in this system without resource growth variation. However, coexistence would not occur (or the self-effects would have to be larger for it to occur) in cases with alternative exclusion outcomes. Thus, one can describe the effects of temporal variation as making coexistence more difficult. (As discussed in Chapter 10, this conclusion may be reversed if the species occur in a spatially structured metapopulation.)

The nonlinear numerical response system with $q = 20$ has four categories of outcomes, including a relatively narrow range of d_2 where coexistence of both species or exclusion of the linear species are alternatives. The latter range of d_2 is even narrower when $q = 30$. Periods of 60 and greater produce wide coexistence bandwidths, and no cases of alternative outcomes. The coexistence bandwidths are all somewhat smaller than those in the corresponding nonlinear functional response models.

In summary of the results in Table 9.2, models with nonlinear numerical responses can exhibit coexistence on a single resource in environments with temporally variable resource growth. However, not all types of environmental variation in resource growth allow some coexistence, and variation with a relatively short environmental period can make coexistence more difficult or impossible. Although this is just one example, analysis of several other systems having large differences in numerical response shape also exhibited a wide range of potential effects of resource-driven variation.

9.5 Other types of environmental variation

Variation in the consumer mortality rates, which is included in eqs (9.1) and (9.2), can also affect coexistence. Such variation could occur in conjunction with nonlinear functional and/or numerical responses. In the linear consumer models treated in Chapter 8, cycles in consumer mortality had no effect on the mean resource abundance, and thus, no effect on coexistence. This is no longer the case when one or both of the consumers' per capita growth rates are nonlinear and this holds for both numerical and functional responses. There are many possibilities for such variation; it can occur in one or both species, and in the latter case it may differ in phase or amplitude between species. If the resource growth rate also varies, there are even more possible combinations of different types of variation. This section will present a few examples to illustrate some of circumstances that have the most obvious effects on coexistence. It will only consider identical sinusoidal variation in one or both consumer species.

9.5.1 Nonlinear numerical responses

I begin by considering correlated mortality in both consumers in the numerical response model, eqs (9.2); both mortality rates vary between zero and twice their mean. The fact that the baseline consumer per capita mortality rates are low relative to the baseline resource per capita growth rate in the system considered here means that a longer period q is required to generate enough variation in resource abundance to have a significant effect on coexistence. Here, coexistence does not occur until q is considerably larger than 60, and the bandwidth is relatively narrow even for $q = 100$. When $q = 60$, not only is coexistence impossible for all d_2, but there are alternative exclusion outcomes for $0.335 < d_2 < 0.426$. When $q = 100$, coexistence is the only outcome for the relatively narrow range of $0.351 < d_2 < 0.424$. A small additional range of mortalities ($0.424 < d_2 < 0.449$) allows alternative outcomes of coexistence or exclusion of consumer 2 by consumer 1. Even if we include this additional range, the total coexistence bandwidth is only about 1/3 that of the comparable model with resource variation (eqs 9.1) in which species 1 has type II functional response (Table 9.1). If the period q is raised to 150, this yields coexistence for $0.313 < d_2 < 0.450$, with exclusion of species 2 above this range, and exclusion of 1 below it; no alternative attractors were observed. If the linear consumer species 2 is the only one that exhibits variable mortality, coexistence is not possible. There are alternative exclusion outcomes depending on initial abundances when $0.150 < d_2 < 0.313$. Mortalities d_2 lower than this range only allow species 2 to persist, while higher d_2 values only allow species 1 to persist. If only the nonlinear species 1 exhibits mortality variation, coexistence occurs for $0.150 < d_2 < 0.451$, with exclusion of species 2 above this range, and exclusion of 1 below it.

The above comparison of single and dual consumer mortality variation can be understood qualitatively based on the general requirement that, in order for two species to coexist, at least one species must have some stabilizing process that decreases its relative competitive ability as its short-term mean abundance increases. If variable mortality is associated with the nonlinear consumer 1 in eqs (9.2), this will lead to greater resource fluctuations as its numbers increase, which will slow its growth and provide an opportunity for the linear species to coexist. Thus, variation in the mortality of consumer 1 produces a relatively wide coexistence bandwidth. This is reduced, but not eliminated by having mortality variation in the linear consumer as well. A numerically dominant linear species 2 produces large amplitude fluctuations in the resource that often give it the ability to exclude species 1.

9.5.2 Nonlinear functional responses

The comparable model with species 1 having a nonlinear functional response (Table 9.1 above) also yields somewhat different results when consumer mortality variation is substituted for resource growth variation. In the absence of any environmentally caused cycles, consumer 1's type II response implies a coexistence bandwidth for the linear species of $d_2 = 0.15$–0.507. Coexistence was examined in this variable consumer mortality model for periods of 15 and 30 to provide a comparison with the case of variation in resource growth rates in Table 9.1. Somewhat less detailed results for $q = 5$ and $q = 100$ are given at the end of this section.

For $q = 15$, coexistence was always observed from $d_2 = 0.286$ through $d_2 = 0.516$. This compares to the much narrower bandwidth of $0.502 < d_2 < 0.539$ for the model considered earlier that only has variation in resource growth. Note that this coexistence attractor in the resource-variation model occurred in conjunction with an alternative attractor characterized by exclusion of consumer 2. Thus, variation driven by fluctuations in both consumers' mortalities was considerably more favourable for coexistence. However, this is largely because the smaller amplitude of consumer mortality cycles compared to resource growth cycles results in less interference of the environmentally driven cycles on the interaction-driven cycles.

If $q = 30$ for the consumer mortality variation, coexistence is possible from $d_2 = 0.450$ to $d_2 = 0.535$. Exclusion of consumer 1 always occurred below this range,

A.   Lower N_1 attractor (Consumer 1 - blue; Consumer 2 - yellow; Resource - black, dashed

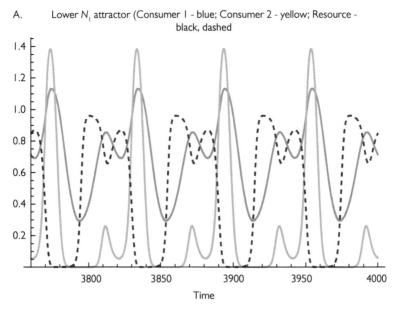

Time

B.   Higher N_1 attractor (Consumer 1 - blue; Consumer 2 - yellow; Resource - black, dashed

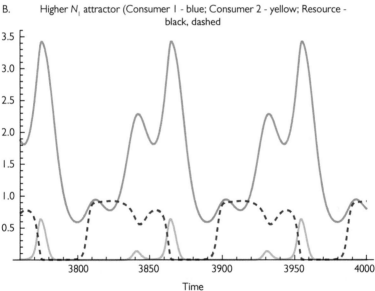

Time

Fig. 9.5 An example of alternative coexistence outcomes that occur with the nonlinear functional response model having variation in both consumer mortality rates. The parameters are: $r = 1$; $k = 1$; $C_1 = 1$, $C_2 = 1$; $h_1 = 10$, $h_2 = 0$; $B_1 = 1$, $B_2 = 1$; $d_1 = 0.06$, $d_2 = 0.5$; $m = 0.0001$; $q = 30$, $\gamma_{p1} = 1$, $\gamma_{p2} = 1$. Note the difference in the scale of the y-axis, due to the much greater maximum abundance of consumer 1.

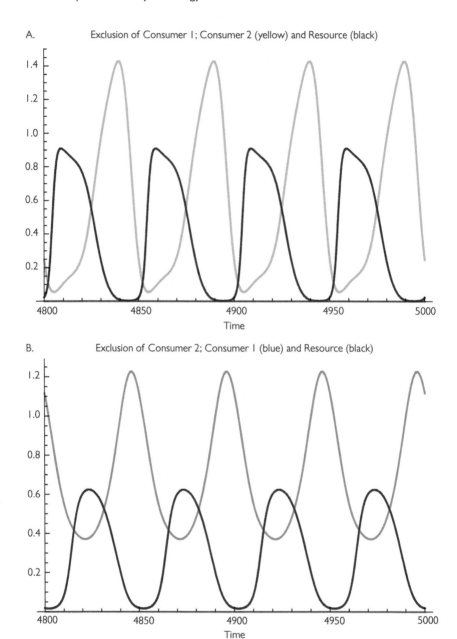

Fig. 9.6 The alternative attractors for eqs (9.2) with variation in both consumers' mortality rates. The parameters are: $r = 1$; $k = 1$; $C_1 = 1$, $C_2 = 1$; $h_1 = 10$, $h_2 = 0$; $B_1 = 1$, $B_2 = 1$; $d_1 = 0.06$, $d_2 = 0.4$; $m = 0.0001$; $q = 50$, $\gamma_{p1} = 1$, $\gamma_{p2} = 1$. Alternative exclusion outcomes exist for values of d_2 between 0.29625 and 0.545.

and exclusion of consumer 2 occurred above this range. This compares to the much wider range of mortalities allowing coexistence (0.340–0.697; see Table 9.1) in the comparable case of variation in resource growth rate with $q = 30$. This is again related to the fact that the environmentally caused variation is of greater amplitude in the case of resource growth variation than in the case of consumer mortality variation. Unlike the preceding example ($q = 15$), $q = 30$ has a very positive effect on coexistence. The bandwidth noted above actually overstates the possibility of coexistence in the consumer mortality variation model because one or more alternative attractors are found for a wide range of different mortalities in the $q = 30$ case. Most significant from the standpoint of coexistence is the fact that exclusion of species 2 by species 1 is a possible outcome for values of d_2 from 0.491 to the maximum of 0.536. Exclusion of species 1 by species 2 occurs as an alternative to coexistence for d_2 values from 0.450 to 0.482. Thus, there is only a narrow range of mortalities of the linear species over which exclusion never occurs for any initial conditions (d_2 from 0.482 to 0.491). In addition, there are parameter ranges over which two different coexistence attractors exist, and parts of these ranges also allow a third (exclusion) attractor. Figure 9.5 illustrates one pair of alternative coexistence attractors that occur in the case of $q = 30$. Alternatives are usually characterized by different periods for the variation in abundances; these are different small integer multiples of the basic period. The period for Figure 9.5A is 60, while that of 9.5B is 90.

Very short periods (e.g. $q = 5$) in this consumer mortality variation model have relatively modest effects on coexistence bandwidth; the cycle driven by consumer 1 dominates the dynamics except when d_2 is low enough to greatly reduce the abundance of consumer 1. Moderately long periods are particularly likely to have alternative exclusion outcomes. Figure 9.6 shows the two exclusion outcomes that are possible if the parameter set described in the previous paragraph has $q = 50$ and $d_2 = 0.4$. This case allows alternative exclusion for a wide range of mortalities between $d_2 = 0.296$ and $d_2 = 0.545$. The alternatives illustrated in Figure 9.6 are characterized by quite different mean resource abundances. Still longer periods (e.g. $q = 100$) have narrower coexistence bandwidths than in the otherwise comparable model with resource-growth driven rather than consumer mortality driven cycles. For the system considered in the previous paragraph coexistence occurs for d_2 between 0.36 and 0.59, which is roughly 1/3 the range for $q = 100$ under resource growth variation (see Table 9.1). This lower bandwidth is largely a consequence of the lower amplitudes of the resource fluctuations caused by consumer mortality rate variation compared to those caused by resource growth variation.

9.5.3 Remaining unknowns

This section definitely does not constitute an adequate analysis of the possible effects on competition between two consumers for a single resource with periodic variation in consumer mortality rates. Many options regarding different amplitudes or phase shifts between consumers, or combinations of consumer- and resource-driven cycles remain to be explored. Obviously, different functional forms or even different parameter values should be examined. However, the examples provided serve to show that the dynamics do not fit well into a framework based purely on invasion, due to the frequent occurrence of alternative attractors. The results also are sufficient to show that differences in linearity do not necessarily promote coexistence in variable environments. Finally, they demonstrate that a considerable diversity of effects on coexistence can arise from relatively small changes in parameter values and from different sources of variation.

Future work should examine: (i) various models of abiotic growth; (ii) biotic resource systems in which the resource's own resource(s) are represented; (iii) models of combined abiotic and biotic growth; (iv) models with two or more potentially interacting resource species/types; and (v) models in which all species exhibit correlated seasonal variation in several parameters, as in most higher-latitude environments. The consequence of different dynamic rates of consumer and resource also needs more study. A wide variety of different functional response forms arise from adaptive behaviour (Chapter 3), and some of these are likely to lead to different impacts on coexistence in variable environments.

9.6 Systems with two or more resources

One of the more basic extensions of theory regarding coexistence in varying environments involves systems with two or more resources (point (iv) in the list in Section 9.5.3). There has been a limited amount of work on coexistence in systems with two resources in which cycles are caused purely by consumer–resource interactions (Egas et al. 2004; Abrams 2006a, b, 2007b). Most of this has been discussed in the context of the coevolution of specialist and generalist species. In spite of having identical ranges of evolutionarily possible functional response parameters, coexistence of three species was possible over a wide range of relative mortality rates at the evolutionary equilibrium. Some of the best possibilities for coexistence arose from features that were not possible in the single-resource model studied here; i.e. adaptive adjustment of relative consumption rates by the consumers (Abrams 2006b), and competitive interactions between biotic resources (Abrams 2007b).

The shared-exclusive version of the 2-consumer–3-resource models considered in previous chapters has not, to my knowledge been extended to consider the impacts of sustained variability affecting either the shared or exclusive resources. Analysing this system should provide another perspective on the range of impacts of sustained variability on the strength of interspecific competition.

9.7 Discussion

A wide variety of traits and mechanisms affect the ability of species to exist and to coexist with others. However, most of these traits and mechanisms can either favour or disfavour coexistence, depending on various specifics regarding functional forms of the per capita growth rates and the nature of the variation. One of the frequent outcomes demonstrated in both this and the previous chapter is that alternative attractors exist in many simple consumer–resource systems undergoing sustained fluctuations. These alternative attractors are generally characterized by different interaction strengths, as measured by the impact of changing a neutral parameter in one consumer species on the two population sizes. This complicates the question of how particular parameter changes affect competitive interactions.

The new and previously published findings reviewed in this chapter bolster several arguments presented in Chapter 8:

1. Multiple outcomes are often possible in systems with seasonal variation that affects one or more parameters of consumers or resources.
2. Inability to invade a system having a second consumer species often does not mean that a focal consumer species is unable to coexist with that second species.
3. The effects of variation on coexistence are difficult to predict without numerical analysis of the system in question.

The shapes of the functional responses of consumers to their resources had been identified as an important determinant of the nature of competitive interactions in stable systems as early as Abrams (1980a). That work, which assumed stable 2-consumer–2-resource models with logistic resource growth, showed that the interspecific effects differed greatly between models with type I (linear) and type II functional responses. The latter were characterized by competition coefficients that changed in magnitude as consumer abundances changed. This was confirmed in Abrams et al. (2003) for the 2-consumer–1-resource model first studied by Armstrong and McGehee (1976a, b). The ability of this model to produce coexistence of two consumers in systems with a single resource has been the subject of most previous and subsequent attention. However, there has been no demonstration that any real or experimental system can be characterized as having a single resource. More attention to the effects of different nonlinearities on the response of population size to parameter changes in multi-resource systems is needed.

Since 1980 there have been mixed opinions on the importance of nonlinearity in responses for influencing competition and/or coexistence. In spite of his 1963 work with Michael Rosenzweig on the effects of saturating functional responses in predator–prey systems, Robert MacArthur never incorporated these responses into his consumer–resource models of competition. In 2005, Facelli et al. (p. 3005) wrote that relative nonlinearity 'has been suggested as applicable to phytoplankton in lakes (Huisman and Weissing 1999) and, despite some recent enthusiasm (Abrams 2004a), does not appear to be a robust promoter of coexistence of many more species than limiting resources (Chesson 1994, Huisman and Weissing 2001)'. Ten years later, Yuan and Chesson (2015, p. 39) revised that opinion, stating that: '. . .relative nonlinearity has a critical role in compensating for a weakened storage effect when there is high correlation between species in their responses to the varying environment. In some circumstances relative nonlinearity is stronger than the storage effect or is even the sole mechanism of coexistence'. It should be noted that my work (Abrams 2004b) did not claim that relative nonlinearity was likely to allow coexistence of many consumers on a single resource. Huisman and Weissing (2001) argue that multi-species coexistence on three resources is a definite possibility, although the required parameters may be evolutionarily unlikely. In any event, the role of differences in the shapes of consumer responses has been recognized in at least one recent empirical study (Letten et al. 2018). The results presented here suggest that nonlinearity should have important effects on competition, regardless of the number of resources. However, it also calls into question the suggestion that such differences always make coexistence more likely.

Future studies of competitive systems in seasonal environments should recognize the potential importance for coexistence of the exact parameters affected by temporal variation, the interaction of different parameter fluctuations, and the wide range of outcomes that may be possible in any given system.

10

Consumers and resources in space

10.1 The nature of spatial competition

Space itself is a resource for many sessile organisms. However, this chapter focuses on space as a 'container' for both resources and consumers. What does spatial variation in physical and chemical conditions mean for competition between consumer species? Answering such a question for any real system requires some understanding of the theoretically possible effects. This in turn requires a family of mathematical models with a range of levels of simplification. The importance of space for interspecific competition was recognized early in the 1970s. Simon Levin (1974, p. 207) stated that, 'The distribution of a species over its range of habitats is a fundamental and inseparable aspect of its interaction with its environment, and no complete study of population dynamics can afford to ignore it'.

Constructing and analysing the simplest consumer–resource models of competition for systems with a small number of patches is a first step in developing an understanding of the range of spatial effects. Surprisingly, this task is still quite incomplete. As a result, many questions remain unanswered. Is a resource item that is present at one spatial location part of the same resource population as a second, physically or biologically identical, resource item at a different location? What amount of spatial separation is sufficient to ensure that two consumer individuals or two resource items belong to distinct populations? These questions cannot be answered in the context of most models of competition discussed thus far in the book, as they do not include space, except as a source of a fixed amount of immigration. In addition, most of the models have assumed that each population is well mixed within the area that is described.

Yet space in reality is a continuous variable, and the level of resolution applied to understand it is always an issue. Both consumer and resource items/individuals are characterized in part by their current location. Even within an area that has uniform physical, chemical, and biological conditions, a consumption event will have its greatest impact on consumers that are located nearby. These close individuals are most likely to have encountered the resource item subsequently if it had not been consumed. Thus, not having an exact and continuous description of space always involves at least some degree of simplification.

Another complication is that most habitats in which species occur are characterized by spatial variation in conditions. These conditions are usually defined as

Competition Theory in Ecology. Peter A. Abrams, Oxford University Press. © Peter A. Abrams (2022).
DOI: 10.1093/oso/9780192895523.003.0010

physical and chemical properties, but they may also be defined to include the presence or abundance of species other than the focal set of competitors. Unfortunately (from a modelling standpoint), different variables and/or different conditions usually do not have identical patterns of spatial variation. This fact may require a high level of spatial resolution to accurately describe the interaction between species. This again causes problems, as representing multiple quantities in continuous space in a tractable mathematical model is generally impossible. Even if the model could be formulated, spatially local parameter values could only be approximated, and the model would be impossible to analyse fully.

As a consequence of these features of space, all ecological analyses have greatly simplified space in one way or another. The approach to spatial structure used here is the most common one, which represents space as a number of discrete 'patches' that are separated by areas that cannot support populations of the species of interest, but that permit travel by one or more of these focal species. This 'metapopulation' (Levins 1969, 1970) or 'metacommunity' (Leibold et al. 2004) approach has been used in studies of interspecific competition since the 1970s (Levins and Culver 1971, Horn and MacArthur 1972, Levin 1974). It is still the approach most often used to study groups of interacting species in spatially subdivided populations. It has the advantage of being a simple extension of non-spatial models. It is also an increasingly accurate representation of space for many species as human modification of the environment continues to fragment areas of semi-natural habitat.

Much of the earliest work on such patch-structured systems followed the framework of Levins (1969, 1970), which does not explicitly model the processes occurring within a patch. This family of models just accounts for extinction and colonization events; abundances within the patch are ignored. Interspecific competition is reflected in a higher extinction probability for a given species within patches that it shares with one or more other competing species. The status of a population is measured by the proportion of patches in which it occurs, rather than the number of individuals or biomass. This leaves out essentially all the details of interactions between species. Hanski (1999) provided a comprehensive summary of this method, which has mostly been used to study the dynamics of single species. This chapter will use a different approach; it will focus on models in which the patches are sufficiently large that stochastic extinction can be ignored on the timescale of interest. The 'within patch' components of these models are provided by the simple resource-based models described in previous chapters.

Adopting a metacommunity framework requires recognition of the differences between patches in conditions that affect the population growth rates of their inhabitants and the nature of movement by the component species/entities between those patches. The definition of coexistence also changes within a metacommunity framework, as has been noted before. Coexistence may be local or global, and local coexistence usually requires that a threshold density be defined, below which the species does not qualify as coexisting in that patch. A metacommunity framework also complicates the issue of invasion, as invasion may occur in a single patch or in many patches more-or-less simultaneously. The framework used here is to consider systems in which all resource types are capable of existing in all patches in the absence of consumers. This simple case makes it possible to attribute differences between spatially subdivided and well-mixed systems to the spatial structure of the former system. In most of this chapter's examples, simultaneous introduction of a small number of individuals into all patches will be assumed when invasion is analysed.

The majority of the analysis here will be based on systems with only two patches. This very simple case is sufficient to illustrate many of the qualitative characteristics that differ between spatially heterogeneous and homogeneous systems. It allows the minimal elements of a spatial model to be identified. The two patches may differ in properties that affect the demographic and/or ecological parameters of either consumer or resource or both. For example, the patches may have different mean temperatures, and this difference is likely to affect a range of processes in both consumer and resource. Regardless of cause, if some property of a patch creates different local dynamics of one or more consumers, this qualifies the resource population in that patch as being a distinct entity from the standpoint of consumer coexistence; this was initially discussed by Haigh and Maynard Smith (1972).

The question whether coexistence at a global scale in a metacommunity qualifies as 'true' coexistence does not have a universally accepted answer. When coexistence is defined locally for a patch that is open to immigration, there is usually some arbitrariness in the decision of what abundance of a locally rare species allows a classification of exclusion. Because consumers and resources can have very different movement abilities, the most general approach would seem to be to define coexistence at a global scale; i.e., species coexist if each is present somewhere in the metacommunity. This allows comparison of spatial resource segregation to other types of resource partitioning. Under this global definition, the quantitative measure of an interaction using the population responses to changes in a neutral parameter should logically assume that the parameter changes in identical ways across all patches. This convention means that, if patches differ in their initial population dynamical parameters, 'quasi-extinction' of a focal consumer population will occur at a different value of an imposed mortality rate for different patches. 'Quasi-extinction' implies that, if the patch were isolated, its consumer population would drop to zero; however, it may have a low abundance sustained by immigration. If each patch is characterized by unique parameter values or species compositions, the proportional rate of decrease of the global equilibrium population size will often change rapidly at a mortality rate that implies quasi-extinction within any one patch. These quasi-extinction events of the resource will generally produce a significant change in the global measures of both inter- and intraspecific competition between consumers.

Each resource/patch combination acts as a resource in that the effect of a consumption event on the consumer's per capita growth rate is dependent on the properties of the space where it occurs. The resources in different patches may also be connected via movement. The nature of movement in both consumers and resources plays a key role in the competitive process. Interest in movement has increased rapidly in recent years (see Lewis et al. 2021), and there is now a journal devoted exclusively to this subject. However, it is fair to say that there is still no consensus on the proper approach to describing movement quantitatively (Gross et al. 2020). This chapter will focus on the different effects of random and adaptive forms of movement.

A more complete account of spatial competition would include an exploration of models using continuous space, and those cases (appropriate for stationary individuals) where adults interact only with a small set of nearest neighbours. Unfortunately, both cases involve more complex mathematics or very large numbers of simulations. Existing results for these interactions cover a much narrower range of biological scenarios than do those for discrete patches, embodied in metacommunity models. See Snyder and Chesson (2004) or Cantrell et al. (2012) for examples of approaches to competition in continuous space.

A major theme of Chapters 8 and 9 was that temporal variation affecting consumers differently could either enable or prevent the coexistence of competitors. The same is true of spatial variation, although this result has been more widely accepted than the corresponding temporal result. Rather than focusing exclusively on coexistence, the present chapter takes a somewhat broader approach and examines how various aspects of movement in both consumer and resource affect the competitive interaction between consumers, measured at the level of the whole metacommunity. The descriptions of within-patch dynamics explored here assume the simple 'continuous time–homogenous population' consumer–resource models used in previous chapters.

The movements of any given consumer or resource type may be independent of the conditions in the occupied patch. However, for biological species that are capable of directed movement, it seems more likely that the movement will be influenced by conditions in the currently occupied patch, and possibly by conditions in other patches. The resource population size and the values of the consumer's own demographic parameters in its current patch are likely to be better known than are those of other patches. As a result, conditions in the currently occupied patch are expected to have a larger effect on consumer movement in most circumstances. If information on resource abundance and patch properties affecting fitness in other patches is available, that information should also affect movement. The local consumer density affects the rate of change of local resource abundance in the near future. Thus, consumer density as well as resource abundance and non-resource related physical/chemical conditions in the currently occupied patch may affect movement, although such purely density-dependent effects are not treated here. It is possible that consumers make exploratory trips to nearby patches that provide information. If such visits are sufficiently short in duration, they can be assumed to be roughly equivalent to remote detection of conditions in other patches. The various possibilities for remote detection and short-term visits suggest that a general approach to consumer (or resource) movement should include the possibility that conditions in both the origin and the destination influence consumer movement rates.

Temporal and spatial variation may interact; for example, the environmental fluctuations experienced in one spatial location may differ from those experienced in another location. A given patch with a temporally constant environment may experience cycles because of consumer–resource interactions within that patch. If so, it

may influence the dynamics of other patches by producing temporally varying numbers of emigrants. Even in cases with identical conditions in each of two patches, the
interaction of consumer–resource cycles with movement between patches can result
in cycles that differ between the two patches.

The rest of this chapter begins (Section 10.2) with a short history of theory regarding competition in subdivided populations. This is followed in Section 10.3 by an
analysis of how spatial subdivision affects the nature of intraspecific competition, as
described by the relationship between a 'neutral' parameter (here, per capita mortality) and population size. Section 10.4 then examines the impact of random movement
by both resources and consumers on their ability to coexist and on the functional form of their interaction. Adaptive movement by consumers and/or resources is
treated in Section 10.5, which uses the simplest possible representation of dynamics
within a patch (MacArthur's consumer–resource model). Section 10.6 broadens this
to consider adaptive movement in the context of consumer–resource models involving some nonlinear components. Possible pathways for advancing our understanding
of competition in metacommunities are treated in Section 10.7.

10.2 A history of metapopulation competition models

Space was already regarded as an important determinant of competitive interactions
early in the modern history of competition theory. The works by Levins (1969, 1970),
Levins and Culver (1971), Horn and MacArthur (1972), and Levin (1974) were particularly influential. Most of these focused on multi-patch systems and ignored the
details of competition within a patch; competition was simply assumed to increase
the extinction rate of each consumer species by a fixed amount in patches that were
occupied by both. Levin (1974) had a treatment of 2-patch systems that included within-patch dynamics rather than just presence/absence. He called attention
to the fact that the alternative-exclusion outcome of the LV model allowed two
competitors to coexist in a landscape in which the different histories of the two
patches led to a different species being dominant in each one. Most experimental
approaches to competition did not explore such subdivided systems until many years
later.

Tilman (1982) stressed that different spatial conditions could allow any number of
competitors to coexist, although he mainly considered the special case in which the
dynamics of two limiting resources varied spatially and he did not include any explicit
models of movement between patches. A variety of spatial models of plant competition were developed in the 1990s (reviewed in Pacala and Levin 1997, and other
chapters in Tilman and Kareiva 1997). Many of these used the individual plant as the
spatial unit, so were dealing with systems in which space itself was the resource. While
these works produced some important insights, they had relatively little influence on
approaches to competition in systems other than terrestrial plants.

Movement has long been recognized as having some counterintuitive effects on total abundance in a metapopulation. The first study to examine this possibility was Holt (1985). He showed that random movement by consumers between two unequal patches altered intraspecific competition and had counter-intuitive effects on equilibrium consumer abundance. The evolutionarily stable movement rate in a constant environment was shown to be zero. Movement rates greater than zero were not favoured at the level of individual selection. However, in the simple constant-environment consumer–resource models that Holt explored, the presence of some consumer movement often reduced overexploitation of the more productive patch(es), and thereby increased the global equilibrium population size.

There were relatively few studies of different types of movement on interspecific interactions until the concept of metacommunities became popular, shortly before 2000. The most common approach at that time was to assume a constant per capita movement rate of individuals out of each patch. Their destination in multi-patch systems was usually assumed to be equally likely to be any other patch, although some works assumed higher probabilities of arriving at closer patches. Subsequent work devoted more attention to the possibility of adaptive movement (Abrams 2000b, 2007a; Amarasekare 2010). As Ronce (2007, p. 233) put it, 'There are good theoretical reasons to believe that informed dispersal decisions would confer an evolutionary advantage over a blind process, unless patterns of variation in habitat quality are totally unpredictable or information acquisition is costly'. The process of developing theory that reflects this generalization has been complicated by the many possible ways of modelling adaptive movement (discussed in Abrams et al. (2007), Amarasekare (2010), and Gross et al. (2020), among others).

10.3 Space and the global shape of intraspecific competition

Interspecific competition is always closely related to intraspecific competition, so it is logical to begin with a treatment of how spatial patch structure affects the nature of intraspecific competition. In general, quantification of interactions defined at the level of a metapopulation (or metacommunity), requires changing a neutral parameter of a focal consumer species across all patches, and then measuring resulting changes in total equilibrium or mean population sizes. One of the most important descriptions of intraspecific competition is the shape of this relationship. If the patches differ from one another (which is normally the case in natural systems), a neutral parameter perturbation of the same magnitude across all patches will likely have quantitatively different effects on consumer abundances in the different patches. Because the per capita competitive effects in most patches change with consumer and resource abundances, this implies that the magnitude of intraspecific effects at the metacommunity level will change as well.

The simplest case to explore is one in which migration is both random and very rare. The within-patch abundances may then be assumed to reach approximately the abundance they would have reached if the patch had been completely isolated.

I consider a multi-patch system having z patches, each obeying the 1-predator–1-prey MacArthur model, with both species potentially present in all patches. This is described by the following equations, where the subscripts denote patch of residence rather than species identity:

$$\frac{dR_i}{dt} = R_i (r_i - k_i R_i) - C_i R_i N_i$$

$$\frac{dN_i}{dt} = N_i (B_i C_i R_i - d_i - D) - m N_i + \sum_{j}^{j \neq i} (m/(z-1)) N_j \qquad (10.1a, b)$$

The parameter definitions are identical to those in the equivalent model in previous chapters. All consumers have equal per capita movement rates (m) out of a patch, and all individuals are assumed to survive the dispersal process. Each disperser has an equal probability of arriving at one of the other patches. D is the perturbation to the mortality rate.

Assuming that the movement rate m is small enough that it does not affect consumer abundance appreciably, the following approximation for the abundance of the consumer in patch i can be applied:

$$\hat{N}_i = \frac{B_i C_i r_i - k_i (d_i + D)}{B_i C_i^2} \qquad (10.2)$$

Note that the consumer population in a patch declines linearly as D increases. This means that the consumer will become extinct (or quasi-extinct) in patch i if

$$D \geq \frac{B_i C_i r_i}{k_i} - d_i \qquad (10.3)$$

The rate of change of the total consumer population with D across all patches that have an appreciable consumer population is:

$$-\sum_{i=1}^{N>0} \frac{k_i}{B_i C_i^2} \qquad (10.4)$$

Note that, as the mortality perturbation, D, increases, consumer extinctions occur in the order given by the value of the right-hand side of inequality (10.3); this means that the patches i having the lowest consumer per capita growth rates when the resource is at its carrying capacity will be the first to drop out of the summation in eq. (10.4). These are patches that are characterized by low values of B, C, and/or r as well as those with high values of k_i and/or d_i. High k, low C, and/or low B within a particular patch are all associated with a large change in N_i per unit change in D. Therefore, the consumer extinctions that occur at relatively low mortality perturbations are likely to reduce the total rate of decline of the total N with D by the largest amount. For a simple example, consider two patches that differ only in their values of C, with patch 2 having the smaller C, equal to 1/3 that of patch 1. The smaller capture rate means that the predator goes extinct in an isolated patch 2 at a lower imposed mortality than in patch 1. Patch 2 has a larger rate of decline of N with D than does patch 1. The result of extinction of N_2 is that there is an abrupt decrease in the magnitude of the negative slope of $N_1 + N_2$ vs D at this point.

The formulas and the 2-patch example just discussed made the simplifying assumption that migration between patches was close to zero. This approximates the case of a very low rate of random movement between patches. In the limiting case of very rapid (and cost-free) movement, the patches merge into a single one having the mean of the two patch-specific parameter values. The effects of most intermediate movement rates must be determined numerically. Figure 10.1 illustrates these effects for two cases with relatively low movement rates, assuming patches are identical except for patch 2 having a C value 1/3 that of patch 1. The blue dots represent a movement rate that is 33.3 times larger than the (otherwise equivalent) system represented by the red dots. Both cases are characterized by a positive second derivative of the relationship between N_{total} and D. This relationship is linear for the MacArthur model in a homogeneous environment, or in a metacommunity model consisting of identical patches with MacArthur-type dynamics in each.

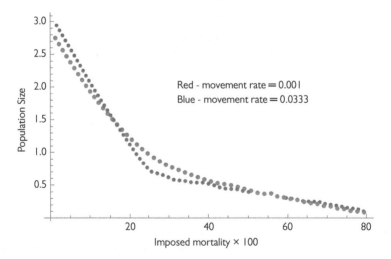

Fig. 10.1 Population size as a function of general mortality applied to the consumers in a 2-patch metacommunity with a single resource and a single consumer species. The parameters are: $B_1 = B_2 = 1$; $C_1 = 1$; $C_2 = 1/3$; $r_1 = r_2 = 1$; $k_1 = 1$, $k_2 = 1$; and $d_1 = d_2 = 0.1$.

If the above approach is expanded to a system having patches that differ in their values of two or more parameters, the same phenomenon of local consumer extinction occurs, and it causes a similar concave shape of N_{total} and D. This phenomenon is similar to what occurs with multi-resource systems in a single patch (see Abrams (2009b) and Chapter 5). If the resources in the individual patches have abiotic growth, the N vs D relationship is concave for a single isolated patch, and the concavity is accentuated by having local consumer extinction with increasing D. If biotic resources have theta-logistic growth with exponents other than 1, the individual segments of the relationship are nonlinear, so the entire relationship becomes more complex in its shape.

Random movement is unlikely in any species that is capable of self-directed travel. Chapter 3 argued that adaptive foraging was prevalent in natural communities, and this is often associated with directed movement. If resources are distinguished both by their physical identity and where they occur, then transfers of both consumers and resources between patches have key roles in determining the strength of indirect interactions among different consumers and among different resources. The next two sections consider the major distinguishing features of such movements; whether they are independent of consumer and resource abundances ('random') or depend on them in a manner that is adaptive for the moving entity.

10.4 Random movement and coexistence

This section assumes random movement of consumer individuals. As above, 'random' movement implies per capita movement rates that are independent of consumer and resource abundances and conditions in both the patch of origin and in all potentially accessible destination patches. All resource items are identical, and there is strictly exploitative competition for this single resource type within each of the two patches. The simple models used here again do not allow alternative equilibria within a patch, unless an interference component to competition is added. However, having different physical conditions in the two patches often allows two consumers to persist in the system as a whole.

The question is how greater movement of consumers and/or resources changes competition between consumers; this requires a measure of competition. Here I measure competitive strength using the change in the 'existence bandwidth' that is produced by the competitor. This is a simple extension of Armstrong's (1976) 'coexistence bandwidth' (see Chapter 6) to purely intraspecific competition. The existence bandwidth is the range of density- and location-independent mortality rates (or the range of another neutral parameter) that allows a consumer to exist in the absence of the competitor. 'Coexistence bandwidth' is the corresponding measure in the presence of the competitor. The strength of competition can be measured by comparing the coexistence bandwidth to the existence bandwidth. This provides a measure that can be used to compare systems with different nonlinear competitive effects. If there is no competitor, the maximum increase in the death rate of one consumer (say the consumer in patch 1) is simply the maximum per capita growth rate from near-zero abundance in the patch where this consumer has its highest per capita growth rate; i.e., $b_1 c_1 r_1 / k_1 - d_1$ in the MacArthur model. Competition is measured by the proportion by which the competitor's presence at the joint equilibrium (or cyclic attractor) reduces this range.

The following two cases are the simplest ones; i.e., those in which either consumer(s) or resource(s) is/are unable to move.

10.4.1 Competition when only one trophic level moves

The first case considered here assumes density-independent resource movement at a fixed per capita rate, with each of the consumer species confined to a different patch. With no resource movement, the two consumers coexist globally. However, resource movement makes coexistence more difficult by making the food supply of the consumer that occupies patch 1 dependent upon consumption by the second consumer species in patch 2. The larger the per capita movement rate of the resource between patches, the greater dependence of the resource in patch 1 on consumption occurring in patch 2. A sufficiently high movement rate means that the resource population is nearly equal across patches, so that even small differences in the zero-growth requirements of the two consumers will result in exclusion of the less efficient one. The ability to coexist can be quantified by examining how the existence bandwidth, defined above, is changed by the presence of the competitor. Again, the MacArthur model provides a useful case for illustrating some numerical results. The question is: how rapid must resource movement be to make coexistence of patch-restricted competitors unlikely? The example discussed here assumes equivalent parameters for the consumers in the two patches, with a per capita resource movement rate of m between the two patches.

Figure 10.2 plots the proportional increase in consumer death rate required to cause extinction as a function of resource movement rate. The two lines illustrate this relationship for two different baseline consumer mortality rates. The red line corresponds to a baseline mortality of 0.25, and the blue line corresponds to a mortality of 0.10. The maximum per capita resource movement rate shown on the figure is four, which is four times larger than the maximum per capita reproductive rate of the resource. At this point, coexistence is possible until the mortality rate disadvantage is approximately 1.3-fold in both cases illustrated; smaller disadvantages allow coexistence. Spatial separation can be effective at generating coexistence even with relatively rapid resource movement.

The second case considered assumes that the resource individuals/units do not move between patches. However, the consumers have random movement between patches and have one or more growth parameters that differ between patches, therefore producing different competitive abilities in the two patches. I assume that a very low level of movement from outside this system ensures that both consumers and

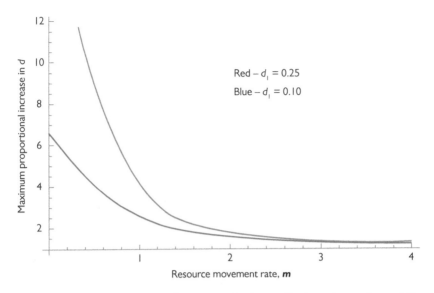

Fig. 10.2 Maximum proportional increase in death rate allowing existence for spatially separated competitors with resource movement at a rate m. Greater movement implies greater competition, reflected in a lower maximal increase in mortality. The common parameters are: $B_1 = B_2 = 1/4$; $C_1 = C_2 = 2/3$; $r_1 = r_2 = 1$; and $k_1 = k_2 = 1/10$. The blue line represents a system with $d_1 = d_2 = 1/10$; the red line is a system with $d_1 = d_2 = 1/4$.

the resource reach both patches. If there is near-zero movement between patches by the consumers, system-wide coexistence at a stable equilibrium only occurs if each consumer is the superior competitor in a different patch. Low rates of random movement of the consumers allow unequal coexistence in the two patches. High rates of random movement can result in exclusion of the less abundant or slower reproducing consumer.

Abrams and Wilson (2004) and Namba and Hashimoto (2004) carried out similar analyses of competition in this system. Both articles showed that robust coexistence was possible in a 2-patch model with random consumer dispersal despite each consumer having identical equilibrium resource densities (R^*) in each patch in isolation. Coexistence was also possible even when the same consumer was the superior competitor in each patch in isolation (i.e., one species had a lower R^* in both patches). This consumer would normally be expected to exclude the other. However, coexistence is possible when the consumer with the exploitation disadvantage (higher R^* values) has a lower movement rate or a faster demographic response to changed resource abundance (e.g., proportionally greater values of both resource conversion efficiency and per capita death rate). Either of these differences results in a greater concentration of the 'inferior' (higher R^*) competitor in the patch with the greater resource density. This means that the consumer that is an inferior exploiter is also an inferior (slower) disperser.

Another mechanism is observed when the system is continually perturbed away from an equilibrium that would otherwise be stable. If the system experiences randomly occurring consumer mortality events, then the slower disperser must usually be a better competitor (lower R^*) in order for both consumers to persist in the system while competing for the same resource within each patch. This is the well-known competition–colonization trade-off, which requires that the patches experience perturbations that keep them from remaining long at near-equilibrium conditions. It does not require directed movement, but the presence of some directed movement is consistent with the mechanism. Rapid resource regeneration following a consumer mortality event usually makes the mechanism more effective in promoting coexistence.

While two species can coexist via the differential movement-rate mechanisms described above, it is unlikely that large numbers of species do so. Most cases of coexistence of multiple consumer species in a metacommunity that contains only one or a small number of physically distinct resources are likely due to different rankings of consumer fitness parameters in different patches. This ranking must imply that each consumer species has some types of patches where it has the lowest R^*. In general, neither of the mechanisms considered above leads to a distribution of consumers characterized by equal fitness in all patches. This outcome requires adaptive movement and is discussed in Section 10.5.

10.4.2 Competition with mobile consumers and interference competition

All the models discussed above have assumed purely exploitative competition. If a consumer's abundance has a direct effect on one or more of fitness parameters of other consumer individuals, this implies an 'interference' component to the competitive process. As noted in previous chapters, within a single patch interference can result in a priority effect, in which an initially more abundant consumer species is able to exclude an initially less abundant one. Alternative exclusion with a priority effect is the standard outcome within a single patch when there is only a single resource type, when attack rates (C) are negatively affected by consumer abundance, and when the magnitude of the effect on the attack rate parameter is greater for heterospecifics than it is for conspecifics. This leads to competition coefficients greater than unity. This may also occur with large interspecific effects on other fitness parameters.

The possibility of priority effects within at least some patches implies the potential for two or more system-wide attractors for metacommunities having alternative attractors in a single-patch setting. The simplest 2-competitor–2-patch system could have species 1 in both patches, species 2 in both patches, and two different configurations having a different species in each patch. The number of alternative outcomes increases with either more patches or more consumer species. Levin (1974) was apparently the first to note this fact. However, at least some of its implications seem to be largely ignored. The primary implication is that having higher levels of interspecific interference often makes coexistence more likely at the metacommunity scale, even though it makes coexistence within a patch less likely. The presence of interspecific interference can be predicted on evolutionary grounds in some systems, and they were observed in some of the earliest laboratory studies of competition between different species of the flour beetle genus, *Tribolium* (Park 1948). If the initial relative abundances of two such competitors differ in two patches, then it is likely that both will persist in the metacommunity, although each patch will consist almost entirely of one species.

The 2-species–2-patch system described in the previous paragraph has a different response of population sizes following a system-wide increase or decrease in one consumer's mortality rate depending on the initial state of the system. It is simplest to describe these relationships using the Lotka–Volterra (LV) model for dynamics within a patch; in this model, alternative attractors within a patch can occur when the product of the two competition coefficients exceeds unity.

Consider a multi-patch system with two competitors and with alternative exclusion outcomes possible in some or all patches before the mortality perturbation. In the LV model this implies that the product of the two competition coefficients is > 1. It is necessary that each species of consumer be present somewhere in the system for changes in species composition to occur as one species experiences a mortality rate perturbation in all patches. Given that this condition is met, the equilibrium population size of the non-perturbed competitor will initially not be affected by increasing mortality of the focal species. However, when mortality on one consumer species produces a sufficiently low population size of that species in a given patch, invasion by the non-manipulated species will become possible, and its equilibrium will jump discontinuously to its carrying capacity within that patch. The focal species' equilibrium in this patch will drop discontinuously to zero. These discontinuities are likely to occur at a different level of general mortality of the manipulated species in different patches because growth parameters of both species will usually differ across patches. The earliest extinctions occur in those patches where the initially present consumer had relatively low equilibrium abundance in the pre-perturbation system. There will then be a series of discontinuous jumps in system-wide equilibrium population size of the competitor as the manipulated species experiences greater mortality across all patches and is displaced successively in the patches where it has a lower ability to exclude its competitor. However, this relationship depends on the initial configuration of the system when the perturbation occurs; this configuration is determined by the initial abundances of consumers and resources in each patch. All these qualitative features are possible for consumer–resource models that also have alternative exclusion outcomes within patches.

10.5 Adaptive movements and their effects on competition

10.5.1 General aspects of adaptive movement

The first treatment of adaptive movement was by Steven Fretwell (1972; also Fretwell and Lucas 1969), who introduced the term 'ideal free distribution' to describe the endpoint of this process. This was a distribution of individuals across patches that was characterized by equal fitness (instantaneous per capita growth rate) in all occupied patches, with occupied patches characterized by greater fitness than a single individual would achieve in an unoccupied patch. 'Ideal' referred to complete and accurate information about conditions, while 'free' referred to the lack of any cost to movement. Both of these defining characteristics are difficult to achieve precisely in real systems, but spatial distributions close to ideal free have often been observed in various behavioural studies (reviewed in Milinski and Parker 1991; Giraldeau and Caraco 2000). If such 'ideal free distributions' (IFDs) are common, it could be because movement at fixed condition-independent rates is a special case that seldom occurs in natural systems. However, there is still considerable debate about the prevalence of IFDs (Matsumura et al. 2010). Lack of IFDs could be due to movement that is not sufficiently rapid, species that are poor at detecting resource abundances, or rapidly fluctuating conditions. General evolutionary considerations suggest that both consumers and living resources should behave adaptively, and it is not obvious whether their combined movements will come to some equilibrium. This implies the need for dynamic models of movement. Fretwell (1972) did not discuss the details of individual movement decisions that would be needed to approximate an ideal free distribution in such a case. Some of the subsequent work on this issue is discussed later in this section. Much work on IFDs has assumed that only a single consumer species was present, but this is unlikely for most species in natural communities.

The mechanistic basis of adaptive movement has received some attention in recent years, but empirical results are scarce, and there is little agreement on a quantitative description of the process. The simplest case is again one with two patches. This scenario has many similarities to adaptive choice between two types of behaviours in a homogeneous system. It is known that adaptive consumer diet choice in a homogeneous (single patch) system often alters intraspecific and interspecific competition (Abrams and Matsuda 2004; Abrams 2006a, b). Adaptive diet choice within a habitat has often been assumed to occur instantaneously with changes in prey or resource abundances; this is true of early models of 'switching' behaviour (Murdoch and Oaten 1975). However, this assumption is biologically unlikely, and it can produce very different dynamics than models with explicit behavioural dynamics (Abrams 1999, 2000b, 2007a; Abrams and Matsuda 2003, 2004). Similarly, understanding movement between patches also requires consideration of the dynamics of the movement process. Adaptive movement by the resources, when they are biological entities, usually depends on the abundances of their own resources, as well as the risk of being eaten by consumers. Understanding spatially structured interactions between the consumers of such living resources is therefore likely to require a model representing three trophic levels explicitly, and having movement rates that can potentially depend on the abundances of all three levels.

Species are likely to exhibit a net movement toward patches with better conditions than their current patch, but it is unclear how often they have information that would enable them to accurately estimate their future demographic parameters in other places. Resource abundance is one of the primary quantities likely to affect movement. Sufficiently low resource abundance for a consumer in its current patch is likely to imply that conditions might be better elsewhere, so adaptive movement could be based solely on the current or recent conditions in the occupied patch. However, long-distance detection of resource abundances in other patches is likely to be possible in some cases, and, as noted above, short visits to a nearby patch may be rapid enough for a model to ignore, provided the species returns to its original patch after a visit to a patch with poorer conditions. Thus, the instantaneous fitness associated with original and potential destination patches should both frequently affect the movement rate.

The models explored below assume that consumers and/or resources move at a rate that is an accelerating function of the fitness difference implied by moving between patches. The accelerating nature of the relationship is based on the fact that both the ability to detect differences, and the reward from moving increase with the fitness difference. Larger fitness differences also imply that the relative ranking of the two patches is likely to persist for a longer time. The specific function used in the numerical analysis here is that proposed in Abrams (2000b): the consumer per capita movement rate from patch i to patch j, M_{ij}, is given by,

$$M_{ij} = mExp[\lambda(W_j - W_i)],\qquad(10.5)$$

where the parameter m gives the between-patch movement rate when patches yield equal fitness change per unit time (and hence, random movement). The parameter λ is a positive constant that scales movement propensity to the difference in fitness, measured by instantaneous per capita growth rate, W. Equation (10.5) implies that there is always some level of movement between patches, but the movement rate from one patch to another is an accelerating function of the fitness gain that results from such a shift. Equation (10.5) with $\lambda = 0$ results in random movement regardless of fitness; i.e., equal per capita movement rates in each direction at a per capita rate, m. Most reasonable models of movement would have properties like those produced by eq. (10.5), and the qualitative conclusions about competitive effects given below do not depend on this exact functional form.

Equation (10.5) could be elaborated by incorporating local consumer abundance. Because local resources will be depleted more rapidly in a patch containing more consumers, one might expect an additional increase in movement out of patch i if the consumer had a high current abundance there (and/or a low abundance in the alternative patch), even when there was no direct impact of consumer abundance on immediate per capita growth rate. Preliminary work (Abrams unpub.) on a model with such consumer dependence suggests that a polymorphism is often produced. The equilibrium consists of some individuals moving according to eq. (10.5), with others being more likely to move when local consumer density is higher.

10.5.2 The shape of competition under adaptive consumer movement

Here I again assume each consumer could maintain a positive population size in each patch in isolation. However, each of the two consumer types is initially the superior competitor (lower R^*) in a different patch. Under constant conditions, accurate and extremely rapid adaptive movement by the consumers leads to approximately the same ultimate outcome as no movement, since it is disadvantageous for consumers to move to the patch where their expected per capita growth rate in the presence of their competitor is negative. However, adaptive movement is likely to affect the transient dynamics of consumers, which implies that it would influence average abundances in a temporally variable system. For example, when a few individuals arrive at a previously unoccupied patch, they are more likely to remain there under adaptive movement than under random movement, because of the high resource abundance. However, when the population growth stops, fitness will be the same as in the other patch and per capita movement rates should be similar in each direction. Those movements should be rare when there is a cost to moving. In the ideal case of movement only to the currently better patch for each consumer individual, there is no competition near equilibrium. If a parameter in one consumer species is changed sufficiently to make it inferior in both patches, then the equilibrium abundance of the inferior species will drop to close to zero, just as with very rare but random consumer movement.

Adaptive movement is almost certain to involve some random component. This is implied by a non-zero value of m in eq. (10.5). It results in some competition between consumers, even when each consumer species is superior in a different patch. When compared to high levels of random consumer movement, adaptive movement is likely to reduce competition between consumers at near-equilibrium conditions, as it results in greater patch segregation between them.

The outcome of competition with spatially restricted resources is usually very different from that of competition with traditional resource utilization differences that are based on physical characteristics of the resources. A simple example is provided by a 2-patch MacArthur system with two consumers and an identical non-moving resource species in each patch. Consumer movement is based on eq. (10.5). The consumers are assumed to each have the advantage in resource exploitation in a different patch. Both are assumed to be capable of sustaining a population in each patch in isolation in the absence of the competitor. The model describes the dynamics of the four consumer subpopulations, N_{ij} for consumer i in patch j, and the dynamics of the resource, R_j, in patch j ($j = 1, 2$):

$$\frac{dN_{11}}{dt} = N_{11}\left(b_{11}c_{11}R_1 - d_{11}\right) - m_1 N_{11} Exp[\lambda_1\left((b_{12}c_{12}R_2 - d_{12}) - (b_{11}c_{11}R_1 - d_{11})\right)]$$
$$+ m_1 N_{12} Exp[\lambda_1\left(-(b_{12}c_{12}R_2 - d_{12}) + (b_{11}c_{11}R_1 - d_{11})\right)]$$

$$\frac{dN_{12}}{dt} = N_{12}\left(b_{12}c_{12}R_2 - d_{12}\right) + m_1 N_{11} Exp[\lambda_1\left((b_{12}c_{12}R_2 - d_{12}) - (b_{11}c_{11}R_1 - d_{11})\right)]$$
$$- m_1 N_{12} Exp[\lambda_1\left(-(b_{12}c_{12}R_2 - d_{12}) + (b_{11}c_{11}R_1 - d_{11})\right)]$$

$$\frac{dN_{21}}{dt} = N_{21}\left(b_{21}c_{21}R_1 - d_{21}\right) - m_2 N_{21} Exp[\lambda_2\left((b_{22}c_{22}R_2 - d_{22}) - (b_{21}c_{21}R_1 - d_{21})\right)]$$
$$+ m_2 N_{22} Exp[\lambda_2\left((b_{21}c_{21}R_1 - d_{21}) - (b_{22}c_{22}R_2 - d_{22})\right)]$$

$$\frac{dN_{22}}{dt} = N_{22}\left(b_{22}c_{22}R_2 - d_{22}\right) + m_2 N_{21} Exp[\lambda_2\left((b_{22}c_{22}R_2 - d_{22}) - (b_{21}c_{21}R_1 - d_{21})\right)]$$

$$- m_2 N_{22} Exp[\lambda_2\left((b_{21}c_{21}R_1 - d_{21}) - (b_{22}c_{22}R_2 - d_{22})\right)]$$

$$\frac{dR_1}{dt} = R_1\left(r_1 - k_1 R_1\right) - c_{11}R_1 N_{11} - c_{21}R_1 N_{21}$$

$$\frac{dR_2}{dt} = R_2\left(r_2 - k_2 R_2\right) - c_{12}R_2 N_{12} - c_{22}R_2 N_{22} \qquad (10.6)$$

If each species is superior (has a lower R^*) in a different patch, the outcome is that the two consumer species are largely, but not completely segregated in space. As a result, changing a neutral parameter (here, d_{ij}) in consumer species i by the same, relatively small amount in both patches, has little effect on the other consumer species; they retain their spatial separation, and the altered consumer decreases when its death rate is increased. Given very accurate habitat selection (very small m and large λ) this means that, as both d_{2j} values are increased, the abundance of species 2 will decrease in both patches, but most of this decrease is in patch 2, where almost all individuals of consumer 2 are located. Species 1 will be nearly unaffected, since it is very rare in patch 2. However, if both d_{2j} values are increased sufficiently, d_{22} will pass a threshold value where species 1 becomes the superior competitor in patch 2. At that point, consumer 2 goes extinct, and consumer 1 achieves a high abundance in patch 2. The increase in species 1 caused by this disappearance may be significantly larger than the original population size of consumer 1 when both species originally have equal population sizes. This will be the case when the species differ in which patch they consume resources at a higher rate, and the consumers over-exploit the resource in their 'preferred' patch (equilibrium $R < r/(2k)$). If so, the facts that $c_{ij} < c_{ii}$ and that the resource growth parameters are equivalent in the two patches imply that, following the loss of consumer 2, the population of consumer 1 in patch 2 will exceed that in patch 1.

In the generic case, there is some movement even when it is disadvantageous. This low level of misdirected movement is expected in natural systems due to errors in estimating resource abundances. Errors are particularly likely to characterize estimates of resources in the patch that is not currently occupied by the estimator. These errors (implied by the random movement component of eq. (10.5)) alter the consequences of mortality perturbations to either of the consumers. The effects of mortality perturbations of different sizes are described for both consumers in Figure 10.3A; this example has relatively accurate movement with $m = 0.25$ and $\lambda = 10$. The two consumers initially have mortality rates of $d = 1$ in both patches, and each has double the c-value in its initially 'preferred' patch compared to the 'non-preferred' one. The figure shows how the total population of each species changes as the per capita mortality of consumer 2 in both patches is increased by the amount given on the x-axis. In contrast to the idealized system described in the previous paragraph, there are some competitive effects at mortality rates lower than that which makes consumer 1 the superior competitor in patch 2.

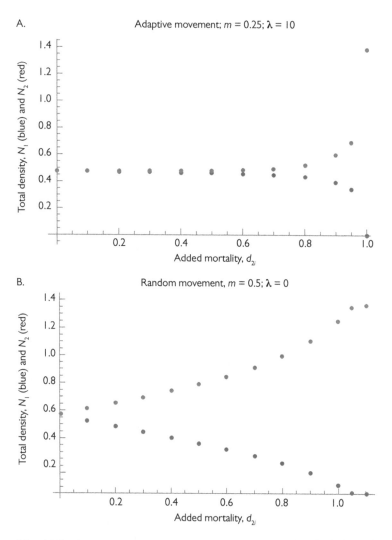

Fig. 10.3 The competitive interaction between two mobile consumers of a single type of resource that is present in both patches, but cannot move between them. Each of the consumers has an advantage in a different patch. The dynamics within a patch are given by the MacArthur model, and the adaptive movement of consumers is based on eq. (10.5). The parameter values for the starting condition of consumer equality are: $b_{11} = b_{12} = b_{21} = b_{22} = 1$; $c_{11} = c_{22} = 2$; $c_{21} = c_{12} = 1$; $d_{11} = d_{22} = d_{12} = d_{21} = 1$; $r_1 = r_2 = 1$; and $k_1 = k_2 = 0.1$. In panel A, movement is adaptive with parameters: $m_1 = m_2 = 0.25$; and $\lambda_1 = \lambda_2 = 10$. In panel B, movement is random, with parameters: $m_1 = m_2 = 0.5$; and $\lambda_1 = \lambda_2 = 0$.

The approximately three-fold increase in the abundance of consumer 1, which follows quasi-extinction of consumer 2, occurs because the resource in patch 1 is overexploited by consumer 1. Consumer 1's lower per capita consumption rate in patch 2 produces a higher, rather than a lower, population size than in patch 1, where it captures resources more rapidly. The pattern shown in Figure 10.3A should be compared to the corresponding case of spatial competition with purely random movement. Figure 10.3B shows such a case where the random movement parameter, m, is double that in panel A. The interaction is still nonlinear, with increasing competitive effects of a fixed increase in d as the initial d becomes larger. A very unequal coexistence is possible even when consumer 2 is inferior in both patches in isolation (e.g., if it has an added mortality of 1.05). This is a result of random movement reducing the abundance of consumer 1 in patch 2, where it is more abundant.

The two consumers may differ in parameters other than c, or in multiple parameters. In all cases, the general pattern shown in Figure 10.3A still applies; uniform, fitness-decreasing changes in the parameters of one consumer will initially have small effects on the other consumer, until the species are close to competitive equality in the patch that is initially better for the manipulated species. At this point, the manipulated species decreases to extinction and its competitor increases rapidly over a relatively narrow range of the manipulated parameter. If the Figure 10.3 example were based on competitors each having a higher b value in a different patch, there would only be a two-fold, rather than a three-fold, increase in the non-perturbed consumer when its competitor became extinct in the system. If the model is modified to have abiotic resources, overexploitation is impossible, again producing smaller changes following extinction of one species. Finally, if eqs (10.6) are modified to have two resource types per patch, coexistence within an isolated patch becomes possible. However, if conditions favour a different consumer in each patch, that coexistence will involve unequal abundances, and adaptive patch movement will produce a similar (but smaller) jump in the density of one consumer when the other consumer becomes extinct.

The above treatment is far from a complete analysis, even of the simple case described by eqs (10.6). A variety of other phenomena occur with different parameter values in that model. However, the example is sufficient to show that competitive interactions in a metapopulation based on the simple MacArthur system with adaptive consumer movement have nonlinear responses quite different from those in the corresponding non-spatial (single patch) system, or a system with random movement. Having minimal spatial overlap in the original system does not imply a lack of competitive effects. The response to uniform changes in density-independent mortality is highly nonlinear, unlike those in an analogous system having differences in resource use within a single patch.

10.5.3 Adaptive movement by the resource

Another possibility is adaptive movement of the resource only, assuming spatially separated consumers. Adaptive movement should often characterize mobile biotic resources, and it might be expected to lead to nearly equal resource fitness in each patch, as predicted by 'ideal free distribution' theory. An increase in the per capita mortality rate of one consumer (say consumer 2) will initially lower its population size, which in the short term produces an increase in the within-patch growth rate of the resource population in patch 2 plus an adaptive movement of resources from patch 1 to patch 2. The movement to patch 2 causes a decrease in the resource abundance in patch 1, which lowers the population of consumer 1. However, in the longer term, both resource populations attain new equilibria, with more resource in patch 2. The effects on the consumer may be seen by again considering a MacArthur system, this time with two patch-restricted consumers and an adaptively moving resource:

$$\frac{dN_1}{dt} = N_1 \left(b_1 c_1 R_1 - d_1 \right)$$

$$\frac{dN_2}{dt} = N_2 \left(b_2 c_2 R_2 - d_2 \right)$$

$$\frac{dR_1}{dt} = R_1 \left(r_1 - k_1 R_1 \right) - c_1 R_1 N_1 + m R_2 Exp[\lambda \left(- \left(r_2 - k_2 R_2 - d_2 - c_2 N_2 \right) \right.$$
$$\left. + \left(r_1 - k_1 R_1 - d_1 - c_1 N_1 \right) \right)] - m R_1 Exp[\lambda \left(\left(r_2 - k_2 R_2 - d_2 - c_2 N_2 \right) \right.$$
$$\left. - \left(r_1 - k_1 R_1 - d_1 - c_1 N_1 \right) \right)]$$

$$\frac{dR_2}{dt} = R_2 \left(r_2 - k_2 R_2 \right) - c_2 R_2 N_2 - m R_2 Exp[\lambda \left(- \left(r_2 - k_2 R_2 - d_2 - c_2 N_2 \right) \right.$$
$$\left. + \left(r_1 - k_1 R_1 - d_1 - c_1 N_1 \right) \right)] + m R_1 Exp[\lambda \left(\left(r_2 - k_2 R_2 - d_2 - c_2 N_2 \right) \right.$$
$$\left. - \left(r_1 - k_1 R_1 - d_1 - c_1 N_1 \right) \right)] \tag{10.7}$$

The equilibrium resource abundances are determined by the two consumer equations in eqs (10.7), unless one consumer has a death rate that is too high for that consumer to exist. The consumer abundances are determined by the resource equations. If the movement had perfect accuracy, there would be no long-term effect of changing the per capita mortality, d, of one consumer on the abundance of the other. When d_2 is high enough to eliminate consumer 2, the resource in patch 2 will approach its carrying capacity. This entails a per capita resource growth rate of close to zero, which is the same growth rate of the resource at equilibrium with consumer 1 in patch 1. Using the movement function in eqs (10.7), the parameter λ must be quite large and m must be quite small to approach zero competitive effect. Otherwise, the random component of movement means that the larger resource population in the patch having few or no consumers effectively fertilizes resource growth in the other patch, leading to a slight increase in the population of consumer 1 in patch 1 when consumer 2 is reduced to zero. The generalization about small competitive effects can be confirmed by examining cases with parameters similar to the adaptive consumer example in the previous section. Purely random resource movement increases competition greatly compared to the zero-competition baseline with no resource movement. However, sufficiently accurate adaptive movement again comes close to eliminating competition at equilibrium when each of the two consumers does better in a different patch. These results have some similarities to a single-patch system in which the resource exhibits adaptive defence with a linear trade-off between avoiding/defending against one or the other type of predator (Matsuda et al. 1993).

10.6 Adaptive movement of both species

The models in Section 10.5 both assumed a MacArthur system within each patch; in this system, the per capita risk of consumption is not directly affected by resource abundance. This independence disappears when the consumers' per capita capture rates change with resource abundance, a characteristic of every nonlinear functional response. Most measurements of functional responses have found nonlinearity (Jeschke et al. 2004). (Note that Jeschke et al. overstate the number of linear responses, by including in the 'linear' category responses with approximate linearity up to a threshold prey/food abundance, beyond which the response is constant.) Nonlinearity raises two types of issues that are important for the dynamics of spatially structured systems. The first is that the expected fitness of a prey individual depends on the number of other individuals in its patch. Such frequency dependence significantly affects the dynamics of traits that affect the vulnerability to consumption (Abrams et al. 1993; Geritz et al. 1998). In most biological systems, either high local prey abundance or high vulnerability of local prey to the predator produce predator satiation (increased total 'handling time'). This satiation, in turn, lowers the risk to prey individuals, and therefore makes it less likely that a prey individual will leave a patch. This can lead to unstable behavioural dynamics, as predators increase in the patch where prey have congregated, eventually producing a rapid exodus of prey individuals. The second issue raised by nonlinear consumer capture rates is their interaction with temporal variation. Saturating functional responses can cause temporal variation via predator–prey cycles in the absence of any adaptive movement by prey. In this case, or in the case of cycles driven by sustained temporal environmental variation, mean population abundances in systems with nonlinear per capita growth rate functions will not equal the equilibrium abundance. Such cycles can even reverse the sign of an interaction between two species (Abrams et al. 2003).

In real communities, consumers have their own predators, which are also likely to have saturating functional responses. Such a functional response in a higher-level predator has the potential to lead to temporal cycles in the movement behaviour of the lower-level predator (consumer). A type II response by the higher-level predator can also generate cycles directly within each patch, even if the lower-level predator does not exhibit adaptive movement. This combination of different sources of instability (both movement and population dynamics) on two trophic levels leads to various potential outcomes, the full range of which has yet to be explored. The rest of this section of the chapter will look at the potential effects of adaptive movement in some of these more complicated scenarios. Unfortunately, relatively little is known about these scenarios in systems with more than one consumer species. Most of the work reviewed below has dealt with a single consumer. This work is directly relevant to intraspecific competition; however, its extension to competition between consumers is largely a task for the future.

Simple predator–prey models having two patches and either adaptive or random movement with nonlinear per capita growth rates were analysed in Jansen (2001), Koelle and Vandermeer (2005), Rowell (2010), Abrams and Ruokolainen (2011), Ruokolainen et al. (2011), Abrams et al. (2012), and Gramlich et al. (2016), among others. All these works have found a wide range of effects of dispersal on mean abundances and the pattern of population variation. Abrams (2007a) discussed the complex cycles in a 2-patch model involving a single species on each of three trophic levels. That model also assumed type II functional responses for both top and middle trophic levels. A wide range of complex cycles occurred in the full model. Even the (unlikely) assumption of fixed abundances of each species led to some unexpected outcomes. In a simple case with fixed population sizes of all species, the type II responses favour prey aggregation in one of the patches, where their higher numbers reduce their per capita risk of being eaten. This results in predators moving to that patch, eventually favouring movement by the prey to the formerly less-occupied patch. This process of cyclic movement can persist over the long term, and cycles are more likely to occur when prey abundances are allowed to vary.

Adaptive habitat choice by both of the top two trophic levels led to exceptionally complex dynamical outcomes. Adaptive movement can lead to asynchrony in many 2-patch systems where consumer–resource cycles would become synchronized in the presence of random movement (Abrams and Ruokolainen 2011; Ruokolainen et al. 2011). Asynchrony usually reduces system-wide variation in consumer abundance. Gramlich et al. (2016) present a more comprehensive analysis of 2-patch systems with a single consumer and resource; they found that antisynchronous oscillations were most common under adaptive dispersal. The variety of outcomes would seem likely to be greater in models that had two or more species per trophic level.

While adaptive movement models have still not been explored in the context of two or more competing consumers in a metacommunity, there have been some analyses of the two species LV competition model in 2-patch systems (Cressman and Křivan 2006; Abrams et al. 2007). The second of these showed that adaptive habitat selection can lead to large amplitude cycles in the locations and population sizes of competing species that would have a stable equilibrium in the absence of adaptive movement.

A variety of different adaptive movement models and food web configurations involving some competition have been investigated. One of the first studies to find that adaptive movement could cause cycles was that of Schwinning and Rosenzweig (1990), who studied a system with intraguild predation. Amarasekare (2010) investigated the role of several types of movement functions in allowing coexistence of two consumers that share a single resource type and a single predator species (a 'diamond food web'). Gross et al. (2020) review several different functional forms of the movement for interacting species in metacommunities. However, it is still unclear to what extent different rules for movement affect the potential for movement-driven cycles.

In spite of the many unknowns, it seems likely that many of the impacts of adaptive consumer movement on the nature of competition in a simple 2-patch–2-consumer system will not depend sensitively on the exact form of the movement function, provided that it results in significant shifts of the consumer population towards better quality patches. Qualitative patterns that do not depend on the exact functional form of movement are particularly likely when neither within-patch dynamics nor movement produce instability. One of the more common effects of adaptive consumer movement in a stable metacommunity is to reduce the extent of competition close to the equilibrium point, by decreasing the spatial overlap of competitors relative to a situation with random movement. However, as shown in the simple 2-patch systems in Section 10.5, large magnitude perturbations can lead to abrupt increase or decrease in population size in this scenario. This is particularly likely when resources are overexploited, and a consumer can achieve a higher density in the patch where it has a lower resource capture rate (Abrams 1998).

10.7 Extending our current understanding of spatial competition

Two of the obvious needs for future work on spatially structured competition are to better understand spatial dynamics in systems with multiple patches and in systems with continuous (or more nearly continuous) spatial structure. Both of these would bring the models closer to reality. Unfortunately, progress in both cases is somewhat limited for technical reasons.

Most studies dealing with multiple-patch systems of interacting species have assumed random movement. This is true of an early analysis of predator–prey systems with type II responses by Jansen and de Roos (2000). They assumed a linear chain of 90 patches, and found that random movement tended to stabilize abundances, as the patches were out of synchrony. One attempt to compare 2- and multiple-patch versions of competition was Wilson and Abrams' (2005) study of two different spatial versions of the Armstrong–McGehee (AM) model, in which the consumer–resource interaction generated cycles in each patch. This was also restricted to random movement. The 2-patch version was based on the continuous time AM model with a single logistic resource and consumers with type II responses. The consumer parameters were identical across patches, but the resource parameters could differ. The 2-patch model typically had a coexistence bandwidth (range of relative mortality rates allowing coexistence) that was approximately 40% greater than that of comparable 1-patch models. Differences in parameter values between patches or the presence of antisynchronized dynamics both resulted in enhanced coexistence. The multi-patch version was based on simulations that attempted to mimic the 2-patch differential equation model with an individual-based approach, but did have several subtle differences. Because the simulations were based on discrete individuals, there was much more scope for random processes to affect dynamics. The cycles generated by the type II functional response with efficient consumers periodically produce very small population sizes, which enhanced the role of stochasticity in determining the outcome of competition. In any event, these explorations suggest that the analysis of multi-patch systems with variation are likely to have some unique features not present in the 2-patch models that have been discussed here.

The study of spatial competition suffers from some of the other characteristics that have commonly been omitted from most published models based on a single homogeneous habitat. Those omissions include: (1) the lack of interactions between resources within a patch, particularly in cases where the resources are biological species; (2) the lack of trophic levels beyond the consumer and resource; and (3) the lack of stage/age structure or intraspecific variation in competition-related traits in both consumers and resources. While much more work is needed, it seems likely that at least some of the qualitative findings regarding random vs adaptive movement revealed by the simple 2-patch models reviewed here will apply to many of these more complex scenarios.

11

Evolution and its ecological consequences

11.1 Evolution's many effects on interspecific competition

'Evolutionary effect' is a phrase that encompasses a wide range of potential initiating factors and responding variables in interacting species. Species can influence each other's evolution via pathways of effect involving changes in population size or changes in any trait affecting the interaction. The initiating event causing traits to change in one species may be a genetic change or environmental shift affecting another species that is already present, or it could be the introduction of a new species. In addition evolutionary change usually alters population sizes of both the evolving species and other species that interact with it. This introductory subsection will consider the processes by which competition affects evolution and vice versa.

Either a change in population size or an evolutionary change in traits related to resource consumption in one consumer can have a wide range of potential effects on the population sizes and resource-use traits of other consumers. This generalization might have been expected because competition is an indirect interaction that has many types of transmitting entities (resources), and species differ widely in their consumption processes. The chain of effects from an initial evolutionary change in one consumer almost always involves alterations in the abundances and in some of the characteristics of its resources. The evolutionary changes in other consumers produced by an altered population size or mean trait value of a focal consumer may depend in magnitude, and possibly sign, on the size of the initial change in the focal consumer. This should not be surprising because we know that the ecological effect on the other consumers usually differs depending on the magnitude of the change.

In contrast to the previous paragraph, most thinking about competitive coevolution has been based on systems without explicit resources. Analyses with resources usually have had just two resources (and two consumer species). In terms of questions that are addressed, most theory has concentrated on the effects produced by addition or loss of one species and has focused on the possibility that evolution causes greater difference (divergence) in traits determining the relative consumption rates of different resources. Divergence does occur, but resource-based models of competing consumers suggest that a much wider range of outcomes are possible, and traits

Competition Theory in Ecology. Peter A. Abrams, Oxford University Press. © Peter A. Abrams (2022).
DOI: 10.1093/oso/9780192895523.003.0011

that do not alter per capita consumption rates can also be affected. Part of the reason for the narrow range of predictions in previous works has been their reliance on theory in which the population dynamics of the competitors are described by a Lotka–Volterra (LV) model. Models that explicitly included resources have raised the possibility that evolution of biotic resources can play a major role in determining the evolutionary impact of one consumer species on another. They suggest that the population dynamical effects of adaptive evolutionary responses to competition may also be quite diverse, and that this type of adaptive change can decrease the population size of the adapting species.

In analysing the evolution of competitors, it is important to realize that competing species in the real world are parts of larger communities. The most common food web model of competition has two consumers and two resources, and most models of this type have assumed that there is no interaction between the two resources. The lack of interaction is reasonable (although not always justified) if the resources are abiotic. However, if the resources themselves are consumers, logical consistency would seem to require consideration of between-resource competition. If the two resources compete with each other, the ecological effects of the two consumers (in the absence of evolution) are often mutually positive, at least over some range of perturbations (Levine 1976; Vandermeer 1980; Abrams and Nakajima 2007). When competition between resources is allowed, mixed-sign (+,−) interactions and hydra effects (which makes sign structure subject to alternative definitions) are both possible for the consumer species, even in very simple models based on linear density dependence in resources and linear functional responses in consumers (Abrams and Cortez 2015a, Cortez and Abrams 2016). Given this wide range of potential ecological effects, it might seem that a similarly wide range of signs and magnitudes of evolutionary effects should also be possible. This topic has not received much theoretical attention.

One of the factors that can influence the evolutionary response to competition is the presence of behaviour or other forms of adaptive plasticity. In many cases, behaviour alters the selective regime to which an individual is exposed. The interaction of rapid adaptation with both evolution and population dynamics may lead to qualitatively different outcomes (e.g., Abrams 2006b), both ecologically and evolutionarily. Behaviour is particularly likely to be important in spatially structured environments with heterogeneous patches. In this scenario, the behaviour determines the selective regime experienced by an individual. Rapid adaptive processes are likely to operate in biotic resources as well as in the consumers, and this could also create opportunities for qualitatively different evolutionary responses of the consumer species to each other's abundances or trait values.

Like many other parts of competition theory, the evolutionary subset has been unduly influenced by both the models and empirical work that appeared early in the history of the field. Empirical studies have concentrated on traits related to food item size in animals. The theoretical literature has followed MacArthur's lead in focusing on the LV model and evolution of a single trait that determines the relative capture rates of many resources. The tradition (again established by MacArthur) has been to assume a 'utilization curve' of a fixed shape (generally Gaussian), whose mean value is the only trait that can evolve. As noted in Chapter 7 and elsewhere, an early empirical examination of these assumptions by David Sloan Wilson found them to be far from reality in the systems that had been studied at that time (Wilson 1975).

Whether or not resources have been included in the model of coevolution, most previous theory has been based on a scenario in which the introduction of a new consumer species and its growth to an ecological equilibrium population size initiate evolutionary change in a resident consumer species. The resident species is assumed to initially be at ecological and evolutionary equilibrium. An example where this scenario may be appropriate is when a new consumer species colonizes an island that has a long existing resident community. This way of conceptualizing the evolutionary response assumes a separation of ecological and evolutionary timescales. However, it is possible for evolution to be rapid enough that both the new and resident species will have evolved during the period required to reach their new 'ecological equilibrium' population sizes (Fussman et al. 2007; Hendry 2017). To use the concept of an ecological equilibrium, it is generally necessary to have a separation of ecological and evolutionary timescales. In the case of purely exploitative competition, both the (one or more) residents' and the invader's responses are affected by the relative and absolute changes in resource abundances that are produced by the invader. Even when the invading consumer's evolution is slow relative to population dynamics, there are often comparably rapid evolutionary changes in the resource populations that may affect the evolutionary change in the focal consumer. These complications have usually been ignored in past field work on coevolving competitors, because the typical study is a natural experiment involving the comparison of one consumer population that occurs 'alone', and a second, spatially distinct population of the same consumer that has existed for a long time in the presence of a second consumer species. Ideally, the environment (including all other species present) should be identical in these two locations, but that is seldom the case in natural systems.

The properties of the resources involved are bound to play an important role in both ecological and evolutionary system dynamics. Even if the resources are abiotic (so cannot evolve), coevolution of the two consumers does not cease after some specified time period, and their evolution will usually affect resource abundances, often altering any quantitative measure of the interaction. When two or more competitors coexist for a significant time span, novel traits are likely to arise at some point in one or both species. If this possibility is realized, it produces longer-term evolutionary change after the initial evolutionary equilibrium, based on pre-existing genetic variation, is attained. It is also possible for either the population interaction itself or the evolutionary change it produces to result in sustained fluctuations in both populations and traits. When this occurs, it usually produces different mean values of traits, as well as different population sizes.

The relative importance of the exploitation and interference components of competition has received little empirical study. This has allowed most of the theoretical work on the evolution of competition to focus on the exploitation component, which is usually easier to quantify. Evolution of intraspecific interference has been the subject of some theoretical work (Abrams and Matsuda 1994; Kisdi 1999), which predicts the existence of equilibria with both a low- and a high-competitive ability phenotype. This could also occur with inter- rather than intraspecific competition, but it has received little attention.

Coevolutionary theory for predator–prey interactions should have important implications for any competitive system with biotic resources. However, the trade-offs assumed in resource-based models of competition are fundamentally different from those assumed in predator–prey models. In addition, much of the work on predator–prey coevolution has followed Dawkins and Krebs (1979), who envisioned the evolutionary interaction strictly as an 'arms race'. However, this is certainly not the only, and quite possibly not the most common, scenario (Abrams 1986b, 2000c; Humphreys and Ruxton 2020). The possibility of evolutionary cycling in models of predator–prey coevolution (Abrams 1992a; Dieckmann et al. 1995) means that similar cycles are also possible in systems having competing predators which are both coevolving with their overlapping sets of prey species.

The importance of evolution and/or adaptive behavioural plasticity in the effect-transmitting species has received more attention in systems with other types of indirect interactions. For example, the impact of evolution in one or more species has been examined in models of three-species food chains (e.g., Abrams and Vos 2003; Loeuille and Loreau 2004), and these models have shown that there are many possible indirect evolutionary effects between top- and bottom-level species that arise from the adaptive change in the middle species. The possibility of coevolution of two prey species that share a common predator was examined in Abrams (2000a) and Abrams and Chen (2002a, b). Here, adaptive foraging by the predator and adaptive evolution in the predator's capture traits are both important determinants of the nature of the indirect interaction between the shared prey species. Given the much larger body of research that has been devoted to interspecific competition than to these two comparably indirect interactions, it is quite surprising how seldom adaptive change in resources is considered in analyses of competitive coevolution. In any interaction chain involving three dynamical entities, change in a property of the middle entity (e.g., adaptive evolution) is expected to be important in determining the population-level effects of one of the indirectly interacting species on the other.

The remainder of this chapter will look at the questions addressed by past work on the evolution of competitors, and what additional questions are most in need of theoretical exploration in the future. It will begin (Section 11.2) with a quick summary of empirical work and a review of early theoretical work. Section 11.3 is a short description of the modelling approach to evolution used here, which is then employed to study several non-standard sets of ecological assumptions (all involving explicit resources) in Section 11.4. Section 11.5 briefly discusses the evolution of apparent competitors. Section 11.6 examines contest/interference traits, while Section 11.7 considers the food web context of evolution. The final section (Section 11.8) looks at evolution and coexistence.

11.2 A brief history of work on the evolutionary responses to competition

11.2.1 *Empirical studies of competitive coevolution*

Empirical work preceded theory dealing with competitive coevolution. Brown and Wilson (1956) coined the term 'character displacement' for what they thought to be the primary evolutionary response to competition; two competing species evolve greater differences in the values of traits determining their use of one or a set of limiting resources. This 'displacement' or 'divergence' was thought to reduce the negative impact of competition on fitness. Divergence was supported by anecdotal observations in Hutchinson (1959) who found some cases where character differences were greater in sympatry than in allopatry. Schoener (1965) presented an extensive analysis of intraspecific variation in bill size in 46 families of bird species and concluded that some of the patterns were consistent with character displacement. Bill dimensions associated with food use diverged in possibly competing species in a number of cases. Grant (1972) argued for the possibility of character convergence of competitors in some cases, but the evidence for such responses was weak, and some convergence could be predicted as a common response to being in a new and different environment. In another review article Schoener (1970) found a significant number of studies that indicated convergence in head and body size measures of pairs of *Anolis* lizard species in the Greater Antilles. The convergence, particularly of larger species towards smaller ones, appeared to be at least partially an effect of different habitat structure rather than the change in competition. In any case, subsequent general theory concentrated on divergence.

The evolutionary responses that have most often been studied using multiple field sites or semi-natural constructed communities are those involving consumption rates of different resources by two or occasionally three species. Studies of communities of two or three Galapagos finches on a variety of islands by Peter and Rosemary Grant (Grant and Grant 2008) and of two stickleback fish subspecies by Schluter and collaborators (Schluter 2000, McKinnon et al. 2004) are probably the best-known examples. Both systems exhibited divergence in response to competition, and a variety of evidence from other systems has strongly suggested that divergence has occurred.

Schluter (2000) provided a summary of the evidence for character displacement at the turn of the millennium; by that time, this set of studies included a larger number of strongly supported examples of character displacement, but most had some caveats. There were 71 cases with 61 of those involving a pair of congeneric species. However, only 14 of the congeneric cases met all six of Schluter's criteria for strong evidentiary support. Most of the 10 additional (not congeneric species) cases had relatively weak evidence. Nevertheless, given the high prevalence of effects of interspecific competition on abundance, the number of strongly supported examples of character divergence in sympatry seems to be rather small (Schluter 2000; Pfennig and

Pfennig 2010). Is this because evolutionary responses are rare, or small in magnitude, or because the responses have forms other than mutual divergence? This question is still unanswered.

11.2.2 A history of theoretical models of competitive coevolution

A proper study of the evolutionary consequences of competitive interactions requires knowledge of the relationship between demographic parameters and the evolving traits. This in turn depends on having some knowledge of the functional and numerical responses of the competing species. Similar requirements apply for studying intraspecific competition and apparent competition. In all of these cases, the evolving parameters and changing population size(s) necessarily influence each other's dynamics via functional and numerical responses. This would seem to make consumer–resource models the logical choice for studying evolution in any of these scenarios. Despite this, most models have been based on modified LV models, in which consumers have direct effects on each other. A problematic aspect (beyond the limited possibility of getting LV dynamics from resource-based models) is that the parameters of the LV model can have a wide variety of relationships to measurable quantities, such as maximum consumption rates of resources. Nevertheless, most models of competitive coevolution continue to be based on a Lotka–Volterra framework, usually assuming MacArthur's proposed formula for the competition coefficient.

The first published model of character displacement was that of Bulmer (1974), although an unpublished Ph.D. thesis treated the topic a decade earlier (Bossert 1963). Bulmer used a one-locus–two-allele model of displacement and focused mainly on intraspecific competition. His underlying ecological model assumed discrete generations of the consumers and used linear effects as an approximation that should be valid near equilibrium. Bulmer's work has not been cited much in the subsequent literature on character displacement, perhaps because his discrete framework was inconsistent with the more popular differential equation models of ecological theory.

A short time later, Lawlor and Maynard Smith (1976) published the first consumer–resource model of character displacement. This study looked at 2-consumer–2-resource models, but adopted MacArthur's assumptions of linear consumer functional and numerical responses. It assumed that a trade-off function specified per capita capture rates of each resource as a function of the trait. This trade-off meant that an evolutionary increase in one capture rate would decrease the other. Depending on whether the trade-off was convex or concave, a single consumer could evolve to be either one of two specialists (concave) or a generalist (convex). This study did not include explicit trait dynamics, but assumed that the system would evolve to an evolutionarily stable state. It predicted that divergence of two competitors would occur, so that they would have a greater difference in trait values in sympatry (living together) than in allopatry (living apart).

The most influential early study was Slatkin's (1980) analysis, which adopted a Lotka–Volterra framework. Some elements of his analysis were present in slightly earlier works by Roughgarden (1976) and Fenchel and Christiansen (1977). Slatkin assumed a continuous trait, with dynamics governed by the 'breeder's equation' of quantitative genetics. This assumes that the mean value of the trait changes at a rate proportional to the fitness gradient (i.e., the slope of the relationship between the trait and fitness). The competition between two consumers having different values of a single trait related to resource use was a decreasing function of their difference in phenotypic values as in MacArthur and Levins (1967). This basic ecological assumption was combined with the breeder's equation model for the evolution of the mean value of the trait determining resource use. Although Slatkin (1980) stated that Lawlor and Maynard Smith's (1976) consumer–resource model produced predictions similar to his, subsequent extensions of the Lawlor and Maynard Smith models (Abrams 1986a; 1987f, g; 1990a, b; Abrams et al. 2008b; Abrams and Cortez 2015b) made predictions that often differ significantly from both Slatkin's work and that of Lawlor and Maynard Smith.

Taper and Case (1985) extended Slatkin (1980) in several ways; one was to numerically study some versions of MacArthur's (1970) consumer–resource model with many explicit resources. However, it appears that their parameter range did not involve resource exclusion, so their results were largely consistent with those derived from the Lotka–Volterra model. Slatkin's (1980) work predicted that convergence of competitors was possible, but this convergence did not represent a difference between sympatric and allopatric traits; it was a case in which the original trait values were not at their allopatric equilibrium values, and both species had their maximum resource capture rate at the same phenotypic value. Taper and Case (1992) and Case and Taper (2000) extended the same coevolutionary model to examine the role of asymmetry and degree of trait variation and to consider the effect of coevolution on species range limits. Taper and Case (1992) stressed the limited real-world applicability of their assumptions of fixed carrying capacity curves, Gaussian utilization curves, and Lotka–Volterra dynamics. Nevertheless, these have remained as the default assumptions in most of the subsequent literature.

The definition of convergence depends on whether it is being applied to the process of character displacement, or just the direction of evolution from some arbitrary starting character value. Character displacement compares the evolutionary equilibrium with and without a competitor. In contrast, invasion by a competitor from another location, potentially having different resources and conditions, can involve evolution that is driven more by the new environment than by the competitor. Both Slatkin (1980) and McPeek (2019b) have described the latter as evolutionary convergence. However, it is not convergent character displacement. It is not surprising that a sufficient inequality in productivities of different resources within a habitat can favour two species evolving to have their resource use concentrated on the most abundant or productive type.

There appeared to be rough consistency of the qualitative predictions of the continuous resource spectrum models growing out of Slatkin's (1980) study and the early

2-consumer–2-resource models of Lawlor and Maynard Smith (1976). Both predicted divergent character displacement (when the sympatric equilibrium is compared to the allopatric one). However, several later studies of resource-based models have examined evolution of characters determining resource capture rates and that produced outcomes other than divergent character displacement (Abrams 1986a, 1987f, g, 1990a, b; Vasseur and Fox 2011). Some of these studies examined non-substitutable resources and/or asymmetric patterns of resource use. Others included the possibility of negative effects of consumer abundance on mortality that were independent of resource consumption rates (i.e., interference). The evolutionary responses in systems with non-substitutable resources are considered further in Section 11.4, and interference is considered in Section 11.6.

Returning to the substitutable resources assumed by all of the earliest theory, both of the early models (Lawlor and Maynard Smith 1976; Slatkin 1980) assumed linear functional responses. Type II responses can give rise to population cycles. Abrams (2006a, b, 2007b) studied evolution in a 2-resource model with type II consumer functional responses. This led to the possibility of up to four coexisting consumers in systems with cycles, even though only two could coexist in the comparable linear functional response models of Lawlor and Maynard Smith (1976). Cases in which addition of the second consumer gave rise to cycles could result in different directions of character response in the original species than in cases without cycles.

Evolutionary responses to purely intraspecific competition are important for both speciation and the regulation of variation within a species. Very early studies examined evolution based on the logistic growth model with two parameters, maximum per capita growth rate, r, and equilibrium population size, K (e.g., Charlesworth 1971). This suggested that populations at equilibrium should maximize population size as a result of evolution of intraspecific competition. This was shown to be inconsistent with consumer resource dynamics by Matessi and Gatto (1984). Most subsequent work has focused on situations that produce disruptive selection, which has the potential to produce two species with distinctly different character states. Understanding disruptive selection almost always requires a resource-based approach. This is because traits determining the relative consumption rates of different resources are usually the characteristics involved in disruptive selection (Abrams et al. 1993; Geritz et al. 1998; Dieckmann et al. 2004). Most work on the evolution of traits involved in intraspecific competition from the past two decades has in fact included explicit resource dynamics (Dieckmann et al. 2004; Rueffler et al. 2006a,b).

Returning to interspecific competition, an important but neglected point is that an evolutionary response to shared resource use with another consumer is likely, even when the shared resource use does not affect the equilibrium population size of a focal consumer (Abrams 2012, Abrams and Cortez 2015b). Consider two consumers, either one or both having a specialist, food-limited predator. In this case, the specialist predator determines the equilibrium abundance of the consumer species it eats. Changing a neutral parameter in this consumer will not change its equilibrium abundance. However, the neutral parameter will generally produce an evolutionary response in a second consumer, because it will result in altered relative abundances of different resources. Figure 11.1 is a food web diagram of such a system with a predator on one of two competing consumers. Assuming a standard 3-level model like that in Chapter 7, a change in the mortality rate of the generalist consumer species 2 (d_2) will not alter the population size of consumer 1 (population N_1), unless the change is a sufficiently large decrease to exclude consumer 2. However, any increase in d_2 decreases N_2, and therefore changes the relative abundances of the two resources at equilibrium. If species 1 has a trait that determines its relative capture rates of the two resources, this trait will change in response to those altered resource abundances, even though species 1's own equilibrium population is unchanged. Whether this response represents convergence or divergence depends upon the consumption rates of the consumers and the trade-off relationship between those rates within each species. The higher-level predator in this example has an impact similar to that of an independent limiting factor, such as nesting sites, for consumer 1.

A question related to the above example is the issue of how evolution of one or both competitors changes their own population size. The usual assumption is that adaptive evolution will increase population size. In fact, simple consumer–resource models suggest that the adaptive evolutionary response to competition may often reduce the population size of the evolving species (Abrams 2012, Abrams and Cortez 2015b). This possibility had not been noted in the earlier theoretical or empirical literature (Schluter 2000). This is primarily because most of the earlier theory lacked explicit resources, and most of the empirical work had not (and still has not) quantified population responses to evolutionary change. Similarly, when two consumers have positive effects on each other's population density (e.g., when there is competition between resources), the divergence brought about by adaptive evolution will frequently reduce the population size(s) of the diverging species (Abrams and Cortez 2015b).

11.3 A simplified approach to evolution

Most of the work on the evolution of competitors discussed here uses the approach now generally known as 'Adaptive Dynamics'. It is an extension of the older 'fitness maximization' approach used in Lawlor and Maynard Smith (1976) and Abrams (1986a). The extension of this approach to situations in which fitness is frequency dependent was pioneered by Eshel (1983), Matsuda (1985), and Wilson and Turelli (1986), and was extended and formalized in Geritz et al. (1998). Frequency dependence arises when the relative fitness of a given phenotype depends on the set of other phenotypes present and their abundances. In the simplest case of a population with a very narrow distribution of trait values, an individual's fitness may be expressed as a function of its own trait value and the mean trait value in the population. The narrow range of variation allows the selection coefficient to be expressed as the derivative of the individual's fitness with respect to its own trait value, evaluated at the population mean. This approach is covered extensively in Dercole and Rinaldi (2008). Frequency

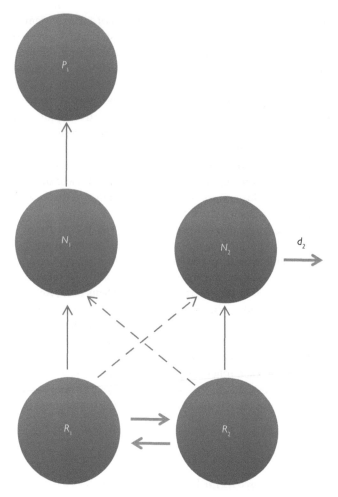

Fig. 11.1 A food web in which one of two competing consumer species (N_1) is controlled by a specialist predator while sharing two resources with the other competitor (N_2). The text discusses the impact of altering the mortality of consumer 2.

dependence requires that the fitness of an individual with a given trait be expressed as a function of that individual's trait and of the mean trait value in the population. The fitness gradient with respect to the individual's trait is what determines the direction and strength of selection, but the sign of this gradient depends on the mean trait value. More generally, the distribution of trait values in the population may be required to determine individual fitness and the fitness gradient. However, dynamics based solely on the mean value can be sufficient if the range of variation is relatively narrow. This framework is often a good approximation in other situations with moderate

variance (Abrams 2001d). While more elaborate models may be required to understand evolution in many scenarios, the simple approximation of Adaptive Dynamics
is surprisingly robust, and it at least provides a minimal set of potential evolutionary
outcomes.

11.4 Examples of non-standard questions and outcomes

The term 'non-standard' in the section heading refers to questions other than the
extent of divergence or to outcomes based on changes in traits other than the sets of
per capita resource capture rates of the consumer.

11.4.1 Evolution in the resource population(s)

Close to 30 years ago a series of papers by Hiroyuki Matsuda and collaborators (Matsuda et al. 1993, 1994 1996; Abrams and Matsuda 1993) explored the impact of
adaptive defence by resources on the interaction between, and coexistence of, competitors and apparent competitors. They concentrated on the case of rapid change
in defence, which could represent either behaviour or rapid evolution. The simplest
system was that considered in Matsuda et al. (1993); this was a system with two strictly food-limited predators consuming a single prey species that exhibited adaptive
change in its relative vulnerability to each of the two predators. Positive short-term
effects of increased abundance of one predator on the short-term per capita growth
rate of the other could occur for two reasons. The first is a direct trade-off, such that
defending against one predator made the prey more vulnerable to the other (a common situation when predators hunt effectively at different locations or times). The
second mechanism for short-term positive effects was an indirect trade-off in which
defending against one predator made it more costly to defend against the other. Both
mechanisms essentially produce 'self-limitation' in each predator. Both mechanisms
therefore allow coexistence of two predators using a common prey, a result that has
been independently and recently rediscovered by Sommers and Chesson (2019) and
by van Velzen (2020). This scenario also has the potential to change the interaction
between the two predators from mutually negative effects to mutually positive effects,
measured either by a neutral parameter perturbation or by changes in equilibrium
population size following addition or removal of one of the predators (Matsuda et al.
1993). It is possible for such adaptive defence to have more complex effects on the
interaction. For example, Figure 4 in Abrams (2000a) considers a case of one prey
type that exhibits evolutionary shifts in defence against two predators; this produces
sustained population cycles, with largely out-of-phase cycles of the two predators. If
such cycles occur, they will usually alter any standard measurement of the interaction
between the two predators.

The type of evolution of the prey species in the simple models of Matsuda et al. (1993) is normally expected to cause some compensating evolutionary responses in one or both predator species. In a system with one predator type, if the prey exhibits higher levels of defence against that type, the resulting low intake rate of the predator would in turn favour predator traits that are advantageous in scenarios with a low maximal food intake rate. The introduction of a second predator that requires a different defence would increase vulnerability of the prey to the first predator, which could favour traits in the first predator that are more appropriate for systems with higher prey availability. The final result could alter the stability of the original equilibrium, and the resulting cycles would likely alter the mean trait value and population size of the original predator.

Adaptive defence by prey also affects apparent competition. Abrams and Matsuda (1993) considered adaptive prey defence in combination with adaptive predator choice in a system with two prey and one predator. The presence of adaptive prey defence often implies a higher relative predation rate on the less common prey because of their lower defence. This would appear to be 'negative switching' behaviour. Nevertheless, the prey still experience a fitness benefit from lower abundance because of their lower defensive costs.

Most of the analyses discussed above assumed relatively rapid adaptive change. While rapid evolution may have the same potential endpoint as adaptive behaviour or plasticity, it does raise the possibility of extinction of the population before sufficient change occurs, or of loss of the theoretically favoured mutant type due to demographic stochasticity. Thus, the applicability of 'rapid adaptive change' models to evolutionary change may depend on mutation rates, selection coefficient magnitudes, and population sizes. In addition, the speed of adaptation may also determine whether population cycles occur. Such cycles can alter the signs as well as the magnitudes of interspecific effects.

The presence of adaptive predator-specific defence traits also affects the dynamics in systems having more than a single prey species. Some of these systems were explored in Matsuda et al. (1994), which argued that predator-specific defences could allow many more predator species than the number of prey species. It also allowed coexistence of predators that had very similar diets, in spite of a large difference in their general efficiency of prey utilization. Matsuda et al. (1996) extended this analysis to larger systems with more than two species on each of two trophic levels.

11.4.2 Evolution with imperfectly or non-substitutable resources

Abrams (1987f) considered the evolution of capture rates in competitive systems with nutritionally essential resources. Fox and Vasseur (2008) and Vasseur and Fox (2011) extended this analysis. Most work on non-substitutable resources has been restricted to the category of perfectly complementary resources (Leon and Tumpson 1975), in which the per capita growth rate is only affected by the resource whose abundance is lowest relative to requirements (given by an ideal ratio where both resources simultaneously limit growth). These models predict that a common response to interspecific competition in a pair of species is convergence in the species' capture rate constants. The reason is that the intake rate of each resource must be above a threshold amount for any growth, regardless of the intake rates of other resources. A new competitor that causes a relatively greater reduction in a given resource type will select for greater consumption rates of that resource in the resident, so that it can maintain its optimal ratio. This evolutionary outcome of convergence may be altered by various factors such as adaptive behaviour, costs of excreting excess nutrients, and the nature of trade-offs that affect other fitness components. It is also possible for the required ratio(s) to evolve. Some of these complications are discussed in Vasseur and Fox (2011). However, all of this work suggests that theory based on perfectly substitutable resources will usually fail to predict the direction of evolutionary change when different resources are both (or all) nutritionally essential.

Substitutable resources were assumed in most theoretical and empirical studies of character displacement. However, a range of mechanisms can prevent perfectly substitutability, and many of these cases have not been recognized as having only partially substitutable resources. For example, otherwise substitutable resources that are characterized by having different ratios of nutritional value to handling time are not really substitutable. This is because handling one type of resource takes away the opportunity to capture the other type(s), a fact that was the basis for much of early optimal foraging theory (Schoener 1971; Stephens and Krebs 1986). Consuming the lower quality food when the higher quality one is above a threshold value actually reduces the rate of nutrient intake and therefore reduces population growth rate.

Situations involving differences between resources in handling or processing requirements can change the qualitative nature of the between-consumer interaction. As noted in the preceding paragraph, early behavioural research on foraging showed that a resource with a lower ratio of energy content (here, B) to handling time (h) should be dropped from the diet when the higher (B/h) resource is sufficiently abundant. If this did not occur, consumption of the lower quality resource would actually reduce the per capita growth rate of that individual. It is known that adaptive diet choice is not perfect, so some consumption of a fitness-decreasing resource is likely to occur in this scenario. If there is intraspecific limitation due to interference or a non-shared resource type, then competition by an otherwise similar species with its own intraspecific interference will result in decreased abundance of both resources. This will favour broadening of the diet to include more low quality (low energy relative to handling time) resource types in both species. Such parallel changes may be unequal in magnitude and may result in either net convergence or net divergence.

Non-substitutability also arises in organisms with digestive capacity limitation when the resources differ in their nutritional content per unit volume. The optimal strategy in such cases is to fill the gut in the time available with as high an average quality of food as possible (Abrams 1990a, b). If there is a high and a low quality food, the optimal strategy is to consume only as much low quality food as needed to fill the gut. In a system with some temporal structure, this strategy implies feeding primarily on the good food in the early part of the potential foraging time period and switching to primarily the poorer food when there is just enough time left to fill the gut. A competitor for the poorer quality food reduces its abundance, meaning that more time is required to fill the gut. This usually favours greater per capita uptake rate of the poorer resource, even at the expense of lower uptake of the better resource. This represents convergence. A specialist competitor for the better resource also causes convergence. However, divergence and parallel change can occur when there are additional, independent sources of intraspecific competition (e.g., habitat with low predation rates) or when the gut size itself evolves (Abrams 1990b). There has been surprisingly little work on evolution under this scenario, and similarly little work on the functional responses it produces (but see Beckerman 2005).

11.4.3 Evolution of other consumer parameters

In theory, any of the parameters of a consumer–resource model should be a candidate for evolutionary change under altered conditions, such as the addition of a new consumer species. In species with nonlinear functional responses, which often involve two or more parameters, any of the parameters could change due to altered relative abundances of different resources. For example, the handling times of different foods are likely to be influenced by many genetic loci and are expected to experience altered selection pressures as the relative abundances of foods change. Unfortunately, handling times have a large degree of plasticity and are difficult to measure (at least relative to traits like external size measurements). Abrams (1986a) briefly considered the evolution of efficiencies of resource conversion in a simple consumer–resource framework. Here again the direction of evolution differed from that predicted for competitors having differences in resource capture rates. In the two-resource framework with a symmetrical trade-off in the two conversion efficiencies, selection favours a higher efficiency (B) on the resource characterized by a larger intake rate (CR). A higher utilization ability, C, may lead to lower intake, due to a lower R, provided the resource is self-reproducing. In a simple 2-consumer–2-resource model with each consumer having an evolutionary trade-off in the two values of B, the pair of B values may exhibit convergence, divergence, or parallel change (Abrams 1986a). In some cases there can be a crossover in the relative values of B. Unfortunately, changes in relative efficiency of competing species seem to have been of little interest to empirical biologists. It is possible that these values are constrained more strongly than are capture rate parameters, but that has not been shown in the vast majority of competitive systems that have been studied.

11.5 Evolution of apparent competitors

There has been a continuing asymmetry in the empirical literature on competition and apparent competition, with many more studies of the former than the latter. Holt and Bonsall's recent review of apparent competition notes that, 'It is puzzling that the issue of enemy-mediated character displacement has not been the focus of more determined empirical inquiry' (Holt and Bonsall 2017, p. 464). They also remark that the possibility of such predator-driven evolution had been mentioned in the article that introduced the term 'apparent competition' (i.e., Holt 1977). Notwithstanding some limited theoretical work (Abrams 2000a, Abrams and Chen 2002a, 2002b) in the interim, the topic has still received very little empirical study.

In the simplest scenario with apparent competition, two otherwise non-interacting species share a common predator. This interaction can itself produce positive or negative effects between the two prey species. For example, if the predator's population density is limited strictly by another factor (e.g., water, nesting sites, or a higher level predator), and the predator has a saturating (type II) functional response, a positive neutral parameter change in prey 1 will always increase the predator's total handling time and thus reduce the per capita predation rate on both prey. Subsequent adaptive change in either prey that affects its vulnerability to the predator will reduce investment in defence, producing parallel decreases in those defensive traits (Abrams 2000a). Parallel changes are also expected in the simplest models with linear functional responses and no direct intraspecific effects in this scenario. The addition of a second prey species that shares the same predator will then increase the predator abundance. This, in turn, favours traits that decrease vulnerability to the predator in both species, which represents parallel displacement. Parallel displacement can be accompanied by either a net increase or a net decrease in the difference between species in their predator-vulnerability traits. These changes in predator vulnerability may also be accompanied by changes in resource uptake rates, since many traits involved in reduced vulnerability also decrease resource intake (Lima and Dill 1990; Abrams 1995; Preisser et al. 2005). More complicated scenarios are possible within a single category of outcome. For example, if the prey species that supports a lower predator population size also has a lower cost to increasing its defence, there may be a crossover in the relative values of the two prey in going from the allopatric to the sympatric state (Abrams 2000a, fig. 1, p. S49).

The presence of two predators that have significantly different foraging behaviours can result in divergent displacement of defensive traits in two prey species (Abrams 2000a), although convergence is also possible here. The primary determinant of outcomes is the nature of the difference in prey traits required to reduce risk from the two predators. Nosil and Crespi (2006) investigated a system with two stick insect species that diverged under predation. The divergence arose because the two insect species specialized to a greater extent on different plant species, with a different colorationbeing more cryptic on each plant species. Marchinko (2009) suggests that different defences against vertebrate and invertebrate predators were involved in divergent selection on threespine stickleback fish. There are surprisingly few examples of evolution driven by apparent competition.

11.6 Evolution with both exploitative and interference competition

When interference is one of the components of a competitive interaction, the addition (or elimination) of a competitor may have a wide variety of effects, depending (inter alia) on the magnitudes of inter- and intraspecific components of interference and what other parameters are affected by the interference. Because interference has been omitted from most competition theory, its evolution is still a topic in need of basic exploration. McPeek (2017, 2019a) has recently stressed the need to include interference in analysing competitive interactions and coevolution. However, some consequences of one very simple form of interference had been considered in Abrams (1986a), and a number of other forms have qualitative effects that can be deduced without detailed models.

In the simplest case, interference is entirely intraspecific, and it only affects per capita mortality rate. Under this scenario, adding a second consumer species (also with intraspecific interference) to a system with a single one usually reduces the abundance of the original consumer. The lower abundance reduces the amount of interference because that interference is purely intraspecific. This lowers each consumer's mortality and therefore reduces the resource intake required for zero population growth. In most cases, those resources consumed at the highest rates will be most affected by sympatry of the consumers, and both consumers will evolve some degree of reduction in their relative consumption of these resources. Assuming an identical trade-off in the relative consumption rates of different resource types, the outcome in a 2-consumer–2-resource model is parallel shifts in the C_{ij} values of both consumers (Abrams 1986a). This qualitative outcome is preserved when the trade-off relationships are not identical, but are sufficiently similar. In this case, the 'preferred' resource of both consumers decreases in relative abundance when the two are in sympatry, so both consumers shift to have somewhat lower consumption of this resource and higher consumption of the resource initially characterized by a lower C_{ij} (Abrams 1986a).

If the interference affecting mortality is purely interspecific, a variety of outcomes are possible. One of these is the outcome of alternative exclusion, originally described by Lotka and Volterra, which would preclude coevolution. If coexistence is still possible, the resource abundances at equilibrium are normally increased in sympatry relative to allopatry. This is again likely to cause parallel shifts in relative consumption rates in the 2-consumer–2-resource case, but in the opposite direction to those produced with intraspecific interference. Interspecific interference is seldom symmetrical (Schoener 1983), so it is possible for very large effects on the abundance of the species that suffers greater interference when the two consumers are sympatric.

In natural systems, interference often has its largest effects on capture rate parameters; this occurs when individuals steal food items or prevent foraging by other individuals. This type of interference may have different effects on the capture rates of different resources. Probably because of this range of mechanisms, capture rate interference has seldom been included in models of competition generally, and I am unaware of its inclusion in any models of competitive coevolution.

Most empirical work has found that both inter- and intraspecific competition is usually asymmetric (Schoener 1983). If there is asymmetric interference between species, it is tempting to assume that the species that is dominant in interactions between individuals is more likely to exclude the subordinate species. However, greater ability in contests usually comes at a cost to some other component of fitness. It is quite possible for a species with low investment in a trait determining ability in competitive contests to replace a species with a high investment (Matsuda and Abrams 1994). The evolution of contest competition has mostly been studied using models of within-species interactions (Abrams and Matsuda 1994; Kisdi 1999). High investment in contest traits is likely to cause a reduction in population size relative to similar scenarios in which such competition is absent. It is also possible to have temporal cycles in the mean competitive ability (Abrams and Matsuda 1994).

11.7 Evolution of competitors in a food web context

Consider a food web in which a consumer of two substitutable resources is itself eaten by a specialist predator or is limited by resources other than food, such as nesting sites. What is the ecological and evolutionary effect of introducing a second consumer of the same two resources? If they can coexist, the second species will have no effect on the first, whose population size is set by its predator. After the introduction, changes in the resource-independent mortality rate of either consumer will not have any effect on the equilibrium abundance of the other. These generalizations hold provided the changes are not so great as to eliminate coexistence or cause extinction of the specialist predator. Regardless of this lack of effect on the equilibrium abundance of species 1, consumer 2 will alter the relative abundances of the resources when it has a different set of consumption rates of the resources than does consumer 1. If consumer 1 has genetic variation for traits influencing the relative capture rates of resources that have changed in relative abundance, then it will evolve. With substitutable resources, this evolution will usually be divergence, and will result in reduced overlap in resource use. If the resources have nutritional interactions, non-divergent displacement may occur. More details are provided in Abrams (2012). The important point in this example is that the evolutionary change occurs independently of the change in population size of the first consumer that is caused by addition of the second.

Chapter 7 discussed a model having two species at each of three trophic levels. This example, with strict resource limitation at each level, exhibited a complete lack of population response of species on the middle level to changes in any parameter that did not directly affect their interaction with the top two predators (e.g., the density-independent mortality rate). However, such changes do alter the abundances of the predators, and are therefore likely to alter traits that affect vulnerabilities to these two predators. In this case, evolutionary responses to an altered mortality in one of the mid-level consumers will reflect apparent competition, but not resource competition.

In larger food webs, and with age- or size- structured populations, it seems likely that a much wider range of evolutionary responses to competition is possible. However, it is also possible that some of the varieties of evolutionary responses observed in models with only a few species are rare in cases with many more species. It remains to be seen how far current results based on simple cases will allow us to understand the evolutionary responses in naturally occurring food webs.

The above treatment of competitive coevolution has been based on an underlying conceptualization that is more appropriate for animals than for plants. It has also concentrated on consumers and resources that have a stable equilibrium. One of the non-equilibrium situations that has been studied theoretically is temporal variation in seedling establishment, which is based on plant competition during the seedling stage in a temporally variable environment (Snyder and Adler 2011; Abrams et al. 2013). The latter work suggests that divergence in environmental sensitivities is likely to evolve under conditions similar to those under which divergence evolves in traditional resource-based models. The pattern of seasonal variation in resource use (as in Chapter 8) is likely to evolve, but this has yet to be explored (although Yamamichi and Letten (2021) consider a related scenario).

11.8 Evolution and coexistence: current theory and the future

It is possible for evolutionary change to make coexistence more or less likely. The predominant view seems to be that the former outcome occurs much more often. This may be true, but it is premature to reach a conclusion. If a new species is in the process of excluding a resident, the two species must start with relatively similar competitive abilities for evolution to be rapid enough compared to the population decline to save the inferior species ('evolutionary rescue'; see e.g., Klausmeier et al. 2020b). Cases of evolution promoting coexistence are more likely to come in the form of making the species less likely to be competitively excluded in the future when the environment changes in a manner that decreases its fitness relative to that of its competitors. However, as noted above, some evolutionarily favoured traits increase competition (e.g., relative capture rates of different essential resources), so they may make a species more susceptible to extinction. The traditional response of divergence under competition for substitutable resources is often associated with a lower range of resources being utilized. Increased specialization increases the probability that the remaining utilized resources of that consumer may all suffer poor conditions simultaneously. Therefore, the effect of evolutionary divergence on long-term coexistence may be negative.

As with a good deal of other recent competition theory, the last few years have seen a focus on how evolutionary responses to competition affect the two requirements for coexistence stressed by Chesson (2000a): niche differences, and differences in general competitive ability. Pastore et al. (2021) was the most recent of these works at the time that this chapter was written. One problem with this two-way classification of coexistence requirements, explained in earlier chapters, is that many traits affect both requirements, and agreement on how to make the bipartite classification in such circumstances is lacking. Another problem is that 'general ability' traits will be selected for in the presence or absence of the competitor, so they are not expected to change significantly because of the addition or removal of a competitor. The fact that a trait influences the relative capture rates of different resources does not allow one to deduce the directional effect of a competitor on the trait unless one knows the nutritional effects of, and interactions between, the resources.

Assume that 'traits' are defined as being the three basic foodweb parameters; capture rates, conversion efficiencies, and mortality/metabolic loss rates. Most commonly measured consumer traits in an animal (e.g., speed or size of mouthparts) are likely to affect both general exploitation ability and the relative abilities to take up different resources. Both general and specific capture rates include the possibility that selectively favoured traits will decrease population size by increasing resource overexploitation (Abrams 2002, 2019). This possibility is almost never considered in analyses of coevolution. In traditional simple models with two or more resources, only the last of the three basic parameters (mortality) is treated as being independent of the niche, and that is due to the often-questionable assumption that mortality is independent of resource intake. A future theory of competitive coevolution will hopefully deal with the consequences of the interconnected, multi-parameter effects of most measurable traits, and will consider their effects on population size.

Given the relative paucity of empirical work on the evolution of competitors, the first order of business for future empirical work should be to understand the direction of trait responses to competition for a wider variety of types of consumers and traits. Evolutionary change is easiest to quantify in species with very short generation times. Studies of phytoplankton or microbes seem particularly promising, given the ability to manipulate resources and dynamic rates in the laboratory.

12

Overview

12.1 A range of viewpoints on theory

A wide range of opinions exists regarding the use of, and usefulness of, competition theory, and of theory in ecology and evolution more broadly. This opening section will explore three diverse statements about this body of theory, all of which are all closely related to the themes of this book:

> Few ecologists are interested now in these misleading equations [Lotka–Volterra models], but mathematicians dote on them and are always trying to foist them off on us—a classic case of the drunkard who loses his watch in the dark but looks for it under the lamp post because that's where the light is.
>
> —Levandowsky 1976, p. 418

> Population ecology as a scientific enterprise . . . is peppered with hypotheses that are not testable and theories that lack or have, at best, weak explanatory relevance.
>
> —Getz 1998, p. 540

> . . .our history of succumbing to enthusiasms is a deep-seated feature of our scientific culture. . . . We have little choice but to seek inspiration from gurus of the newest ideas; sometimes they turn out to be partially right. However, we should never believe them without a struggle.
>
> —Houle 1998, p. 1875

The three quotations all express dissatisfaction with theory in ecology and/or evolutionary biology. Each of the statements has some components that are consistent with the arguments in this book, and others that are not. None is specifically aimed at competition theory; Levandowsky was referring primarily to Lotka's and Volterra's independently proposed predator–prey model. However all of the quotations reflect ideas that are important for understanding the recent history of competition theory.

Levandowsky rightly highlights the combination of popularity and inadequacy with regard to the approach taken by Lotka and Volterra (LV). If he were writing today (46 years later), and were writing about competition, he would be at least partially justified in repeating the same statement. However, it is not mathematicians who are primarily at fault for the continuing popularity of the LV model. Blame must be shared among many parties, and a big share of the responsibility rests with

Competition Theory in Ecology. Peter A. Abrams, Oxford University Press. © Peter A. Abrams (2022).
DOI: 10.1093/oso/9780192895523.003.0012

authors of ecology textbooks. Virtually all modern textbooks focus their treatment of competition theory on the LV model. If they mention other models they usually fail to note their inconsistencies with LV. What has changed since Levandowsky wrote the above statement is that a large body of literature has demonstrated the inability of the Lotka–Volterra model to adequately reflect the consequences of shared resource use. Despite the fact that many of the articles that demonstrated this inability had appeared before the mid-1980s, the LV continues to be the most commonly used model for theoretical studies of multi-species competition and for studies of evolutionary responses to competition.

Getz (1998) was certainly correct in implying that the scientific enterprise needs to have some predictive ability to be useful. However, understanding any complex natural system requires a number of models of intermediate complexity to reveal what outcomes are possible, and what features might affect the outcome of a perturbation that directly alters one or more components of the system. It is not necessary that each model within this array should be assumed to apply well enough to any single natural system to be used directly for any predictive purpose. Developing a predictive model for a particular system requires that we have a large body of models that include different sets of component elements, so that we have some idea of what subset of those elements might be needed in the model, and which processes might be crucial for the specific prediction. In the case of ecological competition, this body of knowledge was basically stalled at a relatively early stage of development. If it is resumed, there should be no requirement that each model produced in this process provide better predictions for any specific real-world system or be 'tested' on any real set of species before exploring other models that build upon it. And if such a test is carried out, its failure does not mean that the 'intermediate' model should be discarded completely. The primary connection to empirical work for these intermediate models is that their component functions should have empirical and/or logical justification. Intermediate models also illuminate aspects of the consumer–resource system that need to be studied further before making a particular prediction (or to improve upon a previous prediction).

Adding more realistic features to a simple model can make some of the predicted dynamics less consistent with natural systems. This does not justify discarding that feature (or set of features) from future models. The development of consumer–resource theory provides a useful example, where a modification that is in the direction of increased realism can produce dynamics that are at least arguably less consistent with what is observed. MacArthur's (1970, 1972) consumer–resource model predicted that competition in 2-species systems never produced cyclic dynamics. Most classically studied laboratory competitive systems are not characterized by cycles. However, MacArthur's underlying assumption of linear functional responses has since been shown to be unlikely (Jeschke et al. 2002, 2004). Functional responses that saturate with increasing food/resource intake are a logical necessity and are almost universally observed. Holling's disc equation type II response is the simplest saturating response, and saturation has been observed in a large fraction of laboratory studies. Type II functional responses are associated with instability at low consumer death rates and/or high consumer attack rates in models of predator–prey systems (Rosenzweig and MacArthur 1963). In many simple models, these cycles are often of such extreme amplitude that they would entail rapid extinction of many real populations. Another empirical problem is that such predator–prey cycles appear to be relatively uncommon (although there is some debate on this; see Kendall et al. 1998). The cycles that have been observed do not have the extreme amplitudes often predicted by the simple model.

These problematic features of models with type II responses do not argue for going back to linear responses in either predator–prey models or models of competition, but rather for modifying the models with type II responses in other ways that reduce instability, and are likely to be present in many systems. Predator-dependence (DeAngelis et al. 1975) of such responses is a mechanism that appears to be very common (Skalski and Gilliam 2001) and has a stabilizing effect. Several other modifications to the Rosenzweig–MacArthur model are known to reduce the probability of cycles, or to eliminate the likelihood of extreme amplitudes; these include stage structure in the prey (Abrams and Walters 1996; Abrams and Quince 2005) and adaptively flexible prey defence behaviour (Abrams 1995). There are a variety of other potential explanations for stabilization, such as direct intraspecific interference in consumers/predators or high levels of immigration of resources/prey. Constructing the predictive models that Getz (1998) calls for a series of models of intermediate complexity incorporating various combinations of these and other features omitted from most extant models.

The third introductory quotation, from Houle (1998), suggests that many of the factors influencing the direction of research have little to do with the ultimate goal of understanding and making predictions about natural systems. Abrupt shifts in the broad topics of primary interest have always been part of the scientific enterprise. And the practice of science is subject to many pressures that have little to do with the optimal advance of knowledge. It is necessary to obtain funding for research, and such funding for academic fields typically requires projects that can be completed in three to five years. Fundable projects must also have significant novel features. In the case of ecology, most of the practitioners are empirical scientists, for whom theory is often viewed as a means of justifying their own work. However, to be an attractive topic for empirical research, a project must be viewed as addressing a currently popular question, and the results must be achievable within the period of a single research grant or the duration of a graduate student's career. And it usually has to be done with limited funding and space. This is likely to eliminate studies of functional responses involving many species. They would not be viewed as sufficiently novel, as functional responses have been measured many times in the past. These have mainly been measurements of responses to a single resource, so it might seem that a study of responses in a multi-species system could be sufficiently novel to be justified. However, investigating response surfaces when three or more species affect the response is usually not possible given constraints of time, money, and space.

The issues raised by the three opening quotations are explored in more detail in the following sections.

12.2 Theory's roles in ecology and competition

12.2.1 The goals of theory

The ultimate goals of theory in ecology should be to understand and predict real-world changes in the abundances of species, or of larger groupings, such as trophic levels. Ecology should also provide the groundwork for understanding and predicting evolutionary change in traits that influence population dynamics. How well practitioners have succeeded in these goals is a matter of debate (Getz et al. 2018). The record may give the appearance of more success than is deserved, as studies have focused on more easily answered questions in a small range of simple systems.

Half a century ago, the indirect interaction of competition seemed to be an ideal case within community ecology for understanding and predicting changes in species abundances. It was, and still is, arguably the simplest, and certainly the most studied, indirect interaction in natural communities, and it is also central to understanding a range of important topics dealing with processes within a species, ranging from evolution to density dependence. MacArthur's well-known 1970 paper and 1972 book clearly reflected his feeling that competition theory should at least be consistent with dynamic processes in both consumers and resources. However, the initial expansion of resource-based competition models following MacArthur's treatments stalled at the end of the 1980s, when competition itself became less popular as an object of study.

The theme of exploring consumer–resource mechanisms seems to be poorly represented (although certainly not absent) in the recent revival of theoretical work on competition (see Chapter 4). Attempts to predict the dynamics of sets of real-world competitors using consumer–resource models seems to have been largely restricted to pairs of very small organisms (e.g., phytoplankton) in the laboratory, a line of research that dates back to Gause (1936). Laboratory research involving phytoplankton as either consumers or resources has also been the subject of some of the more successful attempts to integrate model predictions with experimental results (e.g. Becks et al. 2010; Ellner and Becks 2011). These systems have also been the subjects of some of the relatively few analyses of impacts of non-neutral trait values on species abundances (e.g. Litchman et al. 2010; Klausmeier et al. 2020a), as well as studies of seasonal systems (e.g. Descamps-Julien and Gonzalez 2005).

All models are simplifications, and even overly simplified models often have some value. Before 1970, the 2-species LV model was the main theoretical guide to understanding the interspecific interaction of competition. It provided two important insights that were not widely appreciated before: (1) the importance of the relative amounts of inter- and intraspecific competition in determining the strength of the interaction; and (2) the possibility of alternative exclusion outcomes depending on initial abundance. The multi-species extension showed the possibility of cyclical dynamics (May and Leonard 1975) and of multiple alternative equilibria involving different sets of species (Gilpin and Case 1976). These findings represent the main contributions of the LV model to understanding competition. However, the LV model is too far removed from the mechanisms of competition to provide quantitative predictions regarding population changes in natural systems. More importantly, the LV model is completely inadequate for assessing how other species or resources in a food web should affect the outcome of competition. Chase et al. (2002) illustrate this claim for the question of how predators alter competition between prey species. Letten and Stouffer (2019) is one of the latest in a string of very recent studies to reemphasize the dependence of pairwise competitive effects on other species.

Saavedra et al. (2017) was the one article in the highly cited group reviewed in Chapter 4 that relied entirely on LV models. They end their article by stating that, '...future theory exploring how and whether higher-order interactions in multi-species system influence coexistence would be an important next step'. The problem is that the next step had been taken in the 1970s, before the confusing term 'higher order' had been invented.

12.2.2 The relationship between theory and experiment

The development of all ecological theory has been hampered by a belief that the only role of theory is to make one or more predictions that can immediately be verified or contradicted by a short-term experiment in an empirical system. This seems to underlie the opening quotation by Getz (1998). A prominent example of this 'immediate empirical test' viewpoint is the work of Peters (1991), who essentially argued for reducing ecology to finding significant regression relationships between variables. Such relationships always allow one to make simple (although not necessarily valid) predictions. A very recent article having many prominent authors opens with the claim that, 'Scientific inquiry should operate as a feedback loop in which theory that describes the natural world is developed, tested empirically through carefully articulated hypotheses, modified to better represent reality, and then tested again' (Grainger et al. 2022).

In spite of the views covered in the preceding paragraph, 'testability' is not, or at least should not be, the main criterion for judging all theory for ecological systems (and many other complex systems). Large numbers of theoretical models are needed to develop hypotheses about how complex biological systems might behave under a wide range of circumstances. In ecology, most of the currently popular models are so simple that they cannot be meaningfully used in this way. Slightly more complicated models are needed to identify what earlier assumptions might be misleading for particular circumstances, and what aspects of the population structure and dynamics might be needed to make a better prediction (or a range of potential predictions) for any specific system. Because a wide range of functional forms are known to be possible for most if not all model components, different predictions are likely from assuming different components. Even a single intermediate-complexity model often reveals a much wider range of dynamics and responses to altered conditions than what arises in the simpler model upon which it is based. The intermediate models should use functional forms that have some justification, either from empirical measurement of these relationships, or from purely logical deductions. For the near future, the latter (logical) type of justification will often be more important than empirical measurements in community ecology, given the relative lack of quantitative study of the components of consumer–resource relationships in natural environments.

A common mistake of empirical biologists is to believe that all theory must be empirically tested. The propositions that need to be tested are claims about the causes of observed phenomena or predictions of future population trajectories. These causes or predictions need to be based on an understanding of the range of possible underlying mechanisms. It is this understanding that can be achieved by having a variety of different models with intermediate complexity.

12.2.3 What will a more comprehensive theory look like?

A fully developed ecological theory of competition (or any other ecological interaction) should have multiple levels of complexity. The minimum level of complexity for competition theory is one that includes simple forms of resource dynamics and consumer functional and numerical responses. These constitute the minimum because of the role of resources as the defining feature of the interaction. Part of this minimum was investigated more than 40 years ago, but many of the relatively basic elements—particularly adaptive foraging and the many potential forms of interference competition—have still received very little attention.

Our current knowledge of simple consumer–resource models will probably prove insufficient to make predictions that are likely to apply, even qualitatively, to most ecological communities. The necessary knowledge required includes a greatly expanded set of consumer–resource models that includes more of the food web context. However, this alone will likely not be sufficient. In most real systems, the role of size structure and/or stage structure is an important determinant of the nature of interspecific interactions. Knowing how stages differ in resource use is a key determinant of the effect of stage structure on interspecific interactions. Many of the effects of population structure were reviewed by de Roos and Persson (2013). Genetic and phenotypic variation within species is another aspect of population structure that often needs to be included in competition theory (Schreiber et al. 2011). Similarly, spatial structure has important effects on interactions, some of which were covered in Leibold and Chase (2018), and more briefly reviewed here in Chapter 10. However, the bodies of work on stage-structured or spatially structured competition have an overrepresentation of models with either no resources, or with the simplest possible assumptions about resource growth functions and consumer functional and numerical responses. Furthermore, it is unlikely that population and spatial structure will be the only elaborations required for understanding the dynamics of competing species that are embedded in larger communities. The issue of what species/entities at other trophic levels need to be included to construct a predictive model is another important question that has repeatedly been discussed (Schoener 1993; DeAngelis and Mooij 2003; Loreau 2010). These effects of other levels are treated in Section 12.5.

Some promising developments in the analysis of predator–prey models suggest a movement towards examining the consequences of changing non-neutral parameters in the context of consumer–resource models. Uszko et al. (2017) is an example that focuses on the response of predator–prey dynamics to warming temperatures. Both the inclusion of type III functional responses and explicit resource dynamics produced significant changes in the predicted population-level responses in this study. In general, the push towards 'trait-based' theory (McGill et al. 2006; Klausmeier et al. 2020a) has increased the use of competition models with explicit resources.

Despite some very recent moves towards a broader range of models of interspecific interactions, publications over the last several decades suggest that ecological theory is still playing a minimal role in applied issues. These are the areas where accurate prediction is most important. The applied fields of natural resource management (e.g., fisheries regulation) and conservation biology have largely failed to incorporate models from community ecology. A recent literature analysis of articles published on conservation between 2000 and 2014 (Hintzen et al. 2020, p. 721) suggests that ecology and conservation biology are diverging, '. . .possibly from rising scepticism about the relevance of contemporary ecological theory to practical conservation'. Fisheries biologists have increasingly called for 'ecosystem-based' approaches, which include interacting species. However, thus far few works in fisheries have incorporated the type of consumer–resource models discussed in this book (a problem discussed in Matsuda and Abrams (2006), Abrams (2014), and Abrams et al. (2016). In their analysis of conservation biology, Hintzen et al. (2020, p. 730) conclude with the hope that 'mechanistic models . . . which are still new . . . may permit transparent, general, and useful ecological forecasts'. While I share that hope, the fact that mechanistic models could be regarded as new reflects an unacceptable rate of forgetting the recent past.

12.3 Omitting intermediate entities in models of indirect interactions

The problematic features of leaving resources out of models of competition can be seen by examining theory for other indirect interactions. The primary other indirect interactions involving one intermediate entity are: apparent competition; and top-down or bottom-up effects in systems having three trophic levels. In both cases, there is little if any modern theory that leaves out the intervening species. The predator population(s) has always been present in models of apparent competition following Holt's initial (1977) treatment. It has long been acknowledged that the properties of the predator can change the sign of the transmitted effect (Holt 1977; Holt and Kotler 1987; Abrams 1987d; Holt and Bonsall 2017). In other cases of apparent competition, the magnitudes of the negative effects are greatly altered by different forms of predator functional and numerical responses, without changing their signs. The ultimate reason for including the predator(s) in models of apparent competition is the fact that predators have a wide range of functional and numerical responses and connections to other species, and all of these alter the effect transmitted between two focal prey by a single predator species.

There are at least as wide a range of different properties and configurations of species on the middle trophic level of a 3-level system as in the case of resources in a competitive system. Some very early thought about top-down and bottom-up effects in ecosystems largely ignored the properties of the middle level (Hairston et al. 1960; Oksanen et al. 1981). In this theory positive effects on the top predator(s) would always increase the abundance of the bottom level, and positive effects on the bottom level would increase the top level, producing what Schoener (1993) described as a food chain mutualism. This outcome is the only possibility for simple models with one species per level using MacArthur's (1970) assumptions about dynamics. However, in a 3-trophic-level system, having two rather than one species on the middle level (or adaptive behaviour of a single middle-level species) can completely decouple the abundance of the top species from any effect applied to the bottom level (Leibold 1989; Abrams 1993; Abrams and Vos 2003). Even before this consequence of two species on the middle level was discovered, there was no sentiment that models of 3-level systems directed at top-down or bottom-up effects of separated trophic levels could legitimately ignore the dynamics of the middle level. Using nonlinear functional responses in the model with only one species on each level leads to a much wider range of possibilities due to population cycles and, in some cases, alternative attractors (Abrams and Roth 1994).

The one difference between competition and the two other categories of interaction just discussed is that the transmitting entity in competition can be non-living (i.e. not self-reproducing). This is the case for mineral nutrients for plants. Non-living resources may imply a somewhat narrower range of potential resource dynamics. However, even within the confines of the chemostat dynamics that are often assumed for abiotic resources, the parameter values are important in determining the degree of nonlinearity and the transient dynamics. Transient dynamics, in turn, are related to how the system behaves under periodic environmental variation, as shown in Chapters 8 and 9. And plant populations usually have neglected resources that are not strictly abiotic (Callaway and Walker 1997).

While the seeming consistency of MacArthur's model with the LV model is still used as a justification for leaving resources out of competition models, it has become clear that that seeming consistency is illusory. Even if the unlikely linearity assumptions of that model were generally applicable, the possibility of consumer-caused resource extinction, and the time delay entailed by resource dynamics, imply that very different overall dynamics are possible in systems that otherwise are characterized by MacArthur's assumed functional forms.

One of the common features of consumer–resource models based on living (biotic) resources is that resources can become extinct with a small perturbation to the system. A large enough directional change in a consumer parameter often causes one or more resource extinctions (or quasi-extinctions). Resource extinction in turn frequently causes abrupt (discontinuous) changes in the equilibrium consumer abundances. It is also possible for the addition of a competitor to allow resurrection of a resource that was exclusively used and driven to quasi-extinction by the original consumer. Ratajczak et al. (2018) recently reviewed the phenomenon of abrupt change in ecological communities without once mentioning competition or competitors. Is this because it does not occur, or because it has been ignored or mis-interpreted? Abrams et al. (2008a) documented resource exclusion in multi-resource models of competition between two consumers. A large majority of published papers citing this work either do not acknowledge that resource exclusion was demonstrat-ed in that article or do not consider resource exclusion as a possible outcome in their own analysis of related models. It is basically impossible, on logical grounds, that such exclusion events are absent from all competitive communities, given that apparent competitive exclusion in systems with a single predator has been documented.

Currently we do not know the full range of dynamic behaviours, even for simple models with two consumers and relatively few resources. No mathematical model of a complex biological system at this point in time can include the full range of pro-cesses affecting interspecific interactions, in part because many of the processes are so poorly known. As Schaffer (1981) pointed out long ago, 'abstraction' is common in scientific theories. However, abstraction requires some consideration of what is lost by adopting the implied simplification. Competition models that lack resources make the study of competition inconsistent with other indirect interactions and have very different dynamics and predictions. Even if one is studying the growth of a single species in the laboratory, the nature of the 'density-dependent' feedback that limits growth requires some knowledge of resource dynamics. Indirect interactions trans-mitted by resources are involved in almost all biological communities. It is striking that the recognition of the limitations of the LV model seemed to be much greater four or five decades ago than it is today; for example Gilpin and Case (1976, p. 42), in assessing their own analysis of multi-species LV systems, stated: 'Since these results come from equations that are no more than a caricature of ecological reality, our results must be taken as provisional'.

12.4 Important aspects of consumer–resource relationships

The focus on direct-interaction models of competition (those lacking resource dynamics) has had several biasing effects on our knowledge of that interaction. It has restricted attention to predicting the consequences of altered consumer mortality or immigration (i.e., change in 'neutral' parameters), or predicting effects of the addition or removal of one consumer. The extent and nature of the population-level effects of changes in most consumer fitness parameters has had little theoretical exploration, and little empirical study, in part because those parameters are absent from the direct-interaction models. However, the effects of global climate change will be to alter most of the parameters in any species' functional and numerical responses (Amarasekare 2015). This includes parameters that define the nature of interference competition, those that affect resource capture and conversion rates, and those affecting consumer mortality rates. Higher and lower trophic levels also have largely been ignored in treatments of competition. Levels above the consumer and below the resource typically affect flexible foraging behaviours that respond to risk and reward levels. These processes in higher and lower trophic levels will, in turn, affect the amount and functional form of competition between consumer species.

One aspect of competition that has been of interest throughout the decades is how the similarity of resource use patterns of two consumers is related to the 'tightness' of their population coupling, and consequently, their coexistence bandwidth (see Chapter 7). Greater similarity in resource capture rates is likely to imply tighter coupling, but the exact relationship can vary enormously depending on the nature of resource growth and the details of consumer growth. One important and largely ignored question is how the relative abundance of a particular resource affects its contribution to measures of intraspecific and interspecific competition. This question is clearly not going to arise from a model that lacks resources. In MacArthur's consumer–resource model with perfectly substitutable resources, the abundance of a resource has no effect on how rapidly consumer per capita population growth rate declines with greater consumer population size. As a result, rare resources are as influential as common ones (all other parameters being equal). However, as pointed out in Abrams et al. (2008a), if per capita resource growth is a concave function of resource abundance (and resources do not interact), less abundant resources will have a smaller effect on traditional measures of competition. My impression is that many ecologists believe that less common resources types do in fact contribute less to the overall measure of competition, all else being equal. That density dependence is more often concave is supported by what we know of resource growth. All abiotic growth models suggest a concave relationship, and theta-logistic growth models suggest the same for $\theta < 1$. This inequality is satisfied in a large majority of biotic resources according to Sibly et al.'s (2005) review (discussed in Chapter 5). The shape of the relationship between the abundance of a particular resource and its contribution to a measure of competition is also affected by functional response shape (Abrams 2009b), so this and other factors may alter the predominance of concave relationships. The point here is that knowing the most basic details of consumer–resource relationships is needed to estimate the similarity vs competition relationship, which is still one of the topics of greatest interest to both empirical and theoretical ecologists.

Whether one is interested in the degree of coupling between two competitor populations, or some other question, the first order of business in future studies of competition should be to arrive at a better description of the basic processes involved in the interaction; resource growth and consumer functional/numerical responses. One aspect of these processes that is important and often ignored is the effect of resource and consumer behaviours on their functional responses (see Chapter 3). These behavioural effects should often produce predator interference, which is almost always omitted from models of competition, even though it is the subject of a great deal of attention in predator–prey models (Werner and Peacor 2003; Abrams 1995, 2015). Simple models including adaptive behaviour predict that there should be effects of the abundance of the prey species' resources and the predator's own natural enemies on the functional responses of both species (Abrams 1992c).

Another high-priority topic for future research should be to examine interactions between different resources when studying interactions between their consumers. Early theory by Levine (1976) and Vandermeer (1980) showed the importance of this to understanding the interaction between consumers. After some initial empirical work by Dungan (1987) on a rocky intertidal community, and by Davidson and Brown (e.g. Davidson et al. 1984) on a system based on competition for desert plant seeds between ants and rodents, interactions between resources have routinely been ignored in empirical and theoretical studies of competing consumers. Even for competitive systems with abiotic resources, different resources can affect each other's dynamics; large amounts of resource may either increase or decrease the per capita rate at which it leaves the system. The almost universal assumption of the simple chemostat model has essentially eliminated between-resource interactions from models with abiotic resources. Another widely ignored subject is the fact that many 'resources' have characteristics that have elements of both biotic and abiotic growth. This includes systems in which a prey species is only available for part of its life history or when it is engaged in certain activities or is partially disabled (Abrams and Walters 1996). The extent to which such dynamics differ from those of systems with purely biotic or abiotic dynamics has also received little attention (but see Abrams and Quince 2005; Schreiber and Rudolf 2008).

12.5 Food web structure and its influence on competition

The early theory based on the LV model and MacArthur's interpretation of his consumer–resource model both suggest that the nature of competition is relatively independent of the presence and abundance of species other than the competing consumers. However, the fact that the interaction is highly dependent on resource populations means that the strength of competition (however it is measured) will depend on both the set of resources that are present and other factors that influence their relative abundances. This chapter, and the book as a whole, have stressed the importance of resources in understanding competition. Yet, the interactions between consumers and resources are themselves likely to be influenced by species/entities on trophic levels above the consumers and below the resources if those additional levels are present. Predators of consumers are likely to reduce their resource consumption, and, for biotic resources, lower-level foods/nutrients are likely to influence their foraging, and therefore affect their availability to the consumer. What effects do these processes have on the interaction between consumer and resources? In most systems, we really don't know.

These multi-level effects of adaptive foraging behaviour were explored using food chain models in the 1980s and 1990s (Abrams 1982, 1984b, 1992c, 1995). Higher-level species usually suppress foraging by their prey, and therefore have a positive effect two levels below. Increased resource abundance can increase or decrease foraging; these respectively are likely to have positive and negative effects two levels above. A system with two adjacent levels of a food chain both exhibiting adaptive foraging influenced by predation risk has a variety of potential outcomes (Abrams 1992c). These have not been explored in the context of competition, but there are very likely a wide variety of complex effects that may occur. Given the frequent occurrence of adaptive risk-related foraging, known to be common since Lima and Dill (1990) and Werner and Peacor (2003), it seems highly unlikely that these effects are rare or small in magnitude. The demonstrated presence of behavioural effects on both predators and prey in many systems argues for more research on their effects on interspecific competition.

The results presented in Chapters 8 and 9 show the importance of the time lags in consumer responses to resources in determining the dynamics of temporally variable consumer–resource systems. Time lags are likely to play an even larger role in the dynamics of competitors that are parts of larger food webs, which is to say most competitors in natural systems. The potential role of these lags for interspecific interactions was something that was emphasized by Schoener (1989, 1993). Schoener also noted how difficult they would be to study in real-world food webs.

The argument for a food web approach (i.e., one that includes resources and species on trophic levels above and/or below the ones that define the interaction) is accepted for other indirect interactions. For example, Wise and Farfan (2021, p. 1) recently wrote that, 'Ecologists have long debated how simple mathematical models should be and how much to simplify food webs by lumping taxa into functional groups. For both modellers and empiricists, the food chain is the ultimate simplification'. The lumping of all species on a trophic level into one group is something that would never be considered if the existence or diversity of species on a single level was the topic of interest. However, it is clear that these authors consider lumping species within a level to be more easily justified than making an entire level implicit. Explicit modelling of intermediate species/entities is needed just to be consistent with the approaches used in studying other systems. It is required both because: (1) any single intermediate entity may have a wide range of dynamics, depending on the functional forms of the various components of its growth rate; and (2) there are likely to be a number of dynamically distinct components in the intermediate entity.

12.6 Forces that have biased research on competition

Why have the aspects of consumer and resource dynamics reviewed above been ignored in the vast majority of those systems where competition has been studied? A scientific approach to this question about the direction of research is not possible. However, it is possible to identify some of the factors that are likely to be involved: (1) assumption drag; (2) the economics and sociology of the research environment;

and (3) the urge to oversimplify. To some extent, these factors affect all scientific disciplines. They at least appear to be particularly prominent in the recent history of research on both consumer–resource interactions and competition.

'Assumption drag' is a commonly used term for the persistence of assumptions when they have been used often enough, regardless of evidence against their validity. It is particularly problematic for ecology, given the incredible range of different competitive relationships. The continued frequent use of the LV model and MacArthur's consumer–resource model both reflect assumption drag. The prevalence of functional responses that are functions of the abundance of a single resource is another case of assumption drag within consumer–resource theory. Although specialist consumers are rare, the fact that C. S. Holling originally defined functional responses in terms of a single food/resource type has led to the dominance of such relationships in both the empirical and theoretical literatures. This tendency is reflected in the comprehensive functional response review articles by Jeschke et al. (2002, 2004).

In theoretical biology, using new or difficult mathematical techniques or deriving extremely general results are both important avenues for obtaining recognition. Developing a large body of specific results dependent on simulations of many alternative models usually does not translate into career advancement for theoreticians. In more recent years, the influence of citation counts has introduced a large random component to the direction of research for both theoreticians and empiricists. Thus, some of the types of work required to achieve better understanding of competition, and of consumer–resource interactions more generally, is insufficiently rewarding from a career standpoint. It is unclear how this can be changed quickly, but recognizing the necessity of this type of work for a predictive competition theory is a first step.

Research on the functional responses of consumer species is a good example of how the structure of the research environment discourages a broader perspective. The fact that many measurements of functional responses have already been published makes any new study less likely to attract scientific attention. Because few studies since Abrams (1980a) have analysed the effect of functional response shape on competition, the functional response shapes of competing consumers are almost always unknown. An additional problem is that multi-species functional response measurements require a large number of treatments, which often makes them technically impossible. The fact that competitive systems, particularly those involving biotic resources, often take a substantial amount of time to come to equilibrium has led to a marked lack of studies of organisms with relatively 'slow' life histories. All of this leads to a high level of ignorance regarding the range of functional responses in competitive systems involving multiple resources.

Competitive interactions involving cycling species are also rarely studied. In these cases, standard measures of interspecific effects require calculating an average over the course of a complete cycle, or over a large number of ups and downs in a system with an aperiodic attractor. Keeping a laboratory system going for such periods is seldom possible, and the many years required for field studies of cycling vertebrate populations makes studies of interactions even less likely. Field studies of such systems would also need to account for the environmental change occurring over the course of any experimental study of the interaction. Thus, theory about the characteristics of competitive interactions in cycling systems is not a major component of the literature on competition.

The semi-random process of particular ideas becoming popular was highlighted in the quotation from David Houle in the first paragraph of this chapter. The idea of separating coexistence into the two categories of stabilizing and equalizing effects is currently attracting a huge amount of attention, but it seems unlikely to lead to any useful results for truly understanding competitive processes.

Finally, the urge to oversimplify is a potentially powerful source of bias in most research in most areas of science. Simplification in competition theory finds its ultimate embodiment in the LV model. However, even within the framework of consumer–resource models, the focus on 2-consumer–2-resource models, combined with isocline or invasibility analysis of those models, has narrowed the subject matter of competition studies in a detrimental way. The overuse of simple models has likely been promoted by a seldom-stated idea that simple models are in some sense more general. This applies to the overuse of the MacArthur model within studies of consumer–resource systems. A linear functional response is intermediate between concave and convex responses, so it is thought to potentially apply to some extent to both cases. However, this justification fails if most real responses are strongly nonlinear, and if nonlinear response categories share some dynamical properties that are lacking in the linear case. Both of these conditions happen to be true.

The consumer–resource models used in this book are still much too simple to be used for quantitative predictions for any natural system. However, the diversity of dynamics demonstrated by these models is very likely to be part of the repertoire of potential dynamics in real-world systems. And if they are not, then there must be additional features or restrictions in natural systems that we are currently unaware of, or have not yet studied. Knowing the properties of intermediate-complexity consumer–resource models will be part of the knowledge needed to uncover what those additional features might be.

12.7 Conclusions

Although this book covers many deficiencies in competition theory, it is not an exhaustive review, and other researchers could no doubt come up with other issues. An overlapping set was present in a call for filling gaps in population ecology that appeared more than a decade ago (Agrawal et al. 2007). Following is a list of gaps in the theoretical foundation for understanding competition that covers most of the points made in this book:

1. Lack of an accepted definition of competition, which should be based on its mechanism. This leads to point 2.

2. Neglect of the role of resource properties and dynamics in determining the consequences of mutual resource use for the competitors. Instead, competition is often treated as being solely a function of consumer traits. Among the aspects of resource dynamics left out of most analyses is the key role of resource immigration in determining outcomes when resource exclusion could occur in the absence of such immigration.

3. Concentration on linear relationships for trophic relationships and density dependence, neither of which has been shown to be close to linear in a significant fraction of species/food webs. This is particularly problematic because linear functions uniquely imply that the contribution of a given change in a neutral parameter to a competitive effect is independent of current abundances of both entities, as well as all of the other species in the food web.

4. An excessive focus on 2-consumer–2-resource systems, which are not likely to represent any natural system, and, like all systems with equal numbers of consumers and resources, have many special properties.

5. Scant consideration of adaptive foraging in models of competition.

6. Reliance on oversimplified (and thus often incorrect or inapplicable) methods of analysis, including isocline analysis and the assumption that inability to invade when rare implies failure to coexist.

7. Excessive attention to the outcome of coexistence at the expense of developing a quantitative description of the full range of interspecific effects that can occur with shared resource use.

8. Failure to build a coherent body of competition theory that can integrate with, inform, and be informed by theory about related relationships—i.e., predator–prey relationships, intraspecific competition (i.e., density dependence), apparent competition, and indirect effects in larger food webs.

The above list should be qualified in several ways. None of these criticisms applies to all recent work on competition. There have been many very excellent theoretical and empirical studies of competition during the past four decades, and there are many of those to which none of the above applies. Nevertheless, each of these numbered points characterizes a very large fraction of ecological work on interspecific competition. My own work certainly includes many examples of practices criticized in the above list.

In spite of the qualifications in the preceding paragraph, there is a clear need for restructuring current competition theory. An example of one of the more problematic features of current competition theory is the assumption that an additional consumer species need not influence the interaction of a given pair of consumers. This was discussed in connection with the commentary on Saavedra et al. (2017) earlier in this chapter (in Section 12.2.1). In the most highly cited recent article on competition reviewed in Chapter 4, Levine et al. (2017, p. 59)) state that, '. . . regardless of whether phenomenological or mechanistic models of competition are considered, how much the interaction between two species is dictated by other species in the system remains a relevant question'. It is not clear whether they are referring to resources, other consumers, or predators of the consumers when they refer to 'other species'. They are correct in noting that the effects of additional species on a competitive interaction have not been quantified in the field for most systems, but few such studies have been attempted. In addition, given what we know about consumer–resource interactions, few if any interactions between a given pair of consumers could fail to be significantly influenced by the presence of 'other species'. This was shown to be a general result for systems with nonlinear resource per capita growth rates or nonlinear consumer functional responses in Abrams (1983a). Since then, empirical work has continued to demonstrate nonlinearity to be pervasive in the component functions of consumer–resource interactions. That some other nonlinearity or more complex models would restore linearity is highly unlikely.

Many more examples of issues that could be settled by adopting a more mechanistic, resource-oriented approach to modelling competition were provided in previous chapters. However, even if such an approach is fully adopted, it will never lead to the types of predictive accuracy that are achieved for many systems in physics. There are always too many variables and unknown functional forms in ecological systems. In addition, empirical 'tests' of models have their own limitations. Measuring time-dependent variables without altering system dynamics is likely to remain impossible in many systems for the foreseeable future. Laboratory systems with many species and trophic levels are difficult to maintain and monitor. Repeated censuses of all species in a diverse community over the time span required to estimate interaction strength are beyond the capabilities and/or funding even for most laboratory studies. These problems, and the sheer diversity of ecological communities, mean that predictions will always be approximate. The diversity of consumer–resource models also means that many years will be required to have a reasonably comprehensive body of theory. All of these considerations are likely to have contributed to the title of Getz et al.'s recent (2018) assessment of ecological modelling, 'Making ecological models adequate'. Given the current trends of the Earth and its human population, this goal may be delayed much further by more urgent challenges.

Despite the difficulty of achieving it, the goal of ecology must remain to become better at predicting changes in population sizes and to understand the mechanisms behind observed changes in communities. This was the goal that stimulated the famous mathematician, Vito Volterra, to develop his early models; his son-in-law had piqued Volterra's interest with the problem of why the proportions of different fish species in the Mediterranean Sea changed during the suspension of commercial fishing during the First World War (Hutchinson 1978). Developing better fisheries models is even more relevant today, and natural communities are still facing the coupled threats of human-caused habitat and climate change. Fisheries biology is by no means the only area where competition plays a major role and better predictions are needed. At the time I am writing there is a clear need for a better understanding of competition between different strains of a coronavirus, SARS-CoV2. It presently remains uncertain whether the delta and omicron strains of this virus will coexist.

None of the predictions needed to deal with threats like this can be achieved in the absence of better and more diverse representation of competitive processes involved.

In 1973, near the beginning of the time period covered by this book, Gilpin and Ayala (p. 3590) opened their article on competition by stating that, 'Population ecology is at a Keplerian stage of development. Much of the present theory is based on idealized linear interactions. . .somewhat as pre-Keplerian astronomy was based on idealized circular motions'. In describing 1973 as 'Keplerian' rather than 'pre-Keplerian', they evidently felt that their work and other investigations from that time period would shift the field. A strong case could be made that a great deal of population and community ecology—and certainly that part dealing with interspecific competition—has largely remained in a pre-Keplerian stage for the 49 years following this seminal article analysing nonlinear competition between pairs of *Drosophila* species. That could begin to change if the basic mechanisms underlying competition begin to receive more systematic attention.

References

Abrams, P., and Matsuda, H. (1993). Effects of adaptive predatory and anti-predator behaviour in a two-prey–one-predator system. *Evolutionary Ecology*, 7, 312–326.

Abrams, P., Nyblade, C., and Sheldon, S. (1986). Resource partitioning and competition for shells in a subtidal hermit crab assemblage. *Oecologia*, 69, 429–445.

Abrams, P. A. (1975). Limiting similarity and the form of the competition coefficient. *Theoretical Population Biology*, 8, 356–375.

Abrams, P. A. (1976). Niche overlap and environmental variability. *Mathematical Biosciences*, 28, 357–375.

Abrams, P. A. (1977). Density independent mortality and interspecific competition: a test of Pianka's niche overlap hypothesis. *American Naturalist*, 111, 539–552.

Abrams, P. A. (1980a). Consumer functional response and competition in consumer–resource systems. *Theoretical Population Biology*, 17, 80–102.

Abrams, P. A. (1980b). Are competition coefficients constant? Inductive vs. deductive approaches. *American Naturalist*, 116, 730–735.

Abrams, P. A. (1980c). Resource partitioning and interspecific competition in a tropical hermit crab community. *Oecologia*, 46, 365–379.

Abrams, P. A. (1981a). Alternative methods of measuring competition applied to two Australian hermit crabs. *Oecologia*, 51, 233–240.

Abrams, P. A. (1981b). Competition in an Indo-Pacific hermit crab community. *Oecologia*, 51, 241–249.

Abrams, P. A. (1981c). Shell fighting and competition between two hermit crab species in Panama. *Oecologia*, 51, 84–90.

Abrams, P. A. (1982). Functional responses of optimal foragers. *American Naturalist* 120, 382–390.

Abrams, P. A. (1983a). Arguments in favor of higher order interactions. *American Naturalist*, 121, 887–891.

Abrams, P. A. (1983b). The theory of limiting similarity. *Annual Review of Ecology and Systematics*, 14, 359–376.

Abrams, P. A. (1984a). Variability in resource consumption rates and the coexistence of competing species. *Theoretical Population Biology*, 25, 106–124.

Abrams, P. A. (1984b). Foraging time optimization and interactions in food webs. *American Naturalist*, 124, 80–96.

Abrams, P. A. (1986a). Character displacement and niche shift analyzed using consumer–resource models of competition. *Theoretical Population Biology*, 29, 107–160.

Abrams, P. A. (1986b). Adaptive responses of predators to prey and prey to predators: the failure of the arms race analogy. *Evolution*, 40, 1229–1247.

Abrams, P. A. (1987a). On classifying interactions between populations. *Oecologia*, 73, 272–281.

Abrams, P. A. (1987b). The nonlinearity of competitive effects in models of competition for essential resources. *Theoretical Population Biology*, 32, 50–65.

Abrams, P. A. (1987c). The functional responses of adaptive consumers of two resources. *Theoretical Population Biology*, 32, 262–288.

Abrams, P. A. (1987d). Indirect interactions between species that share a predator: varieties of indirect effects. In W. C. Kerfoot and A. Sih, eds, *Predation: Direct and Indirect Impacts on Aquatic Communities*, pp. 38–54. University Press of New England, Dartmouth, NH.

Abrams, P. A. (1987e). An analysis of competitive interactions between three hermit crab species. *Oecologia*, 72, 233–247.

Abrams, P. A. (1987f). Alternative models of character displacement: I. Displacement when there is competition for nutritionally essential resources. *Evolution*, 41, 651–661.

Abrams, P. A. (1987g). Alternative models of character displacement: II. Displacement when there is competition for a single resource. *American Naturalist*, 130, 271–282.

Abrams, P. A. (1987h). Resource partitioning and competition for shells between intertidal hermit crabs on the outer coast of Washington. *Oecologia*, 72, 248–258.

Abrams, P. A. (1988a). Resource productivity – consumer species diversity: Simple models of competition in spatially heterogeneous environments. *Ecology* 69, 1418–1433.

Abrams, P. A. (1988b). How should resources be counted? *Theoretical Population Biology*, 33, 226–242.

Abrams, P. A. (1989). Decreasing functional responses as a result of adaptive consumer behavior. *Evolutionary Ecology*, 3, 95–114.

Abrams, P. A. (1990a). Mixed responses to resource densities and their implications for character displacement. *Evolutionary Ecology*, 4, 93–102.

Abrams, P. A. (1990b). Adaptive responses of generalist herbivores to competition: convergence or divergence. *Evolutionary Ecology*, 4, 103–114.

Abrams, P. A. (1990c). The effects of adaptive behavior on the type-2 functional response. *Ecology*, 71, 877–885.

Abrams, P. A. (1992a). Adaptive foraging by predators as a cause of predator–prey cycles. *Evolutionary Ecology*, 6, 56–72.

Abrams, P. A. (1992b). Why don't predators have positive effects on prey populations? *Evolutionary Ecology*, 6, 449–457.

Abrams, P. A. (1992c). Predators that benefit prey and prey that harm predators: Unusual effects of interacting foraging adaptations. *American Naturalist*, 140, 573–600.

Abrams, P. A. (1993). Effect of increased productivity on the abundance of trophic levels. *American Naturalist*, 141, 351–371.

Abrams, P. A. (1995). Implications of dynamically variable traits for identifying, classifying, and measuring direct and indirect effects in ecological communities. *American Naturalist*, 146, 112–134.

Abrams, P. A. (1997). Variability and adaptive behavior: Implications for interactions between stream organisms. *Journal of the North American Benthological Society*, 16, 358–374.

Abrams, P. A. (1998). High competition with low similarity and low competition with high similarity: The interaction of exploitative and apparent competition in consumer–resource systems. *American Naturalist*, 152, 114–128.

Abrams, P. A. (1999). The adaptive dynamics of consumer choice. *American Naturalist*, 153, 83–97.

Abrams, P. A. (2000a). Character shifts of species that share predators. *American Naturalist*, 156, S45–S61.

Abrams, P. A. (2000b). The impact of habitat selection on the spatial heterogeneity of resources in varying environments. *Ecology*, 81, 2902–2913.

Abrams, P. A. (2000c). The evolution of predator–prey interactions: Theory and evidence. *Annual Review of Ecology and Systematics*, 31, 359–376.

Abrams, P. A. (2001a). The effect of density independent mortality on the coexistence of exploitative competitors for renewing resources. *American Naturalist*, 158, 459–470.

Abrams, P. A. (2001b). Review of 'The Unified Neutral Theory of Biodiversity and Biogeography'. *Nature*, 412, 858–859.

Abrams, P. A. (2001c). Describing and quantifying interspecific interactions: A commentary on recent approaches. *Oikos*, 94, 209–218

Abrams, P. A. (2001d). Modeling the adaptive dynamics of traits involved in inter- and intraspecific competition: An assessment of three methods. *Ecology Letters*, 4, 166–175.

Abrams, P. A. (2002). Will declining population sizes warn us of impending extinctions? *American Naturalist*, 160, 293–305.

Abrams, P. A. (2003). Effects of altered resource consumption rates by one consumer species on a competitor. *Ecology Letters*, 6, 550–555.

Abrams, P. A. (2004a). Trait initiated indirect effects in simple food webs: Consequences of changes in consumption-related traits. *Ecology*, 85, 1029–1038.

Abrams, P. A. (2004b). When does periodic variation in resource growth allow robust coexistence of competing consumer species? *Ecology*, 85, 372–382.

Abrams, P. A. (2006a). The prerequisites for and likelihood of generalist–specialist coexistence. *American Naturalist*, 167, 329–342.

Abrams, P. A. (2006b). The effects of switching behavior on the evolutionary diversification of generalist consumers. *American Naturalist*, 168, 645–659.

Abrams, P. A. (2007a). Habitat choice in predator–prey systems; spatial instability due to interacting adaptive movements. *American Naturalist*, 169, 581–594.

Abrams, P. A. (2007b). Specialist–generalist competition in variable environments: The consequences of competition between resources. In D. A. Vasseur and K. S. McCann, eds, *The Impact of Environmental Variability on Ecological Systems*, pp. 133–158. Springer, Dordrecht, The Netherlands.

Abrams, P. A. (2009a). When does greater mortality increase population size? The long history and diverse mechanisms underlying the hydra effect. *Ecology Letters*, 12, 462–474.

Abrams, P. A. (2009b). The implications of using multiple resources for consumer density dependence. *Evolutionary Ecology Research*, 11, 517–540.

Abrams, P. A. (2009c). Determining the functional form of density dependence: Deductive approaches for consumer–resource systems with a single resource. *American Naturalist*, 174, 321–330.

Abrams, P. A. (2009d). Adaptive changes in prey vulnerability shape the response of predator populations to mortality. *Journal of Theoretical Biology*, 261, 294–304.

Abrams, P. A. (2010a). Quantitative descriptions of resource choice in ecological models. *Population Ecology*, 52, 47–58.

Abrams, P. A. (2010b). Implications of flexible foraging for interspecific interactions: Lessons from simple models. *Functional Ecology*, 24, 7–17.

Abrams, P. A. (2012). The eco-evolutionary responses of a generalist consumer to resource competition. *Evolution*, 66, 3130–3143.

Abrams, P. A. (2014). How precautionary is the policy governing the Antarctic toothfish (*Dissostichus mawsoni*) fishery? *Antarctic Science*, 26, 3–13.

Abrams, P. A. (2015). Why ratio dependence is (still) a bad model of predation. *Biological Reviews*, 90, 794–814.

Abrams, P. A. (2019). How does the evolution of universal ecological traits affect population size: Lessons from simple models. *American Naturalist*, 193, 814–829.

Abrams, P. A., Ainley, D. G., Blight, L. K., Dayton, P. K., Eastman, J. H., and Jacquet, J. L. (2016). Necessary elements of precautionary management: implications for the Antarctic toothfish. *Fish and Fisheries*, 17, 1152–1174.

Abrams, P. A., Brassil, C. E., and Holt, R. D. (2003). Dynamics and responses to mortality rates of competing predators undergoing predator–prey cycles. *Theoretical Population Biology*, 64, 163–176.

Abrams, P. A., and Chen, X. (2002a). The evolution of traits affecting resource acquisition and predator vulnerability: Character displacement under real and apparent competition. *American Naturalist*, 160, 692–704.

Abrams, P. A., and Chen, X. (2002b). The effect of competition between prey species on the evolution of their vulnerabilities to a shared predator. *Evolutionary Ecology Research*, 4, 897–909.

Abrams, P. A., and Cortez, M. H. (2015a). The many potential interactions between predators that share competing prey. *Ecological Monographs*, 85, 625–641.

Abrams, P. A., and Cortez, M. H. (2015b). Is competition needed for ecological character displacement? Does displacement decrease competition? *Evolution*, 69, 3039–3053.

Abrams, P. A., Cressman. R., and Křivan, V. (2007). The role of behavioral dynamics in determining the patch distributions of interacting species. *American Naturalist*, 169, 505–518.

Abrams, P. A., and Ginzburg, L. R. (2000). The nature of predation: Prey dependent, ratio dependent or neither? *Trends in Ecology and Evolution*, 15, 337–341.

Abrams, P. A., and Holt, R. D. (2002). The impact of consumer–resource cycles on the coexistence of competing consumers. *Theoretical Population Biology*, 62, 281–295.

Abrams, P. A., Holt, R. D., and Roth, J. D. (1998). Apparent competition or apparent mutualism? Shared predation when populations cycle. *Ecology*, 79, 201–212.

Abrams, P. A., and Matsuda, H. (1994). The evolution of traits that determine ability in competitive contests. *Evolutionary Ecology*, 8, 667–686.

Abrams, P. A., and Matsuda, H. (1996). Positive indirect effects between prey species that share predators. *Ecology*, 77, 610–616.

Abrams, P. A., and Matsuda, H. (2003). Population dynamical consequences of switching at low total prey densities. *Population Ecology*, 45, 175–185.

Abrams, P. A., and Matsuda, H. (2004). Consequences of behavioral dynamics for the population dynamics of predator–prey systems with switching. *Population Ecology*, 46, 13–25.

Abrams, P. A., and Matsuda, H. (2005). The effect of adaptive change in the prey on the dynamics of an exploited predator population. *Canadian Journal of Fisheries and Aquatic Science*, 62, 758–766.

Abrams, P. A., Matsuda, H., and Harada, Y. (1993). Evolutionarily unstable fitness maxima and stable fitness minima in the evolution of continuous traits. *Evolutionary Ecology*, 7, 465–487.

Abrams, P. A., Menge, B. A., Mittelbach, G. G., Spiller, D., and Yodzis, P. (1996). The role of indirect effects in food webs. In G. A. Polis and K. O. Winemiller, eds, *Food Webs: Integration of Patterns and Dynamics*, pp. 371–395. Chapman and Hall, New York, NY.

Abrams, P. A., and Nakajima, M. (2007). Does competition between resources reverse the competition between their consumers? Variations on two themes by Vandermeer. *American Naturalist*, 170, 744–757.

Abrams, P. A., and Quince, C. (2005). The impact of mortality on predator population size and stability in systems with stage-structured prey. *Theoretical Population Biology*, 68, 253–266.

Abrams, P. A., and Roth, J. D. (1994a). The responses of unstable food chains to enrichment. *Evolutionary Ecology*, 8, 150–171.

Abrams, P. A., and Roth, J. D. (1994b). The effects of enrichment on three-species food chains with nonlinear functional responses. *Ecology*, 75, 1118–1130.

Abrams, P. A., and Rueffler, C. (2009). Coexistence and limiting similarity of consumer species competing for a linear array of resources. *Ecology*, 90, 812–822.

Abrams, P. A., Rueffler, C., and Dinnage, R. (2008a). Competition–similarity relationships, and the nonlinearity of competitive effects in consumer–resource systems. *American Naturalist*, 172, 463–474.

Abrams, P. A., Rueffler, C. and Kim, G. (2008b). Determinants of the strength of disruptive and/or divergent selection arising from resource competition. *Evolution*, 62, 1571–1586.

Abrams, P. A., and Ruokolainen, L. (2011). How does adaptive consumer movement affect population dynamics in consumer–resource metacommunities? I. Homogeneous patches. *Journal of Theoretical Biology*, 277, 99–110.

Abrams, P. A., Ruokolainen, L., Shuter, B. J., and McCann, K. S. (2012). Harvesting creates ecological traps: consequences of invisible mortality risks in predator–prey metacommunities. *Ecology*, 93, 281–293.

Abrams, P. A., and Shen, L. (1989). Population dynamics of systems with consumers that maintain a constant ratio of intake rates of two resources. *Theoretical Population Biology*, 35, 51–89.

Abrams, P. A., and Schmitz, O. J. (1999). The effect of risk of mortality on the foraging behaviour of animals faced with time and digestive capacity constraints. *Evolutionary Ecology Research*, 1, 285–301.

Abrams, P. A., Tucker, C. M., and Gilbert, B. (2013). Evolution of the storage effect. *Evolution*, 67, 315–327.

Abrams, P. A., and Vos, M. (2003). Adaptation, density dependence, and the abundances of trophic levels. *Evolutionary Ecology Research*, 5, 1113–1132.

Abrams, P. A., and Walters, C. J. (1996). Invulnerable prey and the paradox of enrichment. *Ecology*, 77, 1125–1133.

Abrams, P. A., and Wilson, W. G. (2004). Coexistence of competitors in metacommunities due to spatial variation in resource growth rates: Does R^* predict the outcome of competition? *Ecology Letters*, 7, 929–940.

Abreu, C. I., Woltz, V. L. A., Friedman, J., and Gore, J. (2020). Microbial communities display alternative stable states in a fluctuating environment. *Public Library of Science Computational Biology*, 16(5), e1007934.

Adler, P. B., Ellner, S. P., and Levine, J. M. (2010). Coexistence of perennial plants: an embarrassment of niches. *Ecology Letters*, 13, 1019–1029.

Adler, P. B., Smull, D., Beard, K. H., et al. (2018). Competition and coexistence in plant communities: Intraspecific competition is stronger than interspecific competition. *Ecology Letters*, 21, 1319–1329.

Agrawal, A. A., Ackerly, D. D., Adler, F., et al. (2007). Filling key gaps in population and community ecology. *Frontiers in Ecology and Evolution*, 5, 145–152.

Amarasekare, P. (2010). Effect of non-random dispersal strategies on spatial coexistence mechanisms. *Journal of Animal Ecology*, 79, 282–293.

Amarasekare, P. (2015). Effects of temperature on consumer–resource interactions. *Journal of Animal Ecology*, 84, 665–679.

Amarasekare, P. (2020). The evolution of coexistence theory. *Theoretical Population Biology*, 133, 49–51.

Amarasekare, P., and Simon, M. W. (2020). Latitudinal directionality and ectotherm invasion success. *Proceedings of the Royal Society B*, 287, 20191411.

Armstrong, R. A. (1976). Fugitive species: Experiments with fungi and some theoretical considerations. *Ecology*, 57, 953–963.

Armstrong, R. A., and McGehee, R. (1976a). Coexistence of two competitors on one resource. *Journal of Theoretical Biology*, 56, 499–502.

Armstrong, R. A., and McGehee, R. (1976b). Coexistence of species competing for shared resources. *Theoretical Population Biology*, 9, 317–328.

Armstrong, R. A., and McGehee, R. (1980). Competitive exclusion. *American Naturalist*, 115, 151–170.

Arnqvist, G., and Rowe, L. (2005). *Sexual Conflict*. Princeton University Press, Princeton, NJ.

Ayala, F. J. (1969). Experimental invalidation of the principle of competitive exclusion. *Nature*, 224, 1076–1079.

Ayala, F. J., Gilpin, M. E., and Ehrenfeld, J. G. (1973). Competition between species: Theoretical models and experimental tests. *Theoretical Population Biology*, 4, 331–356.

Barabás, G., and D'Andrea, R. (2020). Chesson's coexistence theory: Reply. *Ecology*, 101, e03140.

Barabás, G., D'Andrea, R., and Stump, S. M. (2018). Chesson's coexistence theory. *Ecological Monographs*, 88, 277–303.

Barabás, G., Michalska-Smith, M. J., and Allesina, S. (2016). The effect of intra- and interspecific competition on coexistence in multi-species communities. *American Naturalist*, 188, E1–E12.

Barabás, G., Meszéna, G., and Ostling, A. (2012). Community robustness and limiting similarity in periodic environments. *Theoretical Ecology*, 5, 265–282.

Barraquand, F., Louca, S., Abbott, K. C., et al. (2017). Moving forward in circles: Challenges and opportunities in modeling population cycles. *Ecology Letters*, 20, 1074–1092.

Bazykin, A. D. (1974). *Structural and Dynamic Stability of Model Predator–Prey Systems*. Institute of Animal Resource Ecology, University of British Columbia.

Bazykin, A. D. (1998). *Nonlinear Dynamics of Interacting Populations*. World Scientific Publishers, Singapore.

Beckerman, A. P. (2005). The shape of things eaten: The functional response of herbivores foraging adaptively. *Oikos*, 110, 591–601.

Becks, L., Ellner, S. P., Jones, L. E., and Hairston, Jr.,N. G. (2010). Reduction of adaptive genetic diversity radically alters eco-evolutionary community dynamics. *Ecology Letters*, 13, 989–997.

Begon, M., Townsend, C. R., and Harper, J. L. (2006). *Ecology: From Individuals to Ecosystems*. 4th edition. Wiley-Blackwell, Oxford, UK.

Bender, E. A., Case, T. J., and Gilpin, M. E. (1984). Perturbation experiments in community ecology: Theory and practice. *Ecology*, 65, 1–13.

Berec, L., Kramer, A. M., Bernhauerová, V., and Drake, J. M. (2018). Density-dependent selection on mate search and evolution of Allee effects. *Journal of Animal Ecology*, 87, 24–35.

Bertness, M. D. (1981). Interference and exploitation competition in some tropical hermit crabs. *Journal of Experimental Marine Biology and Ecology*, 49, 189–202.

Bertness, M. D., and Callaway, R. (1992). Positive interactions in communities. *Trends in Ecology and Evolution*, 9, 191–193.

Billick, I., and Case, T. J. (1994). Higher order interactions in ecological communities: What are they and how can they be detected? *Ecology*, 75, 1529–1543.

Birch, J. D., Simard, S. W., Beiler, K., and Karst, J. (2020). Beyond seedlings: Ectomycorrhizal networks and growth of mature Pseudotsuga menziesii. *Journal of Ecology*, 109, 806–818.

Birch, L. C. (1957). The meanings of competition. *American Naturalist*, 91, 5–18.

Bolker, B., Holyoak, M., Křivan, V., and Schmitz, O. J. (2003). Connecting theoretical and empirical studies of trait-mediated effects. *Ecology*, 84, 1001–1114.

Bolker, B. M. (2008). *Ecological Models and Data in R*. Princeton University Press, Princeton, NJ.

Bolnick, D. I., and Preisser, E. L. (2005). Resource competition modifies the strength of trait-mediated predator–prey interactions: A meta-analysis. *Ecology*, 86, 2771–2779.

Bossert, W. H. (1963). *Simulation of character displacement in animals*. Ph.D. Dissertation, Harvard University.

Bowman, W. D., Hacker, S. D., and Cain M. L. (2017). *Ecology*. 4th edition. Sinauer Associates, Sunderland, MA.

Broeckman, M. J. E., Muller-Landau, H. C., Visser, M. D., Jongejans, E., Wright, S. J., and de Kroon, H. (2019). Signs of stabilisation and stable coexistence. Ecology Letters, 22, 1957–1975.

Bronstein, J. L. (1994). Conditional outcomes in mutualistic interactions. *Trends in Ecology and Evolution*, 9, 214–217.

Brook, B. W., and Bradshaw, C. J. A. (2006). Strength of evidence for density dependence in abundance time series of 1198 species. *Ecology* 87, 1445–1451.

Brooker, R. W., Maestre, F. T., Callaway, R. M., et al. (2008). Facilitation in plant communities: The past, the present, and the future. *Journal of Ecology*, 96, 18–34.

Brown, J. S., Laundre, J. W., and Gurung, M. (1999). The ecology of fear: Optimal foraging, game theory and trophic interactions. *Journal of Mammalogy*, 80, 385–399.

Brown, W. L., Jr., and Wilson, E. O. (1956). Character displacement. *Systematic Zoology*, 5, 49–64.

Bulmer, M. G. (1974). Density-dependent selection and character displacement. *American Naturalist*, 108, 45–68.

Burke, N. W., and Holwell, G. I. (2021). Male coercion and female injury in a sexually cannibalistic mantis. *Biology Letters*, 16, 20200811.

Burson, A., Stomp, M., Mekkes, L., and Huisman, J. (2019). Stable coexistence of equivalent nutrient competitors through niche differentiation in the light spectrum. *Ecology*, 100, e02873.

Callaway, R. M., Brooker, R. W., Choler, P., et al. (2002). Positive interactions among alpine plants increase with stress. *Nature*, 417, 844–848.

Callaway, R. M., and Walker, L. R. (1997). Competition and facilitation: A synthetic approach to interactions in plant communities. *Ecology*, 78, 1958–1965.

Cantrell, R. S., Cosner, C., and Lou, Y. (2012). Evolutionary stability of ideal free dispersal strategies in patchy environments. *Journal of Mathematical Biology*, 65, 943–965.

Cardinaux, A., Hart, S. P., and Alexander, J. M. (2018). Do soil microbes influence the outcome of novel interactions between competitors? *Journal of Ecology*, 106, 1853–1863.

Case, T. J. (1990). Invasion resistance arises in strongly interacting species-rich model competition communities. *Proceedings of the National Academy of Sciences USA*, 87, 9610–9614.

Case, T. J. (1995). Surprising behavior from a familiar model and implications for competition theory. *American Naturalist*, 146, 961–966.

Case, T. J. (2000). *An Illustrated Guide to Theoretical Ecology*. Oxford University Press, New York, NY.

Case, T. J., and Taper, M. L. (2000). Interspecific competition, environmental gradients, gene flow, and the coevolution of species borders. *American Naturalist*, 155, 583–605.

Charlesworth, B. (1971). Selection in density-regulated populations. *Ecology*, 52, 469–474.

Chase, J. M., Abrams, P., Grover, J., Diehl, S., Chesson, P., Holt, R., Richards, S., Nisbet, R., and Case, T. J. (2002). The effects of predators on competition between their prey. *Ecology Letters*, 5, 302–313.

Chase, J. M., and Leibold, M. A. (2003). *Ecological Niches; Linking Classical and Contemporary Approaches*. University of Chicago Press, Chicago, IL.

Chesson, P. (1990). MacArthur's consumer–resource model. *Theoretical Population Biology*, 37, 26–38.

Chesson, P. (1994). Multispecies competition in variable environments. *Theoretical Population Biology*, 45, 227–276.

Chesson, P. (2000a). Mechanisms of maintenance of species diversity. *Annual Review of Ecology and Systematics*, 31, 343–366.

Chesson, P. (2000b). General theory of competitive coexistence in spatially-varying environments. *Theoretical Population Biology*, 58, 211–287.

Chesson, P. (2003). Quantifying and testing coexistence mechanisms arising from recruitment fluctuations. *Theoretical Population Biology*, 64, 345–357.

Chesson, P. (2018). Updates on mechanisms of maintenance of species diversity. *Journal of Ecology*, 106, 1773–1794.

Chesson, P. (2020a). Species coexistence. In K. S. McCann and G. Gellner, eds, *Theoretical Ecology: Concepts and Applications*, pp. 5–27. Oxford University Press, Oxford, UK.

Chesson, P. (2020b). MacArthur (1970) and mechanistic coexistence theory. *Theoretical Population Biology*, 133, 13–14.

Chesson, P. (2020c). Chesson's coexistence theory: Comment. *Ecology*, 101(11), e02851.

Chesson, P., and Huntly, N. (1997). The roles of harsh and fluctuating conditions in the dynamics of ecological communities. *American Naturalist*, 150, 519–553.

Chesson, P. L., and Warner, R. R. (1981). Environmental variability promotes coexistence in lottery competitive systems. *American Naturalist*, 117, 923–943.

Clark, J. S., Bell, D., Chu, C., et al. (2010). High-dimensional coexistence based on individual variation: a synthesis of evidence. *Ecological Monographs*, 80, 569–608.

Cody, M. L. (1973). *Competition and the Structure of Bird Communities*. Princeton University Press, Princeton, NJ.

Cody, M. L., and Diamond, J. M., eds (1975). *Ecology and Evolution of Communities*. Harvard University Press, Cambridge, MA.

Cohen, J. E., Briand, F., and Newman, C. M. (1990). *Community Food Webs: Data and Theory*. Springer-Verlag, New York, NY.

Connell, J. H. (1961). The influence of interspecific competition and other factors on the distribution of the barnacle, *Chthamalus stellatus*. *Ecology*, 42, 710–723.

Connell, J. H. (1975). Some mechanisms producing structure in natural communities: a model and evidence from field experiments. In M. L. Cody and J. M. Diamond, eds, *Ecology and Evolution of Communities*, pp. 460–490. Harvard University Press, Cambridge, MA.

Connell, J. H. (1983). On the prevalence and relative importance of interspecific competition: Evidence from field experiments. *American Naturalist*, 122, 661–696.

Cortez, M. H., and Abrams, P. A. (2016). Hydra effects in stable communities and their implications for system dynamics. *Ecology*, 97, 1135–1145.

Cressman, R., and Křivan, V. (2006). Migration dynamics for the ideal free distribution. *American Naturalist*, 168, 384–397.

Cushing, J. M. (1980). Two species competition in a periodic environment. *Journal of Mathematical Biology*, 10, 385–400.

Cushing, J. M. (1986). Periodic Lotka–Volterra competition equations. *Journal of Mathematical Biology*, 24, 381–403.

D'Andrea, R., and Ostling, A. (2016). Challenges in linking trait patterns to niche differentiation. *Oikos*, 125, 1369–1385.

Davidson, D. W., Inouye, R. S., and Brown, J. H. (1984). Granivory in a desert ecosystem: experimental evidence for indirect facilitation of ants by rodents. *Ecology*, 65, 1780–1796.

Dawkins, R., and Krebs, J. F. (1979). Arms races between and within species. *Proceedings of the Royal Society of London B*, 205, 489–511.

Day, T. (2000). Competition and the form of spatial resource heterogeneity on evolutionary diversification. *American Naturalist*, 155, 790–803.

de Roos, A. M., and Persson, L. (2013). *Population and Community Ecology of Ontogenetic Development*. Princeton University Press, Princeton, NJ.

DeAngelis, D. L., Goldstein, R. A., and O'Neill, R. V. (1975). A model for trophic interaction. *Ecology*, 56, 881–892.

DeAngelis, D. L., and Mooij, W. M. (2003). In praise of mechanistically-rich models. In C. D. Canham, J. J. Cole, and W. K. Lauenroth, eds, *Models in Ecosystem Science*, pp. 63–82. Princeton University Press, Princeton, NJ.

Denno, R. R., McClure, M. S., and Ott, J. R. (1995). Interspecific interactions in phytophagous insects: Competition reexamined and resurrected. *Annual Review of Entomology*, 40, 297–331.

Dercole, F., and Rinaldi, S. (2008). *Analysis of Evolutionary Processes: The Adaptive Dynamics Approach and Its Applications*. Princeton University Press, Princeton, NJ.

Descamps-Julien, B., and Gonzalez, A. (2005). Stable coexistence in a fluctuating environment: An experimental demonstration. *Ecology*, 86, 2815–2824.

Dieckmann, U., Doebeli, M., Metz, J. A. J., and Tautz, D., eds (2004). *Adaptive Speciation*. Cambridge University Press, Cambridge, UK.

Dieckmann, U., Marrow, P., and Law, R. (1995). Evolutionary cycling in predator–prey interactions; population dynamics and the Red Queen. *Journal of Theoretical Biology*, 176, 91–102.

Doncaster, C. P. (2006). Comment on 'On the regulation of populations of mammals, birds, fish, and insects' III. *Science*, 311, 1100c.

Doncaster, C. P. (2008). Non-linear density dependence in time series is not evidence of non-logistic growth. *Theoretical Population Biology*, 78, 483–489.

Dungan, M. L. (1987). Indirect mutualism: Complementary effects of grazing and predation in a rocky intertidal community. In W. C. Kerfoot and A. Sih, eds, *Predation: Direct and*

Indirect Impacts on Aquatic Communities, pp. 188–200. University Press of New England, Hanover, NH.

Edmunds, J., Cushing, J. M., Costantino R. F., Henson, S. E., Dennis, B., and Desharnais, R. A. (2003). Park's *Tribolium* competition experiments: a non-equilibrium species coexistence hypothesis. *Journal of Animal Ecology*, 72, 703–712.

Edwards, K. F., Klausmeier, C. A., and Litchman, E. (2013). A three-way trade-off maintains functional diversity under variable resource supply. *American Naturalist*, 182, 786–800.

Egas, M., Dieckmann, U., and Sabelis, M. W. (2004). Evolution restricts the coexistence of specialists and generalists: the role of tradeoff structure. *American Naturalist*, 163, 518–531.

Ellner, S. P., and Becks, L. (2011). Rapid prey evolution and the dynamics of two-predator food webs. *Theoretical Ecology*, 4, 133–152.

Ellner, S. P., and Guckenheimer, J. (2006). *Dynamic Models in Biology*. Princeton University Press, Princeton, NJ.

Ellner, S. P., Snyder, R. E., Adler, P. B., and Hooker, G. (2019). An expanded modern coexistence theory for empirical applications. *Ecology Letters*, 22, 3–18.

Eshel, I. (1983). Evolutionary and continuous stability. *Journal of Theoretical Biology*, 103, 99–111.

Facelli, J. M., Chesson, P., and Barnes, N. (2005). Differences in seed biology of annual plants in arid lands: a key ingredient of the storage effect. *Ecology*, 86, 2998–3006.

Falster, D. S., Kunstler, G., FitzJohn, R. G., and Westoby, M. (2021). Emergent shapes of trait-based competition functions from resource-based models: A Gaussian is not normal in plant communities. *American Naturalist*, 198, 253–267.

Fenchel, T. M., and Christiansen, F. B. (1977). Selection and interspecific competition. In F. B. Christiansen and T. M. Fenchel, eds, *Measuring selection in natural populations*, pp. 477–498. Lecture notes in biomathematics, Volume 19. Springer-Verlag, New York, NY.

Fox, J. W., and Vasseur, D. A. (2008). Character convergence under competition for nutritionally essential resources. *American Naturalist*, 172, 667–680.

Fretwell, S. D. (1972). *Populations in a Seasonal Environment*. Princeton University Press, Princeton, NJ.

Fretwell, S. D., and Lucas, H. L. (1969). On territorial behaviour and other factors influencing habitat distributions in birds. *Acta Biotheoretica*, 19, 16–36.

Friedman, J., Higgins, L. M., and Gore, J. (2017). Community structure follows simple assembly rules in microbial microcosms. *Nature Ecology and Evolution*, 1, 0109.

Fryxell, J. M., and Lundberg, P. (1994). Diet choice and predator–prey dynamics. *Evolutionary Ecology*, 8, 407–421.

Fryxell, J. M., and Lundberg, P. (1998). *Individual Behavior and Community Dynamics*. Chapman and Hall, New York, NY.

Fussmann, G. F., Loreau, M., and Abrams, P. A. (2007). Community dynamics and evolutionary change. *Functional Ecology*, 21, 465–477.

Gause, G. (1936). *The Struggle for Existence*. Williams and Wilkins, Baltimore, MD.

Geritz, S. A. H., Kisdi, É., Meszéna G., and Metz, J. A. J. (1998). Evolutionarily singular strategies and the adaptive growth and branching of the evolutionary tree. *Evolutionary Ecology*, 12, 35–57.

Germain, R. M., Williams, J. L., Schluter, D., and Angert, A. L. (2018). Moving character displacement beyond characters using contemporary coexistence theory. *Trends in Ecology and Evolution*, 33, 75–84.

Getz, W. M. (1993). Metaphysiological and evolutionary dynamics of populations exploiting constant and interactive resources; r–K selection revisited. *Evolutionary Ecology*, 7, 287–305.

Getz, W. M. (1998). An introspection on the art of modeling in population ecology. *BioScience*, 48, 540–552.

Getz, W. M., and Lloyd-Smith, J. O. (2006). Comment on 'On the regulation of populations of mammals, birds, fish, and insects' I. *Science*, 311, 1100a.

Getz, W. M., Marshall, C. R., Carlson, C. J., et al. (2018). Making ecological models adequate. *Ecology Letters*, 21, 153–166.

Gilbert, B., Tunney, T. D., McCann, K. S., et al. (2014). A bioenergetic framework for trophic interactions. *Ecology Letters*, 17, 902–914.

Gilpin, M. E. (1975). Limit cycles in competition communities. *American Naturalist*, 109, 51–60.

Gilpin, M. E., and Ayala, F. J. (1973). Global models of growth and competition. *Proceedings of the National Academy of Sciences USA*, 70, 3590–3593.

Gilpin, M. E., Carpenter, M. P., and Pomerantz, M. J. (1986). The assembly of a laboratory community: Multi-species competition in *Drosophila*. In J. M. Diamond and T. J. Case, eds, *Community Ecology*, pp. 23–40. Harper and Row, New York, NY.

Gilpin, M. E., and Case, T. J. (1976). Multiple domains of attraction in competition communities. *Nature*, 261, 40–42.

Gilpin, M. E., and Justice, K. E. (1972). Re-interpretation of the invalidation of the principle of competitive exclusion. *Nature*, 236, 273–274.

Giraldeau, L. -A., and Caraco, T. (2000). *Social Foraging Theory*. Princeton University Press, Princeton, NJ.

Godoy, O., Bartomeus, I., Rohr, R. P., and Saavedra, S. (2018). Towards the integration of niche and network theories. *Trends in Ecology and Evolution*, 33, 287–300.

Golubski, A. J., and Abrams, P. A. (2011). Modifying modifiers: What happens when interspecific interactions interact? *Journal of Animal Ecology*, 80, 1097–1108.

Gómez-Llano, M., Germain, R. M., Kyogoku, D., McPeek, M. A., and Siepielski, A. M. (2021). When ecology fails: How reproductive interactions promote species coexistence. *Trends in Ecology and Evolution*, 36, 610–622.

Gotelli, N. J. (1991). *A Primer of Ecology*. Sinauer Associates, Sunderland, MA.

Grainger, T. N., Levine, J. M., and Gilbert, B. (2019). The invasion criterion: A common currency for ecological research. *Trends in Ecology and Evolution*, 34, 925–935.

Grainger, T. N., Senthilnathan, A., Ke, P.-J., Barbour, M. A., et al. (2022). An empiricist's guide to using ecological theory. *American Naturalist*, 199, 1–20.

Gramlich, P., Plitzko, S. J., Rudolf, L., Drossel, B., and Gross, T. (2016). The influence of dispersal on a predator–prey system with two habitats. *Journal of Theoretical Biology*, 398, 150–161.

Grant, P. R. (1972). Convergent and divergent character displacement. *Biological Journal of the Linnean Society*, 4, 39–69.

Grant, P. R., and Grant, B. R. (2008). *How and Why Species Multiply: The Radiation of Darwin's Finches*. Princeton University Press, Princeton, NJ.

Grenny, W. J., Bella, D. A., and Curl, Jr.,H. C. (1973). A theoretical approach to interspecific competition in phytoplankton communities. *American Naturalist*, 107, 405–425.

Grime, P. (1979). *Plant Strategies and Vegetation Processes*. John Wiley & Sons, New York, NY.

Gross, K. (2008). Positive interactions among competitors can produce species-rich communities. *Ecology Letters*, 11, 929–936.

Gross, T., Allhoff, K. T., Blasius, B., et al. (2020). Modern models of trophic metacommunities. *Philosophical Transactions of the Royal Society B*, 375, 20190455.

Grover, J. P. (1997). *Resource Competition*. Chapman and Hall, London, UK.

Gurney, W. S., and Nisbet, R. M. (1998). *Ecological Dynamics*. Oxford University Press, New York, NY.

Gyllenberg, M., and Yan, P. (2009). On the number of limit cycles for three-dimensional Lotka–Volterra systems. *Discrete and Continuous Dynamical Systems, Series B*, 11, 347–352.

Gyllenberg, M., Yan, P., and Wang, Y. (2006). A 3D competitive Lotka–Volterra system with three limit cycles: A falsification of a conjecture by Hofbauer and So. *Applied Mathematics Letters*, 19, 1–7.

Haigh, J., and Maynard Smith, J. (1972). Can there be more predators than prey? *Theoretical Population Biology*, 3, 290–299.

Hairston, N. G., Smith, F. E., and Slobodkin, L. B. (1960). Community structures, population control, and competition. *American Naturalist*, 91, 421–425.

Hallam, T. G., Svoboda, L. J., and Gard, T. C. (1979). Persistence and extinction in three species Lotka–Volterra competitive systems. *Mathematical Biosciences*, 46, 117–124.

Hanski, I. (1999). *Metapopulation Ecology*. Oxford University Press, Oxford, UK.

Hardin, G. (1960). The competitive exclusion principle. *Science*, 131, 1292–1297.

Harpole, W. S., Ngai, J. T., Cleland, E. E., et al. (2011). Nutrient co-limitation of primary producer communities. *Ecology Letters*, 14, 852–862.

Hart, S. P., Freckleton, R. P., and Levine, J. M. (2018). How to quantify competitive ability. *Journal of Ecology*, 106, 1902–1909.

Hart, S. P., Usinowicz, J., and Levine, J. M. (2017). The spatial scales of species coexistence. *Nature Ecology and Evolution*, 1, 1066–1073.

Hastings, A., and Powell, T. (1991). Chaos in a three species food chain. *Ecology*, 72, 896–903.

Hazlett, B. A. (1978). Shell exchanges in hermit crabs: Aggression, negotiation, or both? *Animal Behavior*, 26, 1268–1279.

Hazlett, B. A. (2013). Shell exchange behavior in the Hawaiian hermit crab, *Calcinus hazletti* (Decapoda, Diogenidae). *Crustaceana*, 86, 253–257.

Hendry, A. (2017). *Eco-evolutionary Dynamics*. Princeton University Press, Princeton, NJ.

Hening, A., and Nguyen, D. H. (2020). The competitive exclusion principle in stochastic environments. *Journal of Mathematical Biology*, 80, 1323–1351.

Herrando-Pérez, S., Delean, S., Brook, B. W., and Bradshaw, C. J. A. (2012). Density dependence: An ecological Tower of Babel. *Oecologia*, 170, 585–603.

Hilborn, R., and Stokes, K. (2010). Defining overfished stocks: Have we lost the plot? *Fisheries*, 35, 113–120.

HilleRisLambers, J., Adler, P. B., Harpole, W. S., Levine, J. M., and Mayfield, M. M. (2012). Rethinking community assembly through the lens of coexistence theory. *Annual Review of Ecology, Evolution, and Systematics*, 43, 227–248.

Hintzen, R. E., Papadopoulou, M., Mounce, R., et al. (2020). Relationship between conservation biology and ecology shown through machine reading of 32,000 articles. *Conservation Biology*, 34, 721–732.

Hirsch, M. W., Smale, S., and DeVaney, R. L. (2004). *Differential Equations, Dynamical Systems, and an Introduction to Chaos*. 2nd edition. Elsevier, New York, NY.

Holling, C. S. (1959). The components of predation as revealed by a study of small mammal predation of the European pine sawfly. *Canadian Entomologist*, 91, 293–320.

Holling, C. S. (1965). The functional response of predators to prey and its role in mimicry and population regulation. *Memoirs of the Entomological Society of Canada*, 455, 1–60.

Holling, C. S., and Buckingham, S. (1976). A behavioral model of predator–prey functional responses. *Behavioral Science*, 21, 183–195.

Holt, R. D. (1977). Predation, apparent competition, and the structure of prey communities. *Theoretical Population Biology*, 12, 197–229.

Holt, R. D. (1984). Spatial heterogeneity, indirect interactions, and the coexistence of prey species. *American Naturalist*, 124, 377–406.

Holt, R. D. (1985). Population dynamics in two-patch environments: Some anomalous consequences of optimal habitat distribution. *Theoretical Population Biology*, 28, 181–208.

Holt, R. D. (2008). Habitats and seasons. *Israel Journal of Ecology and Evolution*, 54, 279–285.

Holt, R. D., and Barfield, M. (2013). Direct food–predator interactions as determinants of food chain dynamics. *Journal of Theoretical Biology*, 339, 47–57.

Holt, R. D., Barfield, M., and Gonzalez, A. (2003). Impacts of environmental variability in open populations and communities: 'inflation' in sink environments. *Theoretical Population Biology*, 64, 315–330.

Holt, R. D., and Bonsall, M. B. (2017). Apparent competition. *Annual Review of Ecology, Evolution, and Systematics*, 48, 447–471.

Holt, R. D., and Kotler, B. P. (1987). Short-term apparent competition. *American Naturalist*, 130, 412–430.

Holt, R. D., and Lawton, J. H. (1994). The ecological consequences of shared natural enemies. *Annual Review of Ecology and Systematics*, 25, 495–520.

Holyoak, M., Leibold, M. A., and Holt, R. D., eds (2004). *Metacommunities: Spatial Dynamics and Ecological Communities*. University of Chicago Press, Chicago, IL.

Horn, H. S., and MacArthur, R. H. (1972). Competition among fugitive species in a harlequin environment. *Ecology*, 53, 749–752.

Houle, D. (1998). High enthusiasm and low r-squared. *Evolution*, 52, 1872–1876.

Hsu, S. B., and Hubbell, S. (1979). Two predators competing for two prey species: An analysis of MacArthur's model. *Mathematical Biosciences*, 47, 143–171.

Hubbell, S. P. (2001). *The Unified Neutral Theory of Biodiversity and Biogeography*. Princeton University Press, Princeton, NJ.

Huisman, J., and Weissing, F. J. (1999). Biodiversity of plankton by species oscillations and chaos. *Nature*, 402, 407–410.

Huisman, J., and Weissing, F. J. (2001). Biological conditions for oscillations and chaos generated by multi-species models of competition. *Ecology*, 82, 2682–2695.

Humphreys, R. K., and Ruxton, G. D. (2020). The dicey dinner dilemma: Asymmetry in predator–prey risk-taking, a broadly-applicable alternative to the life–dinner principle. *Journal of Evolutionary Biology*, 33, 377–383.

Hutchinson, G. E. (1959). Homage to Santa Rosalia, or why are there so many kinds of animals? *American Naturalist*, 93, 145–159.

Hutchinson, G. E. (1961). The paradox of the plankton. *American Naturalist*, 95, 137–145.

Hutchinson, G. E. (1965). *The Ecological Theatre and the Evolutionary Play*. Yale University Press, New Haven, CT.

Hutchinson, G. E. (1978). *An Introduction to Population Ecology*. Yale University Press, New Haven, CT.

Jansen, V. A. A. (2001). The dynamics of two diffusively coupled predator–prey populations. *Theoretical Population Biology*, 59, 119–131.

Jansen, V. A. A., and de Roos, A. M. (2000). The role of space in reducing predator–prey cycles. In U. Dieckmann, R. Law, and J. A. J. Metz, eds, *The Geometry of Ecological Interactions: Simplifying Spatial Complexity*, pp. 183–201. Cambridge University Press, Cambridge, UK.

Jeschke, J. M., Kopp, M., and Tollrian, R. (2002). Predator functional responses: Discriminating between handling and digesting prey. *Ecological Monographs*, 72, 95–112.

Jeschke, J. M., Kopp, M., and Tollrian, R. (2004). Consumer–food systems; Why type-I functional responses are exclusive to filter-feeders. *Biological Reviews*, 79, 337–349.

Johst, K., Berryman, A., and Lima, M. (2008). From individual interactions to population dynamics: Individual resource partitioning simulation exposes the causes of nonlinear intra-specific competition. *Population Ecology*, 50, 79–90.

Kamran-Disfani, A. R., and Golubski, A. J. (2013). Lateral cascade of indirect effects in food webs with different types of adaptive behavior. *Journal of Theoretical Biology*, 339, 58–69.

Kang, Y., and Chesson, P. (2010). Relative nonlinearity and permanence. *Theoretical Population Biology*, 78, 26–35.

Kaplan, I., and Denno, R. F. (2007). Interspecific interactions in phytophagous insects revisited: a quantitative assessment of competition theory. *Ecology Letters*, 10, 977–994.

Ke, P.-J., and Wan, J. (2020). Effects of soil microbes on plant competition: a perspective from modern competition theory. *Ecological Monographs*, 90(1), e01391.

Keddy, P. A. (2000). *Competition*. 2nd edition. Kluver Academic Publishers, Dordrecht, Netherlands.

Kendall, B. A., Prendergast, J., and Bjornstad, O. (1998). The macroecology of population dynamics: Taxonomic and biogeographic patterns of population cycles. *Ecology Letters*, 1, 160–164.

King, A. A., and Schaffer, W. M. (1999). The rainbow bridge: Hamiltonian limits and resonance in predator–prey dynamics. *Journal of Mathematical Biology*, 39, 439–469.

Kingsland, S. E. (1995). *Modeling Nature*. 2nd edition. Univeristy of Chicago Press, Chicago, IL.

Kisdi, É. (1999). Evolutionary branching under asymmetric competition. *Journal of Theoretical Biology*, 197, 149–162.

Klausmeier, C. A., Kremer, C. T., and Koffel, T. (2020a). Trait-based ecological and eco-evolutionary theory. In K. S. McCann and G. Gellner, eds, *Theoretical Ecology: Concepts and Applications*, pp. 161–193. Oxford University Press, Oxford, UK.

Klausmeier, C. A., and Litchman, E. (2012). Successional dynamics in the seasonally forced diamond food web. *American Naturalist*, 180, 1–16.

Klausmeier, C. A., Osmond, M. M., Kremer, C. T., and Litchman, E. (2020b). Ecological limits to evolutionary rescue. *Philosophical Transactions of the Royal Society B*, 375, 20190453.

Klironomos, J. N. (2002). Feedback with soil biota contributes to plant rarity and invasiveness in communities. *Nature*, 417, 67–70.

Koelle, K., and Vandermeer, J. (2005). Dispersal-induced desynchronization: from metapopulations to metacommunities. *Ecology Letters*, 8, 167–175.

Kraft, N. J. B., Godoy, O., and Levine, J. M. (2015). Plant functional traits and the multidimensional nature of coexistence. *Proceedings of the National Academy of Sciences USA*, 112, 797–802.

Krebs, C. J. (1972). *Ecology: The Experimental Analysis of Distribution and Abundance*. Harper and Row, New York, NY.

Krebs, C. J. (1995). Two paradigms of population regulation. *Wildlife Research*, 22, 1–10.

Krebs, C. J. (2001). *The Experimental Analysis of Distribution and Abundance*. 5th edition. Benjamin Cummings, San Francisco, CA.

Kremer, C. T., and Klausmeier, C. A. (2017). Species packing in eco-evolutionary models of seasonally fluctuating environments. *Ecology Letters*, 20, 1158–1168.

Kuang, J. J., and Chesson, P. (2008). The interaction between predation and competition. *Nature*, 436, 235–238.

Laird, R. A., and Schamp, B. S. (2006). Competitive intransitivity promotes species coexistence. *American Naturalist*, 168, 182–193.

Laird, R. A. and Schamp, B. S. (2015). Competitive intransitivity, population interaction structure, and strategy coexistence. *Journal of Theoretical Biology*, 365, 149–158.

Lawlor, L. R. (1979). Direct and indirect effect of n-species competition. *Oecologia*, 43, 355–364.

Lawlor, L. R., and Maynard Smith, J. (1976). The coevolution and stability of competing species. *American Naturalist*, 110, 79–99.

Leibold, M. A. (1989). Resource edibility and the effects of predators and productivity on the outcome of trophic interactions. *American Naturalist*, 134, 922–949.

Leibold, M. A. (1996). A graphical model of keystone predation in food webs: trophic regulation of abundance, incidence, and diversity patterns in communities. *American Naturalist*, 147, 784–812.

Leibold, M. A., and Chase, J. M. (2018). *Metacommunity Ecology*. Princeton University Press, Princeton, NJ.

Leibold, M. A., Holyoak, M., Mouquet, M., et al. (2004). The metacommunity concept: A framework for multi-scale community ecology. *Ecology Letters*, 7, 601–613.

Leibold, M. A., and McPeek, M. A. (2006). Coexistence of the niche and neutral perspectives in community ecology. *Ecology*, 87, 1399–1410.

Leibold, M. A., Urban, M. C., De Meester, L., Klausmeier, C. A., and Vanoverbeke, J. (2019). Regional neutrality evolves through local adaptive niche evolution. *Proceedings of the National Academy of Sciences USA*, 116, 2612–2617.

Leimar, O., Sasaki, A., Doebeli, M., and Dieckman, U. (2013). Limiting similarity, species packing, and the shape of competition kernels. *Journal of Theoretical Biology*, 339, 3–13.

Lekberg, Y., Bever, J. D., Bunn, R. A., et al. (2018). Relative importance of competition and plant–soil feedback, their synergy, context-dependency and implications for coexistence. *Ecology Letters*, 21, 1268–1281.

Leon, J. A., and Tumpson, D. B. (1975). Competition between two species for two complementary or substitutable resources. *Journal of Theoretical Biology*, 50, 185–201.

Letten, A. D., Dhani, M. K., Ke, P.-J., and Fukami, T. (2018). Species coexistence through simultaneous fluctuation-dependent mechanisms. *Proceedings of the National Academy of Sciences USA*, 115, 6745–6750.

Letten, A. D., Ke, P.-J., and Fukami, T. (2017). Linking modern coexistence theory and contemporary niche theory. *Ecological Monographs*, 87, 161–177.

Letten, A. D., and Stouffer, D. B. (2019). The mechanistic basis for higher-order interactions and non-additivity in competitive communities. *Ecology Letters*, 22, 423–436.

Levandowsky, M. (1976). The cats in Zanzibar. *Quarterly Review of Biology*, 51, 417–419.

Levi, T., Barfield, M., Barrantes, S., Sullivan, C., Holt, R. D., and Terborgh, J. (2019). Tropical forests can maintain hyperdiversity because of enemies. *Proceedings of the National Academy of Sciences USA*, 116, 581–586.

Levin, S. A. (1970). Community equilibria and stability, and an extension of the competitive exclusion principle. *American Naturalist*, 104, 413–423.

Levin, S. A. (1974). Dispersion and population interactions. *American Naturalist*, 108, 207–228.

Levine, J. M., Bascompte, J., Adler, P. B., and Allesina, S. (2017). Beyond pairwise mechanisms of coexistence in complex communities. *Nature*, 546, 56–64.

Levine, S. H. (1976). Competitive interactions in ecosystems. *American Naturalist*, 110, 903–910.

Levins, R. (1968). *Evolution in Changing Environments*. Princeton University Press, Princeton, NJ.

Levins, R. (1969). Some demographic and genetic consequences of environmental heterogeneity for biological control. *Bulletin of the Entomological Society of America*, 15, 237–240.

Levins, R. (1970). Extinction. In M. Gerstenhaber, ed., *Some Mathematical Questions in Biology*, pp. 75–107. Lecture Notes in Mathematics. American Mathematical Society, Providence, RI.

Levins, R., and Culver, D. (1971). Regional coexistence of species and competition between rare species. *Proceedings of the National Academy of Sciences USA*, 68, 1246–1248.

Lewis, M. A., Fagan, W. F., Auger-Méthé, M., et al. (2021). Learning and animal movement. *Frontiers in Ecology and Evolution*, 9, 681704.

Lewontin, R. C. (1974). *The Genetic Basis of Evolutionary Change*. Columbia University Press, New York, NY.

Li, L., and Chesson, P. (2016). The effects of dynamical rates on species coexistence in a variable environment: The paradox of the plankton revisited. *American Naturalist*, 188, E46–E58.

Lima, S. L., and Dill, L. M. (1990). Behavioral decisions made under the risk of predation: A review and prospectus. *Canadian Journal of Zoology*, 68, 619–640.

Litchman, E., de Tezanos Pinto, P., Klausmeier, C. A., Thomas, M. K., and Yoshiyama, K. (2010). Linking traits to species diversity and community structure in phytoplankton. *Hydrobiologia*, 653, 15–28.

Litchman, E., and Klausmeier, C. A. (2001). Competition of phytoplankton under fluctuating light. *American Naturalist*, 157, 170–187.

Litchman E., and Klausmeier, C. A. (2008). Trait-based community ecology of phytoplankton. *Annual Review of Ecology, Evolution, and Systematics*, 39, 615–639.

Loreau, M. (1992). Time scale of resource dynamics and coexistence through time partitioning. *Theoretical Population Biology*, 41, 401–412.

Loreau, M. (2010). *From Populations to Ecosystems: Theoretical Foundations for a New Ecological Synthesis*. Princeton University Press, Princeton, NJ.

Loeuille, M., and Loreau, M. (2004). Nutrient enrichment and food chains: can evolution buffer top-down control? *Theoretical Population Biology*, 65, 285–298.

Lundberg, S., and Stenseth, N. C. (1985). Coevolution of competing species: Ecological character displacement. *Theoretical Population Biology*, 27, 105–119.

Lynch, M., Conery, J., and Bürger, R. (1995a). Mutational meltdowns in sexual populations. *Evolution*, 49, 1067–1080.

Lynch, M., Conery, J., and Bürger, R. (1995b). Mutation accumulation and the extinction of small populations. *American Naturalist*, 146, 489–518.

Ma, B. O., Abrams, P. A., and Brassil, C. E. (2003). Dynamic versus instantaneous models of diet choice. *American Naturalist*, 162, 668–684.

MacArthur, R. H. (1970). Species packing and competitive equilibria for many species. *Theoretical Population Biology*, 1, 1–11.

MacArthur, R. H. (1972). *Geographical Ecology*. Princeton University Press, Princeton, NJ.

MacArthur, R. H., and Levins, R. (1967). The limiting similarity, convergence and divergence of coexisting species. *American Naturalist*, 101, 377–385.

Manlick, P. J., and Pauli, J. N. (2020). Human disturbance increases trophic niche overlap in terrestrial carnivore communities. *Proceedings of the National Academy of Sciences USA*, 117, 26842–26848.

Marchinko, K. (2009). Predation's role in repeated phenotypic and genetic divergence of armor in threespine stickleback. *Evolution*, 63, 127–138.

Matessi, C., and Gatto, M. (1984). Does K-selection imply prudent predation? *Theoretical Population Biology*, 25, 347–363.

Matsuda, H. (1985). Evolutionary stable strategies for predator switching. *Journal of Theoretical Biology*, 115, 351–366.

Matsuda, H., and Abrams, P. A. (1994). Runaway evolution to self-extinction under asymmetrical competition. *Evolution*, 48, 1764–1772.

Matsuda, H., and Abrams, P. A. (2006). Maximal yields from multi-species fisheries systems: Rules for systems with multiple trophic levels. *Ecological Applications*, 16, 225–237.

Matsuda, H., and Abrams, P. A. (2013). Is feedback control effective for ecosystem-based fisheries management? *Journal of Theoretical Biology*, 339, 122–128.

Matsuda, H., Abrams, P. A., and Hori, M. (1993). The effect of adaptive anti-predator behavior on exploitative competition and mutualism between predators. *Oikos*, 68, 549–559.

Matsuda, H., Hori, M., and Abrams, P. A. (1994). Effects of predator-specific defense on community complexity. *Evolutionary Ecology*, 8, 628–638.

Matsuda, H., Hori, M., and Abrams, P. A. (1996). Effects of predator-specific defense on biodiversity and community complexity in two-trophic-level communities. *Evolutionary Ecology*, 10, 13–28.

Matsumura, S., Arlinghaus, R., and Dieckmann, U. (2010). Foraging on spatially distributed resources with sub-optimal movement, imperfect information, and travelling costs: Departures from the ideal free distribution. *Oikos*, 119, 1469–1483.

May, R. M. (1973). *Stability and Complexity in Model Ecosystems*. Princeton University Press, Princeton, NJ.

May, R. M. (1974). On the theory of niche overlap. *Theoretical Population Biology*, 5, 297–332.

May, R. M., and Leonard, W. J. (1975). Nonlinear aspects of competition between three species. *SIAM Journal of Applied Mathematics*, 29, 243–253.

May, R. M., and MacArthur, R. H. (1972). Niche overlap as a function of environmental variability. *Proceedings of the National Academy of Sciences USA*, 69, 1109–1113.

Mayfield, M. M., and Levine, J. M. (2010). Opposing effects of competitive exclusion on the phylogenetic structure of communities. *Ecology Letters*, 13, 1085–1093.

Mayfield, M. M., and Stouffer, D. B. (2017). Higher-order interactions capture unexplained complexity in diverse communities. *Nature Ecology and Evolution*, 1, 0062.

McCallen, E., Knott, J., Nunez-Mir, G., Taylor, B., Jo, I., and Fei, S. (2019). Trends in ecology: Shifts in ecological research themes over the past four decades. *Frontiers in Ecology and the Environment*, 17, 109–116.

McCann, K. S. (2012). *Food Webs*. Princeton University Press, Princeton, NJ.

McCann, K. S., Hastings, A., and Huxel, G. R. (1998). Weak trophic interactions and the balance of nature. *Nature*, 395, 794–798.

McGill, B. J., Enquist, B. J., Weiher, E., and Westoby, M. (2006). Rebuilding community ecology from functional traits. *Trends in Ecology and Evolution*, 21, 178–185.

McKinnon, J. S., Mori, S., Blackman, B. K., et al. (2004). Evidence for ecology's role in speciation. *Nature*, 429, 294–298.

McMeans, B. C., McCann, K. S., Guzzo, M. M., et al. (2020). Winter in water: Differential responses and the maintenance of biodiversity. *Ecology Letters*, 23, 922–938.

McPeek, M. A. (2017). The ecological dynamics of natural selection: Traits and the coevolution of community structure. *American Naturalist*, 189, E91–E117.

McPeek, M. A. (2019a). Mechanisms influencing the coexistence of multiple consumers and multiple resources: resource and apparent competition. *Ecological Monographs*, 89(1), e01328.

McPeek, M. A. (2019b). Limiting similarity? The ecological dynamics of natural selection among resources and consumers caused by both apparent and resource competition. *American Naturalist*, 193, E92–E115.

McPeek, M. A., and Gomulkiewicz, R. (2005). Assembling and depleting species richness in metacommunities: Insights from ecology, population genetics and macroevolution. In M. Holyoak, M. A. Leibold, and R. D. Holt, eds, *Metacommunity Ecology*, pp. 355–373. University of Chicago Press, Chicago, IL.

McPeek, M. A., and Siepielski, A. M. (2019). Disentangling ecologically equivalent from neutral species: The mechanisms of population regulation matter. *Journal of Animal Ecology*, 88, 1755–1765.

Meszéna, G., Gyllenberg, M., Pasztor, L., and Metz, J. A. J. (2006). Competitive exclusion and limiting similarity: a unified theory. *Theoretical Population Biology*, 69, 68–87.

Milinski, M., and Parker, G. A. (1991). Competition for resources. In J. R. Krebs and N. B. Davies, eds, *Behavioural Ecology: An Evolutionary Approach*, 3rd edition, pp. 137–168. Blackwell Scientific, Oxford, UK.

Miller, E. T., and Klausmeier, C. A. (2017). Evolutionary stability of coexistence due to the storage effect in a two-season model. *Theoretical Ecology*, 10, 91–103.

Mittelbach, G. G., and McGill, B. J. (2019). *Community Ecology*. 2nd edition. Oxford University Press, New York, NY.

Morosov, A., and Petrovskii, S. (2013). Feeding on multiple sources: Towards a universal parameterization of the functional response of a generalist predator. *PLoS One*, 8(9), e74586.

Murdoch, W. W. (1969). Switching in generalist predators: Experiments on predator specificity and stability of prey populations. *Ecological Monographs*, 39, 345–354.

Murdoch, W. W., Briggs, C. J., and Nisbet, R. M. (2003). *Consumer–Resource Dynamics*. Princeton University Press, Princeton, NJ.

Murdoch, W. W., and Oaten, A. (1975). Predation and population stability. *Advances in Ecological Research*, 9, 1–131.

Murdoch, W. W., and Walde, S. J. (1989). Analysis of insect population dynamics. In P. J. Grubb and J. B. Whittaker, eds, *Toward a More Exact Ecology*, pp. 113–140. Blackwell Scientific, Oxford, UK.

Mylius, S. D., and Diekmann, O. (2001). The resident strikes back: Invader-induced switching of resident attractor. *Journal of Theoretical Biology*, 211, 297–311.

Namba, T. (1984). Competitive coexistence in a seasonally fluctuating environment. *Journal of Theoretical Biology*, 111, 369–386.

Namba, T., and Hashimoto, C. (2004). Dispersal-mediated coexistence of competing predators. *Theoretical Population Biology*, 66, 53–70.

Namba, T., and Takahashi, S. (1993). Competitive coexistence in a seasonally fluctuating environment. II. Multiple stable states and invasion success. *Theoretical Population Biology*, 44, 374–402.

Neill, W. E. (1974). The community matrix and the interdependence of the competition coefficients. *American Naturalist*, 108, 399–408.

Neill, W. E. (1975). Experimental studies of microcrustacean competition, community composition, and efficiency of resource utilization. *Ecology*, 56, 809–826.

Nicholson, A. J. (1937). The role of competition in determining animal populations. *Journal of the Council for Scientific and Industrial Research (Australia)*, 10, 101–106.

Noonburg, E., and Abrams, P. A. (2005). Transient dynamics and prey species coexistence with shared resources and predator. *American Naturalist*, 165, 322–335.

Nosil, P., and Crespi, B. J. (2006). Experimental evidence that predation promotes divergence in adaptive radiation. *Proceeding of the National Academy of Sciences USA*, 103, 9090–9095.

Nyblade, C. F. (1974). *Coexistence in Sympatric Hermit Crabs*. Ph.D. thesis, University of Washington, Seattle.

O'Dwyer, J. P. (2018). Whence Lotka–Volterra? Conservation laws and integrable systems in ecology. *Theoretical Ecology*, 11, 441–452.

Oaten, A., and Murdoch, W. W. (1975). Functional response and stability in predator–prey systems. *American Naturalist*, 109, 289–298.

Ohgushi, T. (2005). Indirect interaction webs: Herbivore-induced effects through trait-changes in plants. *Annual Review of Ecology, Evolution, and Systematics*, 36, 81–105.

Oksanen, L., Fretwell, S. D., Arruda, J., and Niemela, P. (1981). Exploitation ecosystems in gradients of primary productivity. *American Naturalist*, 118, 240–261.

Otto, S. P., and Day, T. (2007). *A Biologist's Guide to Mathematical Modeling in Ecology and Evolution*. Princeton University Press, Princeton, NJ.

Ousterhout, B. H., Serrano, M., Bried, J. T., and Siepielski, A. M. (2019). A framework for linking competitor ecological differences to coexistence. *Journal of Animal Ecology*, 88, 1534–1548.

Pacala, S. W., and Levin, S. A. (1997). Biologically generated spatial pattern and the coexistence of competing species. In D. Tilman and P. Kareiva, eds, *Spatial Ecology: The Role of Space in Population Dynamics and Interspecific Interactions*, pp. 204–232. Princeton University Press, Princeton, NJ.

Park, T. (1948). Experimental studies of interspecies competition. I. Competition between populations of the flour beetles, *Tribolium confusum* Duval and *Tribolium castaneum* Herbst. *Ecological Monographs*, 18, 265–307.

Pastore, A. I., Barabás, G., Bimler, M. D., Mayfield, M. M., and Miller, T. E. (2021). The evolution of niche overlap and competitive differences. *Nature Ecology and Evolution*, 5, 330–337.

Pásztor, L., Botta-Dukát, Z., Magyar, G., Czáeán, T., and Meszéna, G. (2016). *Theory-based Ecology: A Darwinian Approach*. Oxford University Press, Oxford, UK.

Pásztor, L., Meszéna, G., and Barabás, G. (2020). Competitive exclusion and evolution: Convergence almost never produces ecologically equivalent species. *American Naturalist*, 195, E112–E117.

Peacor, S. D. (2002). Positive effects of predators on prey through induced modifications of prey behavior. *Ecology Letters*, 5, 77–85.

Pearse, I. S., LoPresti, E., Schaeffer, R. N., et al. (2020). Generalising indirect defence and resistance of plants. *Ecology Letters*, 23, 1137–1152.

Pella, J. S., and Tomlinson, P. K. (1969). A generalized stock-production model. *Bulletin of the InterAmerican Tropical Tuna Commission*, 13, 421–496.

Perälä, T., and Kupareinen, A. (2017). Detection of Allee effects in marine fishes: Analytical biases generated by data availability and model selection. *Proceedings of the Royal Society Series B*, 284, 20171284.

Peters, R. H. (1991). *A Critique for Ecology*. Cambridge University Press, Cambridge, UK.

Pfennig, D. W., and Pfennig, K. S. (2010). Character displacement and the origins of diversity. *American Naturalist*, 176, S22–S44.

Picoche, C., and Barraquand, F. (2019). How self-regulation, the storage effect, and their interaction contribute to coexistence in stochastic and seasonal environments. *Theoretical Ecology*, 12, 489–500.

Pimm, S. L. (1991). *The Balance of Nature*. University of Chicago Press, Chicago, IL.

Pither, J. W., Simard, S. W., Ordonez, A., and Williams, J. W. (2018). Below-ground biotic interactions moderated the postglacial range dynamics of trees. *New Phytologist*, 220, 1148–1160.

Polis, G. A., and Winemiller, K. O., eds (1996). *Food Webs: Integration of Patterns and Dynamics*. Chapman and Hall, New York, NY.

Pomerantz, M. J. (1981). Do 'higher-order interactions' in competition systems really exist? *American Naturalist*, 117, 583–592.

Pomerantz, M. J., Thomas, W. R., and Gilpin, M. E. (1980). Asymmetries in population growth regulated by intraspecific competition: Empirical studies and model tests. *Oecologia*, 47, 311–322.

Preisser, E. L., Bolnick, D. I., and Benard, M. F. (2005). Scared to death? Behavioral effects dominate predator–prey interactions *Ecology*, 86, 501–509.

Pulliam, H. R. (1975). Coexistence of sparrows: A test of community theory. *Science*, 189, 474–476.

Pulliam, H. R. (1983). Ecological community theory and the coexistence of sparrows. *Ecology*, 64, 45–52.

Pyke, G. H., Pulliam, H. R., and Charnov, E. L. (1977). Optimal foraging: A selective review of theory and tests. *Quarterly Review of Biology*, 52, 137–154.

Ratajczak, Z., Carpenter, S. R., Ives, A. R., et al. (2018). Abrupt change in ecological systems: Inference and diagnosis. *Trends in Ecology and Evolution*, 33, 513–526.

Reiners, W. A., Lockwood, J. A., Reiners, D. S., and Prager, S. D. (2017). 100 years of ecology: what are our concepts and are they useful? *Ecological Monographs*, 7, 260–277.

Reynolds, S. A., and Brassil, C. E. (2013). When can a single-species density dependent model capture the dynamics of a consumer–resource system? *Journal of Theoretical Biology*, 339S1, 70–83.

Ricklefs, R. E., and Miller, G. L. (2000). *Ecology*. 4th edition. W. H. Freeman, New York, NY.

Rinaldi, S., Muratori, S., and Kuznetsov, Y. (1993). Multiple attractors, catastrophes and chaos in seasonally perturbed predator–prey communities. *Bulletin of Mathematical Biology*, 1, 15–35.

Ronce, I. (2007). How does it feel to be like a rolling stone? Ten questions about dispersal evolution. *Annual Review of Ecology, Evolution, and Systematics*, 38, 231–253.

Rosenzweig, M. L., and MacArthur, R. H. (1963). Graphical representation of stability conditions of predator–prey interactions. *American Naturalist*, 97, 209–223.

Ross, J. V. (2006). Comment on 'On the regulation of populations of mammals, birds, fish, and insects' II. *Science*, 311, 1100b.

Rossberg, A. G. (2013). *Food Webs and Biodiversity: Foundations, Models, Data*. Wiley, Hoboken, NJ.

Roughgarden, J. (1974). Species packing and the competition function with illustrations from coral reef fish. *Theoretical Population Biology*, 5, 163–186.

Roughgarden, J. (1976). Resource partitioning among competing species—a coevolutionary approach. *Theoretical Population Biology*, 9, 388–424.

Rowell, J. T. (2010). Tactical population movements and distributions for ideally motivated competitors. *American Naturalist*, 176, 638–650.

Royama, T. (1992). *Analytical Population Dynamics*. Chapman and Hall, London, UK.

Rueffler, C., Egas, M., and Metz, J. A. J. (2006a). Evolutionary predictions should be based on individual-level traits. *American Naturalist*, 168, E148–E162.

Rueffler, C., Van Dooren, T. J. M., and Metz, J. A. J. (2006b). The evolution of resource specialization through frequency-dependent and frequency-independent mechanisms. *American Naturalist*, 167, 81–93.

Ruokolainen, L., Abrams, P. A., McCann, K. S., and Shuter, B. J. (2011). Spatial coupling of heterogeneous consumer–resource systems: The effect of adaptive consumer movement on synchrony and stability. *Journal of Theoretical Biology*, 291, 76–87.

Saavedra, S., Rohr, R. P., Bascompte, J., Godoy, O., Kraft, N. J. B., and Levine, J. M. (2017). A structural approach for understanding multi-species coexistence. *Ecological Monographs*, 87, 470–486.

Sauve, A. M. C., Taylor, R. A., and Barraquand, F. (2020). The effects of seasonal strength and abruptness on predator–prey dynamics. *Journal of Theoretical Biology*, 491, 110–175.

Schaffer, W. M. (1981). Ecological abstraction: the consequences of reduced dimensionality in ecological systems. *Ecological Monographs*, 51, 383–401.

Scheffer, M. (2009). *Critical Transitions in Nature and Society*. Princeton University Press, Princeton, NJ.

Schippers, P., Verschoor, A. M., Vos, M., and Mooij, W. M. (2001). Does 'supersaturated coexistence' resolve the 'paradox of the plankton'? Ecology Letters, 4, 404–407.

Schluter, D. (2000). *The Ecology of Adaptive Radiation*. Oxford University Press, Oxford, UK.

Schoener, T. W. (1965). The evolution of bill size differences among sympatric species of birds. *Evolution*, 19, 189–213.

Schoener, T. W. (1970). Size patterns in West Indian *Anolis* lizards: II. Correlations with the sizes of particular sympatric species: Displacement and covergence. *American Naturalist*, 104, 155–174.

Schoener, T. W. (1971). Theory of feeding strategies. *Annual Review of Ecology and Systematics*, 11, 369–404.

Schoener, T. W. (1973). Population growth regulated by intraspecific competition for energy or time: Some simple representations. *Theoretical Population Biology*, 4, 56–84.

Schoener, T. W. (1974a). Some methods for calculating competition coefficients from resource utilization spectra. *American Naturalist*, 108, 332–340.

Schoener, T. W. (1974b). Resource partitioning in ecological communities. *Science*, 185, 27–39.

Schoener, T. W. (1974c). Competition and the form of the habitat shift. *Theoretical Population Biology*, 6, 265–307.

Schoener, T. W. (1976). Alternatives to Lotka–Volterra competition: Models of intermediate complexity. *Theoretical Population Biology*, 10, 309–333.

Schoener T. W. (1978). Effects of density restricted food encounter on some simple single-level competition models. *Theoretical Population Biology*, 13, 365–381.

Schoener, T. W. (1982). The controversy over interspecific competition. *American Scientist*, 70, 586–595.

Schoener, T. W. (1983). Field experiments on interspecific competition. *American Naturalist*, 122, 240–285.

Schoener, T. W. (1986). Mechanistic approaches to community: A new reductionism? *American Zoologist*, 26, 81–106.

Schoener, T. W. (1989). Food webs from the small to the large. *Ecology*, 70, 1559–1589.

Schoener, T. W. (1993). On the relative importance of direct versus indirect effects in ecological communities. In H. Kawanabe, J. E. Cohen, and K. Iwasaki, eds, *Mutualism and Community Organization: Behavioural, Theoretical, and Food-web Approaches*, pp. 365–411. Oxford University Press, Oxford, UK.

Schreiber, S., and Rudolf, V. H. W. (2008). Crossing habitat boundaries: Coupling dynamics of ecosystems through complex life cycles. *Ecology Letters*, 11, 576–587.

Schreiber, S. J. (2021). Positively and negatively autocorrelated environmental fluctuations have opposing effects on species coexistence. *American Naturalist*, 197, 405–414.

Schreiber, S. J., Bürger, R., and Bolnick, D. I. (2011). The community effects of phenotypic and genetic variation within a predator population. *Ecology*, 92, 1582–1593.

Schreiber, S. J., Yamamichi, M., and Strauss, S. Y. (2019). When rarity has costs: Coexistence under positive frequency-dependence and environmental stochasticity. *Ecology*, 100, e02664.

Schwinning, S., and Rosenzweig, M. L. (1990). Periodic oscillations in an ideal free predator-prey distribution. *Oikos*, 59, 85–91.

Scranton, K., and Vasseur, D. A. (2016). Coexistence and emergent neutrality generate synchrony among competitors in fluctuating environments. *Theoretical Ecology*, 9, 353–363.

Shurin, J. B., Borer, E. T., Seabloom, E. W., et al. (2002). A cross-ecosystem comparison of the strength of trophic cascades. *Ecology Letters*, 5, 785–791.

Sibly, R. M., Barker, D., Denhan, M. C., Hone, J., and Pagel, M. (2005). On the regulation of populations of mammals, birds, fish, and insects. *Science*, 309, 607–610.

Sibly, R. M., Barker, D., Denhan, M. C., Hone, J., and Pagel, M. (2006). Response to comments on 'On the regulation of populations of mammals, birds, fish, and insects'. *Science*, 311, 1100d.

Sibly, R. M., and Hone, J. (2002). Population growth rate and its determinants: An overview. *Philosophical Transactions of the Royal Society B*, 357, 1153–1170.

Siepielski, A. M., and McPeek, M. A. (2010). On the evidence for species coexistence: A critique of the coexistence program. *Ecology*, 91, 3153–3164.

Skalski, G. T., and Gilliam, J. W. (2001). Functional responses with predator interference: viable alternatives to the Holling Type II model. *Ecology*, 82, 3083–3092.

Slatkin, M. (1980). Ecological character displacement. *Ecology*, 61, 163–177.

Smith, D. J., and Amarasekare, P. (2018). Toward a mechanistic understanding of thermal niche partitioning. *American Naturalist*, 191, E57–E75.

Snyder, R. E., and Adler, P. B. (2011). Coexistence and coevolution in fluctuating environments; Can the storage effect evolve? *American Naturalist*, 178, E76–E84.

Snyder, R. E., and Chesson, P. (2004). How spatial scales of dispersal, competition, and environmental heterogeneity interact to affect coexistence. *American Naturalist*, 164, 633–650.

Soliveres, S., Maestre, F. T., Ulrich, W., et al. (2015). Intransitive competition is widespread in plant communities and maintains their species richness. *Ecology Letters*, 18, 790–798.

Sommer, U. (1984). The paradox of the plankton: Fluctuations of phosphorus availability maintain diversity of phytoplankton in flow-through cultures. *Limnology and Oceanography*, 29, 633–636.

Sommer, U., Adrian, R., Domis, L. D. S., et al. (2012). Beyond the plankton ecology group (PEG) model: Mechanisms driving plankton succession. *Annual Review of Ecology, Evolution, and Systematics*, 43, 429–448.

Sommers, P., and Chesson, P. (2019). Effects of predator avoidance behavior on the coexistence of competing species. *American Naturalist*, 193, E132–E148.

Song, C., Barabás, G., and Saavedra, S. (2019). On the consequences and interdependence of stabilizing and equalizing mechanisms. *American Naturalist*, 194, 627–639.

Stephens, D. W., and Krebs, J. R. (1986). *Foraging Theory*. Princeton University Press, Princeton, NJ.

Stewart, F. M., and Levin, B. R. (1973). Partitioning of resources and the outcome of interspecific competition: A model and some general considerations. *American Naturalist*, 107, 171–198.

Strobeck, C. (1973). N species competition. *Ecology*, 54, 650–654.

Strogatz, S. H. (1994). *Nonlinear Dynamics and Chaos*. Addison Wesley, Reading, MA.

Strong, D. R., Simberloff, D., Abele, L. G., and Thistle, A. B., eds (1984). *Ecological Communities: Conceptual Issues and the Evidence*. Princeton University Press, Princeton, NJ.

Suraci, J. P., Clinchy, M., Zanette, L. Y., and Wilmers, C. C. (2019). Fear of humans as apex predators has landscape-scale impacts from mountain lions to mice. *Ecology Letters*, 22, 1578–1586.

Taper, M. L., and Case, T. J. (1985). Quantitative genetic models for the evolution of character displacement. *Ecology*, 66, 355–371.

Taper, M. L., and Case, T. J. (1992). Models of character displacement and the theoretical robustness of taxon cycles. *Evolution*, 46, 317–333.

Terborgh, J. W. (2015). Toward a trophic theory of species diversity. *Proceedings of the National Academy of Sciences USA*, 112, 11415–11422.

Tilman, D. (1980). Resources: A graphical-mechanistic approach to competition and predation. *American Naturalist*, 116, 362–393.

Tilman, D. (1982). *Resource Competition and Community Structure*. Princeton University Press, Princeton, NJ.

Tilman, D. (1987). The importance of the mechanisms of interspecific competition. *American Naturalist*, 129, 769–774.

Tilman, D., and Karieva, P. M., eds (1997). *Spatial Ecology: The Role of Space in Population Dynamics and Interspecific Interactions*. Princeton University Press, Princeton, NJ.

Tredennick, A. T., Adler, P. B., and Adler, F. R. (2017). The relationship between species richness and ecosystem variability is shaped by the mechanism of coexistence. *Ecology Letters*, 20, 958–968.

Turchin, P. (2003). *Complex Population Dynamics*. Princeton University Press, Princeton, NJ.

Turelli, M. (1977). Random environments and stochastic calculus. *Theoretical Population Biology*, 12, 140–178.

Turelli, M. (1978). Does environmental variability limit niche overlap? *Proceedings of the National Academy of Sciences USA*, 75, 5085–5089.

Turelli, M. (1981). Niche overlap and invasion of competitors in random environments: 1. Models without demographic stochasticity. *Theoretical Population Biology*, 20, 1–56.

Uszko, W., Diehl, S., Englund, G., and Amarasekare, P. (2017). Effects of warming on predator-prey interactions: A resource-based approach and a theoretical synthesis. *Ecology Letters*, 20, 513–523.

van Leeuwen, E., Jansen, V. A. A., Dieckmann, U., and Rossberg, A. G. (2013). A generalized functional response for predators that switch between multiple prey species. *Journal of Theoretical Biology*, 328, 89–98.

van Velzen, E. (2020). Predator coexistence through emergent fitness equalization. *Ecology*, 101(5), e02995.

Vance, R. R. (1972). Competition and mechanism of coexistence in three sympatric species of intertidal hermit crabs. *Ecology*, 53, 1062–1074.

Vandermeer, J. H. (1969). The competitive structure of communities: An experimental approach with Protozoa. *Ecology*, 50, 362–371.

Vandermeer, J. H. (1980). Indirect mutualism: Variations on a theme by Stephen Levine. *American Naturalist*, 116, 441–448.

Vandermeer, J. H. (2004). Coupled oscillations in food webs: Balancing competition and mutualism in simple ecological models. *American Naturalist*, 163, 857–867.

Vandermeer, J. H. (2006). Oscillating populations and biodiversity maintenance. *Bioscience*, 56, 967–975.

Vasseur, D. A., and Fox, J. W. (2011). Adaptive dynamics of competition for nutritionally complementary resources: Character convergence, displacement, and parallelism. *American Naturalist*, 178, 501–514.

Velland, M. (2010). Conceptual synthesis in community ecology. *Quarterly Review of Biology*, 85, 183–206.

Velland, M. (2016). *The Theory of Ecological Communities*. Princeton University Press, Princeton, NJ.

Velland, M., Srivastava, D. S., Anderson, K. M., et al. (2014). Assessing the relative importance of neutral stochasticity in ecological communities. *Oikos*, 123, 1420–1430.

Verhulst, P.-F. (1838). Notice sur la loi que la population poursuit dans son accroissement. *Correspondance Mathématique et Physique*, 10, 113–121.

Vincent, R. L., and Brown, J. S. (2005). *Evolutionary Game Theory, Natural Selection, and Darwinian Dynamics*. Cambridge University Press, Cambridge, UK.

Volterra, V. (1931). Variations and fluctuations of the number of individuals in animal species living together. In R. N. Chapman, ed., *Animal Ecology*, pp. 409–448. McGraw-Hill, New York, NY.

Warren, B. H., Simberloff, D., Ricklefs, R. E., et al. (2015). Islands as model systems in ecology and evolution: Prospects fifty years after MacArthur–Wilson. *Ecology Letters*, 18, 200–217.

Wei, N., Kaczorowski, R. L., Arceo-Gómez, G., O'Neill, E. M., Hayes, R. A., and Ashman, T.-L. (2021). Pollinators contribute to the maintenance of flowering plant diversity. *Nature*, 597, 688–692.

Werner, E. E., and Peacor, S. D. (2003). A review of trait-mediated indirect interactions in ecological communities. *Ecology*, 84, 1083–1100.

Williams, F. M. (1972). Mathematics of microbial populations, with an emphasis on open systems. *Transactions of the Connecticut Academy of Arts and Sciences*, 44, 397–426.

Wilson, D. S. (1975). The adequacy of body size as a niche difference. *American Naturalist*, 109, 769–784.

Wilson, D. S., and Turelli, M. (1986). Stable underdominance and the evolutionary invasion of empty niches. *American Naturalist*, 127, 835–861.

Wilson, E. O., and Bossert, W. H. (1971). *A Primer of Population Biology*. Sinauer Associates, Sunderland, MA.

Wilson, W. G., and Abrams, P. A. (2005). Coexistence of cycling and dispersing consumer species: Armstrong and McGehee in space. *American Naturalist*, 165, 193–205.

Wise, D. H., and Farfan, M. A. (2021). Effects of enhanced productivity of resources shared by predators in a food-web module: Comparing results of a field experiment to predictions of mathematical models of intra-guild predation. *Ecology and Evolution*, 2021, 1–11.

Wollrab, S., Ismest'yeva, L., Hampton, S. E., Silow, E. A., Litchman, E., and Klausmeier, C. A. (2021). Climate change-driven regime shifts in a planktonic food web. *American Naturalist*, 197, 281–295.

Yamamichi, M., and Letten, A. D. (2021). Rapid evolution promotes fluctuation-dependent species coexistence. *Ecology Letters*, 24, 812–818.

Yamamichi, M., Yoshida, T., and Sasaki, A. (2014) Timing and propagule size of invasion determine its success by a time-varying threshold of demographic regime shift. *Ecology*, 95, 2303–2315.

Yodzis, P. (1988). The indeterminacy of ecological interactions as perceived through perturbation experiments. *Ecology*, 69, 508–515.

Yodzis, P. (1989). *Introduction to Theoretical Ecology*. Harper and Row, New York, NY.

Yodzis, P. (1996). Food webs and perturbation experiments: Theory and practice. In G. A. Polis and K. O. Winemiller, eds, *Food Webs: Integration of Patterns and Dynamics*, pp. 192–200. Chapman and Hall, New York, NY.

Yuan, C., and Chesson, P. (2015). The relative importance of relative nonlinearity and the storage effect in the lottery model. *Theoretical Population Biology*, 105, 39–52.

Index